D0854039

A QUESTION OF POWER

ALSO BY ROBERT BRYCE:

Pipe Dreams

Cronies

Gusher of Lies

Power Hungry

Smaller Faster Lighter Denser Cheaper

A QUESTION OF POWER

Electricity and the Wealth of Nations

ROBERT BRYCE

PUBLICAFFAIRS
New York

PublicAffairs
Hachette Book Group
1290 Avenue of the Americas, New York, NY 10104
www.publicaffairsbooks.com
@Public_Affairs

Printed in the United States of America

First Edition: March 2020

Published by PublicAffairs, an imprint of Perseus Books, LLC, a subsidiary of Hachette Book Group, Inc. The PublicAffairs name and logo is a trademark of the Hachette Book Group.

Print book interior design by Jeff Williams.

Library of Congress Cataloging-in-Publication Data
Names: Bryce, Robert, 1960– author.
Title: A question of power: electricity and the wealth of nations / Robert Bryce.
Description: First edition. | New York: PublicAffairs, [2020] | Includes bibliographical references and index. |
Identifiers: LCCN 2019044917 | ISBN 9781610397490 (hardcover) | ISBN 9781610397506 (ebook)
Subjects: LCSH: Electric industries. | Electric industries—History. | Electricity—Social aspects.
Classification: LCC HD9695.A2 B79 2020 | DDC 333.793/2—dc23

LC record available at https://lccn.loc.gov/2019044917

ISBNs: 978-1-61039-749-0 (hardcover); 978-1-5417-5714-1 (international paperback); 978-1-61039-750-6 (ebook)

LSC-C

10 9 8 7 6 5 4 3 2 1

For Lorin

Then there is electricity!—the demon, the angel, the mighty physical power, the all-pervading intelligence! . . . Is it a fact—or have I dreamt it—that, by means of electricity, the world of matter has become a great nerve, vibrating thousands of miles in a breathless point of time?

—NATHANIEL HAWTHORNE,

The House of the Seven Gables, 1851

CONTENTS

List of Illustrations, xi
Acknowledgments, xiii
Introduction: Barrio Antón Ruíz, xvii

PART ONE. ELECTRICITY MEANS MODERNITY

1 Electricity 101 3
2 The Transformative Power of Electricity 10
3 The Vertical City 22
4 The New (Electric) Deal 36
5 Wiring the Superpower 48
6 Women Unplugged 60

PART TWO. WHY ARE BILLIONS STILL STUCK IN THE DARK? AND WHAT ARE THEY DOING ABOUT IT?

7 My Refrigerator Versus the World 73
8 The Power Imperatives: Integrity, Capital, and Fuel 80
9 The American Way of War 87
10 Beirut's Generator Mafia 96
11 It's Not Possible to Keep the Lights on Without Coal 105

PART THREE. THE VIEW FROM ON HIGH-WATT

12 The New (Electric) Economy 119
13 Electrified Cash 133

14 Watts Into Weed 142
15 The Blackout Will Not Be Televised 152

PART FOUR. TWENTY-FIRST-CENTURY TERAWATTS

16 The Terawatt Challenge 165
17 The All-Renewable Delusion 171
18 This Land Is My Land 189
19 The Nuclear Necessity 217
20 Future Grid 236

Conclusion 245

Appendix A, 249
Appendix B, 251
Notes, 253
Bibliography, 305
Index, 309

ILLUSTRATIONS

GRAPHICS

Global GDP and Global Electricity Use: 1980 to 2014 21
Expected Global Growth in Christian and Muslim Populations, 2015 to 2060 64
The High-Watt, Low-Watt, and Unplugged Worlds, 2012 76
Corruption and Per-Capita Electricity Use, 2012 83
Lebanon's Electricity Generated from Oil, 1971 to 2014 103
Share of Global Electricity Generation, by Fuel, 1985 to 2017 109
Electricity Consumption Increases and Market Capitalization Increases of the Giant Five, 2012 to 2017 122
Comparing the Monopolies of Old with the New Tech Giants 125
Electricity Consumption of the Giant Five in 2017 127
Renewable and Conventional Electricity Generation Capacity of the Giant Five, 2018 132
High-Watt Weed 149
California Electricity Prices Compared to US Average, 2011 to 2017 177
Per-Capita Renewable-Energy Spending Around the World, 2016 183
Powering New York City 220
The Terawatt Challenge 243
SI Numerical Designations and Power Units 249–250
Dollar Value of Electricity Stored in Common Batteries 251

PHOTOS

Frank Sprague, 1892 26
Architect rendering of the Postal Telegraph Building in 1893 32
The Postal Telegraph Building in 2017 32
US Senator Burton Wheeler of Montana 44
US Senator George Norris of Nebraska 45
Speaker of the House of Representatives Sam Rayburn and Senator Lyndon Johnson, January 8, 1956 50
Rural electrification, San Joaquin Valley, California, 1938 54
Rehena Jamadar and Joyashree Roy, West Bengal, India, 2016 61
Women in Majlishpukur, 2016 63
Reddy Kilowatt 66
Electric and telecommunications wires above a market street, Delhi, India, 2016 85
Hussein Mousl, driver, Beirut, Lebanon, 2017 97
Fatmagül Sultan, a powership owned by the Turkish firm Karadeniz Holding 104
Street scene, Kolkata, India, 2016 111
Black-market marijuana grow operation, Denver, 2017 143
Anti-wind-energy sign near Denmark, Wisconsin, 2016 209
Rose and Dave Enz, Wrightsville, Wisconsin, August 4, 2016 211
The Indian Point Energy Center, Buchanan, New York, 2018 228

TABLES

1. Top Twenty Engineering Achievements of the Twentieth Century 18
2. America's Crazy Quilt Electric Grid 56
3. Top Ten Countries with the Highest Rates of Child Marriage Are All Electricity Poor 68
4. The Top Ten Problems Facing the World 165

ACKNOWLEDGMENTS

A gang of people helped me produce this book. I must first thank the team at PublicAffairs—Clive Priddle, Peter Osnos, Kaitlin Carruthers-Busser, Liz Dana, and Jaime Leifer—who challenged me to make this book as good as it could possibly be. I owe particular thanks to my dear friend and editor, Lisa Kaufman. As she has done five times before, Lisa helped me convert a bunch of affiliated ideas into a coherent manuscript. Without her insight and patient guidance, this book would not have happened. Thanks to my agent, Dan Green, who over the past nineteen years has always been able to talk me through times of distress while encouraging me to look forward and get back to work.

Thanks also to the team that helped produce the documentary *Juice: How Electricity Explains the World*, which was created simultaneously with this book. My good friend and collaborator Tyson Culver has been a loyal business partner and sounding board since we first began talking about making a movie back in 2016. In addition to Tyson, the *Juice* production team included Deanna DeHaven, James Treakle, John Moody, Dino Maglaris, Matthew L. Wallis, and Ted Powers. Other key people who were instrumental in the making of the film and therefore deserve my sincere thanks: Chris Wright, Liz Wright, Steven R. Anderson, Bud Brigham, Roland Pritzker, Rachel Pritzker, Ray Rothrock, Ed Schweitzer, Stephanie Schweitzer, David Costello, Schweitzer Engineering Laboratories, and Arthur Smith.

My former colleagues at the Manhattan Institute were supportive throughout the long process of writing this book. In particular, I want to acknowledge Howard Husock, Troy Senik, Vanessa Mendoza, Bernadette Serton, and the institute's former president, Larry Mone, for their patience and encouragement.

Dr. Joyashree Roy—to borrow the title of Austin-based singer-songwriter Ruthie Foster's most famous album—is a phenomenal woman. When Lorin, Tyson, and I visited Kolkata in late 2016, Joyashree, an economics professor at Jadavpur University, was gracious and supportive at every turn. She arranged our visit to Vidyasagar University in Midnapore, as well as our visits to rural villages in West Bengal. She also arranged for us to obtain Indian rupees, which was no mean feat in the weeks after Prime Minister Narendra Modi imposed demonetization, which, at a stroke, made illegal all 500- and 1,000-rupee notes, or roughly 80 percent of all the currency in India. She set up meetings, patiently explained the local geography, and even did simultaneous translations. I hope one day to repay her kindness. I must also thank Joyashree's colleagues in Kolkata, Nandini Das and Suman Dutta, for their help in facilitating our travel, as well as Dr. Sebak Jana, who not only hosted us in Midnapore, but took us to several rural villages in West Bengal.

Jimmy Nassour, my friend and fellow board member at Parkside Community School in Austin, was extraordinary. He and his always-smiling brother-in-law, Simon Najm, shepherded us around Beirut, set up meetings, and taught us key Arabic slang. The love that Jimmy and Simon showed for Lebanon made us love Lebanon too.

My friend J. Paul Oxer helped me formulate some of the ideas in this book and, just as important, introduced me to Hlynur Guðjónsson, the consul general for Iceland in New York. Hlynur, in turn, connected me with numerous contacts in Reykjavík. Tinna Traustadóttir and Nanna Baldvinsdóttir from Landsvirkjun were charming tour guides who showed us around the Búrfell hydro project. Gísli Katrínarson explained why Iceland has become a hotspot for high-performance computing, and Helmut Rauth walked us through Genesis Mining's operations and the cryptocurrency business.

Thanks to the Breakthrough Institute for including me in the annual Breakthrough Dialogue. That event allowed me to meet many of the people I interviewed for this book. My particular thanks to Ted Nordhaus, Alex Trembath, and Jessica Lovering, all of whom I consider my friends. Thanks, too, to Michael Shellenberger and the people at Environmental Progress for allowing me to conduct interviews at their office in Berkeley.

My friend and graphics wizard Seth Myers was, once again, patient and diligent. This book is better because of his skill at turning numbers into meaningful graphics. Jude Clemente and Yevgeniy Feyman also provided research assistance and encouragement during the early days of this project. Thanks to the punctilious and pulchritudinous Mimi Bardagjy, who has fact-checked all of my books with humor and precision.

I must also thank the people who read various versions of the book manuscript. My pal Robert Elder Jr., once again, was a patient and careful reader. My father-in-law, Paul Rasmussen, an emeritus chemistry professor at the University of Michigan, has read—and provided insight and feedback on—all my book manuscripts. He provided valuable perspective again on this one. Other friends who provided helpful comments include: Peter Z. Grossman, Joe Cunningham, John Sennett, Rex Rivolo, Chris Pedersen, Steve Brick, Stan Jakuba, Omar Kader, Chris Cauthon, and Bryan Shahan. Thanks also to my friend Jonathan Lesser, who was a valuable consultant on all grid-related matters as well as a patient reader.

Given this book's focus on electricity, a pair of technical notes may be of interest. My laptop (a MacBook Air) uses about 50 watts of power when being charged. The monitor that I use with the laptop, a Thunderbolt display, draws another 100 watts.

Finally, I must thank my wife, Lorin. We had our first date on Halloween 1982. We have been together ever since. Her gentle love and steadfast support saw me through some challenging times as I worked through the many iterations of this book. Thanks, too, to our children: Mary, Michael, and Jacob. They have endured more than a few lectures on energy, power, and electricity. I may be wrong, but I think they are starting to enjoy my disquisitions.

Any errors in this book are mine. If you spot a mistake, please let me know so it can be corrected for the paperback edition.

June 11, 2019
Austin, Texas

INTRODUCTION

Barrio Antón Ruíz

There's no such thing as a low-energy, high-income country.

—**TODD MOSS**, Energy for Growth Hub[1]

Electricity has transformed humanity like no other form of energy. Since the dawn of the Electric Age less than 140 years ago, electricity has changed how we live, communicate, learn, and eat. In doing so, it has fueled an unprecedented period of human flourishing. Never in human history have so many people lived in such wealth and prosperity. And electricity continues to change and enrich our lives. From our ability to navigate foreign cities with maps on our iPhones to the staggering quantities of information available to us on the Internet, we use electricity without a second thought. Nearly every technology we use requires reliable flows of electricity. And yet, as we become ever more connected, ever more wired, billions of people are being left behind.

The vast disparity between the rich and the poor is, in large part, defined by the disparity between those who have electricity and those who scrape by on small quantities of juice or none at all. People in wealthy countries assume that reliable electricity is akin to a birthright. We seldom think about the relationship between electric power and human empowerment. But to bring home the implications of our dependency on electricity—and our vulnerability to the lack of sufficient supplies—all we need to do is spend some time with our neighbors in Puerto Rico and see what happened to them after Hurricane Maria shattered the island's electric grid. That deadly storm left thousands of *Puertoriqueños* in the dark. Among them were Wilfredo Roque, Iris Ortiz, and their three girls.

The first thing I saw as I drove down the steep driveway next to Wilfredo and Iris's modest home in Barrio Antón Ruíz was the orange extension cord. The cord, which was intertwined with a piece of brown rope, was suspended less than two meters over the surface of the couple's driveway. The line stayed aloft thanks to a two-by-four piece of lumber on the left side of the driveway that was secured to the ground near a 4,500-watt gasoline-fueled generator perched next to the fence. On the right side of the driveway, the rope and electric cord were secured to a railing on the house by a couple of knots.

Wilfredo, a slightly built, energetic man, came to greet me right away. He and Iris were both eager to talk. Hurricane Maria had been far more powerful than they had expected. "We weren't prepared for the devastation," Wilfredo told me in Spanish, as he pulled shut the heavy rolling metal gate that separated his driveway from the narrow asphalt street in the neighborhood, which is located amid a set of green rolling hills about one hour's drive southeast of Puerto Rico's capital, San Juan.

As I looked at his generator, a Chinese-made Black Max, he quickly volunteered the numbers. "We spend $100 to $125 per week on gasoline for the generator." In the wake of Hurricane Maria, which devastated Puerto Rico on September 20, 2017, the family's only source of electricity had been the generator. In the first few weeks after they got the machine, they ran it ten or twelve hours per day. "We run it less now. We were spending too much money on fuel. So now we only run the generator five hours per day," Wilfredo said. "No more than that."

After he gave me the figures for the generator, I asked him to repeat them. I did so for two reasons. The first: my tourist-level

Spanish is, well, tourist level. I've traveled a fair amount in Latin America and can navigate and order dinner at the café, but in several of my conversations with Puertoriqueños, I knew I was missing key details. The other reason: it was hard to hear Wilfredo due to the eardrum-shattering racket coming from his neighbor's generator. A short distance from where we were standing, on the other side of the fence, the machine was running full blast. Like Wilfredo's generator, it lacked sound insulation. Both machines sat near the ground, protected from the sun, wind, and rain by a few too-small pieces of roofing material and plywood.

As Wilfredo repeated the numbers, I wrote them in my notebook, in Spanish: "$100 to $125 *por semana*, 5 *horas por dia*." I then showed the notes to Wilfredo for confirmation. He nodded and said, "*Correcto*."

Before Hurricane Maria devastated Puerto Rico with winds that hit 180 miles (290 kilometers) per hour, Wilfredo, Iris, and their three young girls, Alannis (thirteen), Arianna (ten), and Ayamie (five), were paying the state-run grid operator, Puerto Rico Electric Power Authority (PREPA), about $90 per month for their electricity. But that money didn't buy them reliable power and the family's modest home was regularly hit by blackouts. Around the time Alannis was a baby, Iris told me, the blackouts happened several times per day. After she and Wilfredo complained, PREPA workers switched out a transformer in the barrio; things got better and the blackouts were reduced to several times per week. The electricity service still wasn't great, but their connection to PREPA's grid was good enough that they didn't need to run a generator. After the deadly hurricane pummeled the island, they couldn't rely on Puerto Rico's electric grid for anything. So Wilfredo began searching for a small generator to buy. It took him two months.

Even with the four or five hours of electricity per day that was being provided by the generator, life was suddenly much harder.[2] The generator was smelly and loud. So were all their neighbors' generators. The nearly constant noise made it hard to sleep at night. Iris was having to wash some of the family's clothes by hand. Doing the laundry, which used to take just a few minutes, was suddenly taking hours. The children's schoolwork was suffering because they were not getting enough time on the Internet. "We are being left behind," Iris told me. "We have returned to the time of my grandmother, or my great-grandmother."

About the time I visited Puerto Rico, the Rhodium Group, a US-based consulting firm, published a report that found that the island had endured the largest blackout in US history.[3] The firm reported that "more customer-hours have been lost in Puerto Rico due to Hurricane Maria than in the rest of the US over the past five years due to all causes combined." Not only that, Iris, Wilfredo, their three girls, and other Puertoriqueños were enduring the second-largest blackout in world history.[4]

Imagining such energy hardship is almost beyond our ken. We flip the switch, we plug in our phones, laptops, and AirPods, and we expect the power to be on. Every time. And it almost always is. But the Roque Ortiz family's electricity predicament could befall pretty much anyone in the United States or any other country. The risk of an extended blackout—and the societal upheaval that would come with it—is real.[5] Such a blackout could be caused by extreme weather such as a hurricane, tornado, or snowstorm. It could also be caused by extraplanetary forces. In 2017, the American Geophysical Union estimated that an extreme solar storm could cause blackouts that would affect two-thirds of the US population and that "daily domestic economic loss could total $41.5 billion plus an additional $7 billion loss through the international supply chain."[6]

Saboteurs are constantly probing for weaknesses in the electric grid. In 2018, the Department of Homeland Security warned that Russian hackers had infiltrated numerous US energy companies, including electric utilities.[7] If hackers succeed in bringing down all or part of the American electric grid, they could cause billions of dollars in damage without having to leave the comfort of their computer keyboards. In addition to threats from weather and cyberspace, electric grids are also vulnerable to physical sabotage. Well-prepared saboteurs could disable key transformer stations or transmission lines and in doing so cause blackouts across significant swaths of the American grid. Millions, or perhaps even tens of millions, of Americans could be blacked out and plunged into the very same predicament Wilfredo, Iris, and other residents of Barrio Antón Ruíz were enduring. Instead of having cheap, abundant, and reliable electricity, Americans could be faced with a situation in which electricity is expensive, scarce, and intermittent. Unable to rely on the electric grid, they would have to get some of their electricity from small, inefficient, diesel- or gasoline-fired generators. That, in turn, would require having plentiful motor fuel at local service stations, which themselves would need electricity in order to pump the fuel into customers' tanks.

We could also sabotage ourselves. Numerous environmental groups and politicians have claimed that we can completely eliminate the use of hydrocarbons (coal, oil, and natural gas) and nuclear, and instead rely mainly on solar and wind energy. While those policies are intended to slow or stop climate change, they are little more than wishful thinking. Advocates for an all-renewable economy ignore the myriad downsides of attempting to rely on intermittent sources of energy, as well as the vast amounts of land, concrete, steel, copper, and other commodities that would be required to make those projects work at the scale our modern society demands. Politically popular proposals like the Green New Deal claim that if only we adopt a warlike approach to our energy and power systems we can completely eliminate all greenhouse gas emissions from our economy and do so in just two or three decades.[8] Electric cars like the Tesla have gained an almost cult-like following, with little understanding of the fact that we have to get the electricity to charge them from somewhere. Further, making those cars requires mining and smelting megatons of ore to produce the lithium, cobalt, dysprosium, neodymium, and other elements that are used in the vehicles' batteries and motors. In short, the production and consumption of electricity always comes with a cost. Forsaking our existing electricity-generation systems for ones that rely solely on renewables could make our grid less stable and less reliable.

Super-reliable electricity is essential to the Information Age. America's biggest and richest companies have spent billions of dollars building their own electric grids to make sure their computer networks never go dark. Retailing and computing giant Amazon alone controls about 4,700 megawatts of electricity-generation capacity; that's as much as entire countries like Croatia or Laos.[9] At the same time that megacorporations are able to effectively secede from our electric grid, billions of people around the world today are *disempowered*.

The numbers of the disempowered are staggering: About one billion people on the planet today have no access to electricity at all. Another two billion or so are using only tiny amounts. Furthermore, the electricity that the world's energy poor use often resembles the expensive, smelly, intermittent power that Puertoriqueños like Wilfredo and Iris had after Hurricane Maria. Unable to rely on the electric grid, these billions of people routinely plan their days around electricity—when they will have it and when they won't. They often have no choice but to get their electricity from generators similar to the Black Max that Wilfredo was refueling every day or two. If they don't own a generator themselves, the electricity poor often pay subscription fees to local businesspeople who own generators that supply power to customers in their neighborhoods.

Put short, when it comes to electricity, we don't know how good we have it or just how important electricity is. We take it for granted. But nearly everything we touch—almost everything we read, eat, or wear—has, in one way or another, been electrified. Electricity is the world's most important and fastest-growing form of energy.[10] It's also the most difficult to supply and do so reliably. That paradox has shaped and will continue to shape global politics. It underlies the chasm between the rich and the poor, the educated and the uneducated.

That leads to the thesis of this book: electricity is the fuel of the twenty-first century. Electricity makes modern life possible. And yet, some three billion people around the world are still stuck in the dark. Their opportunities, their potential to develop lives beyond the backbreaking work of subsistence farming and day labor, their possibilities for economic and social development, depend on increasing their access to reliable electricity. Electricity is the ultimate poverty killer. No matter where you look in the world, as electricity use has increased, so have personal incomes. Having electricity doesn't guarantee wealth. But its absence almost always means poverty. How we empower the powerless while meeting soaring global electricity demand will be the key factor in addressing some of the world's biggest challenges, including women's rights, climate change, and inequality.

I am also focused on electricity because it is the world's second-largest industry, trailing only the oil and gas sector in overall revenue.[11] Global electricity sales total some $2.4 trillion per year.[12] That means that the electricity business is bigger than the global automobile business and twice as big as the pharmaceutical sector.[13] In the United States alone, electricity sales total about $400 billion per year.[14] If the US electric sector were a single stand-alone business, its revenues would nearly equal those of Ford Motor Company, General Electric, and General Motors combined.[15]

Electricity production matters to climate change because it accounts for the biggest single share of global carbon dioxide emissions: about 25 percent.[16] Furthermore, countries that have vibrant electric sectors—places where electricity is abundant and reliable—are leading the global economy. Countries that are hindered by expensive, intermittent power are being left behind. The nineteenth century was the age of coal and steam. The twentieth century was dominated by oil and engines. The twenty-first century is about electrons and bits. Big data, robotics, and artificial intelligence are the hottest technologies of the moment, and all of them depend on electricity.

In the pages ahead, I will look at the world through the lens of electricity. My lens will be wide-angle. I will look at everything from how electricity improves the lives of women and girls to the enormous amount of electricity used by the marijuana business to the mechanics of creating, fueling, and maintaining a functioning electric grid.

In looking at the world through the lens of electricity, I seek answers to several questions, including: Why are countries like the United States, Germany, and France electricity rich, while billions of people around the world are still stuck in the dark? Which industries are showing the biggest growth in electricity demand? How secure is the electric grid? Which fuels will be used to meet future electricity demand, and how will that demand growth affect the efforts to fight climate change? I will share insights from the journey I took to answer these questions—a journey that brought me to India, Lebanon, Iceland, Puerto Rico, New York, and Colorado and involved discussions with dozens of people, including engineers, politicians, activists, academics, and authors, as well as Bitcoin miners, cab drivers, cannabis growers, and others whose lives are shaped by their access, or lack of access, to electricity.

In the first section, I will show why electricity means modernity. To do that, I will take you on a quick jog through Electricity 101, so that you can tell your watts from your watt-hours. I'll then explain why electricity has had such a transformative effect on humanity and, in particular, for women and girls. I will travel back in time to the early days of the Electric Age to show how electricity changed the shape and height of our cities and the lives of farmers. I'll introduce you to the small group of New Dealers who liberated electricity from the grip of self-interested trusts, passed the legislation that assured rural electrification, and thus set the stage for the economic boom that assured America's emergence as an economic superpower.

In the second section, I will illuminate the vast disparity in electricity use around the world today and explain why so many people are stuck in energy poverty, with implications for human rights, economic and cultural development, military strategy, and geopolitics. I will then show what various societies and countries are doing to get the electricity they need and discuss the hard reality about electrification: when forced to choose between energy poverty and access to electricity, consumers and policymakers will always choose electricity, and they will always make it as cheaply as they can so they can provide it to the greatest number of people, regardless of the environmental impacts.

In the third section, I will focus on the electricity rich, to show how and why electricity demand continues to increase, as well as the growing interdependence of electricity, information, money, and the economy. I'll also examine the dark side of this development: an increased vulnerability to a shutdown of the grid, whether the culprit is squirrels, hackers, or nuclear devices.

Finally, I will look at the future of electricity and discuss how electricity demand in both rich and poor nations is likely to be met. Over the next few decades, global electricity generation will double. The electric grids that will be built over the next twenty to thirty years will have significant impacts on global prosperity and on efforts to address climate change. I will explain why renewables alone cannot meet soaring global electricity demand. I will explore the most promising nuclear-energy technologies, discuss why solar, natural gas, and nuclear will play prominent roles, and explain why I continue to be idiotically optimistic about the future of our high-energy world.

Energy politics are tribal. Everyone, it seems, has their favorite. Me, I'm a proponent of what I call N2N, or natural gas to nuclear. Some people say we will need more coal, while others tout geothermal, hydro, wind, and solar. The hard reality is that there are no quick or easy solutions. Energy transitions take decades.[17] Sure, we can desire decisive action on climate change. We can want more rights for women and push for an end to global poverty. But we must be discerning. My hope is that this book, by showing you how the world looks through the lens of electricity, will help you see energy and power systems as they are, not how you may want them to be. We have to separate the glib rhetoric that dominates many of today's energy discussions from the reality. Only then can we understand the stakes and consequences of our energy policies, as well as the fuels and technologies that will help bring more people out of the dark and into the bright lights of modernity.

Before delving into all of those issues, though, it's important to take a few minutes to understand what electricity is, why it's so difficult to supply reliably, and why it has been so transformative.

PART ONE
Electricity Means Modernity

1

ELECTRICITY 101

> I've found out so much about electricity that I've reached the point where I understand nothing and can explain nothing.
>
> —**PIETER VAN MUSSCHENBROEK**, Dutch scientist

Electricity lingo pervades our everyday speech. We want to get amped up, flash high-wattage smiles, and deliver electrifying orgasms. We idolize human dynamos who can produce high-voltage performances. We get wired until we blow a fuse. After that, we unplug and recharge.

We've electrified our vernacular for a simple reason: human history can be divided into two epochs, the Electric Age, and everything that came before it. Sure, the Renaissance gave us Michelangelo. Electricity gave us Elvis.

Electricity means modernity. While we have grown accustomed to having cameras that can take high-definition video on our mobile phones, it's easy to forget just how short the Electric Age is when compared to the rest of human history. Archeological records show that humans (or rather, our hominid ancestors) first used fire about one million years ago, but it didn't become common until about 400,000 years ago.[1] By contrast, we have only been putting electricity to work since the 1880s. Therefore, if we could compress the 400,000 years that humans have been using fire into one twenty-four-hour period, the Electric Age would span only the last thirty seconds before midnight.[2]

Electricity means modernity because we are harnessing forces we can't see or feel. For millennia, we could only corral energy from things like wood, dung, coal, oil, rivers, horses, the sun, and wind. With electricity, we are exploiting energy forces invisible to the eye with stunning precision and ever-greater efficiency. Over the past century and a half, we've gone from harnessing animals—and enduring all the shit they shat—to harnessing the subatomic motion of electrons. The more we can control flows of electrons, the more work we can do. The more work we do, the more work we want to do. And here's the really good news: we are getting better and better at wringing more work out of those electrons.

To get an idea of the staggering number of electrons we are harnessing, consider this: making a single cup of tea with an electric kettle requires about 4.9 sextillion electrons.[3] In scientific notation that is 4.9×10^{21}. When typed out, 4.9 sextillion looks like this: 4,900,000,000,000,000,000,000. And remember, that's what's required for a single cup of tea. Running an air conditioner and a full-size refrigerator will require adding a grocery bag filled with zeros to that number. If you are planning to energize a skyscraper, or fire up an electric-arc furnace to cook up a batch of steel I-beams, you'll need a couple shipping containers loaded with zeros to type out the number of electrons you'll be using.

While we can calculate the number of electrons needed to make a cup of tea, it's still hard to grasp the thingness of electricity. It's a force that propels our lives while being both ubiquitous and invisible.

Benjamin Franklin, the publisher, writer, diplomat, and raconteur, pioneered our understanding of electricity. In 1752, he conducted his famous kite experiment, which featured a piece of metal attached to the top of a kite and a metal key tied to the earthbound end of the string. The key, in turn, was connected to a Leyden jar, which was a primitive type of battery. Franklin controlled the kite with a dry piece of silk fabric, which insulated him from being shocked. Franklin's experiment proved that the lightning in the sky was the same as the static electricity that could be obtained by rubbing amber with fabric. Franklin's work provided the foundation for the other great electricity pioneers and entrepreneurs who followed him. Philadelphia's founding father coined a spate of electricity terms, including battery, charge, conductor, and condenser. He was also one of the authors and signatories of the Declaration of Independence, and he called electricity a "common element," which he termed "electric fire."[4] He also thought electricity was a fluid that flowed from one body to another.

Electricity is not a fluid. But given the ineffability and complexities of electricity, it helps to think of it in that way. Franklin's take on electricity became known as the single-fluid theory. He held that objects having a negative charge lost electrical fluid and those with a positive charge gained it. If an object lost or gained electrical fluid, and thereby became unbalanced, it would become charged. Objects with similar charges repelled each other. Keeping that electricity-as-a-fluid analogy in your head, imagine that the electricity in your house is being delivered through a garden hose. To further grasp the analogy, it will help if you understand this simple equation:

$$\text{watts} = \text{current} \times \text{voltage} = \text{amps} \times \text{volts}$$

In other words, the number of watts delivered is the product of amperes multiplied by volts. Now consider the amount of electricity coming into your house through that same garden hose. The amount of electric power (which is measured in watts, but in this case think liters) that can be pushed through that hose is the product of the current, or flow rate (amperes) multiplied by the water pressure (voltage).[5] The more pressure applied to the water in the garden hose, the greater the flow rate that can be pushed through it. The higher the pressure and flow rate, the more liters of water (watts) get delivered to your house.

To bring this analogy home, let's assume your house has caught fire. You immediately call the fire department, because you want to save your Cézanne paintings and Beanie Baby collection from the growing inferno. But instead of using fire hydrants, high-pressure pumps, and large-diameter hoses to douse the blaze, the firefighters try to extinguish the flames with a pair of leaky garden hoses that they've attached to spigots on your neighbor's house. The firefighters wouldn't have much firefighting capability. Why not? The number of liters (watts) of water they could put on the raging fire would be restricted by a low-pressure (voltage) garden hose that was delivering liquid at a low rate (amperage).[6]

The water analogy also helps when thinking about the generation of electricity. Just as the local water utility uses its pumps to deliver tons of water at high pressure and volume to its customers, the electric utility uses spinning generators—think of them as electron pumps—to push huge volumes of electrons, at high pressure, into the local grid. The key difference between the water grid and the electric grid is that the water grid is far simpler. For instance, if the pressure in the water grid drops, it only means that customers must spend a little more time filling up their coffee pots or swimming pools. On the electric grid, voltage (again, think water pressure) must be kept stable regardless of how many customers are using electricity. Further, that voltage must be kept steady day and night, 24/7, regardless of whether a customer needs a few watts for lighting or hundreds of megawatts to coax aluminum out of bauxite. The grid must be continually tuned so that electricity production and electricity usage match. Matching generation and consumption helps assure that voltage on the grid stays at near-constant levels. If voltage fluctuates too much, blackouts can occur.

Electricity means modernity because—as my son Michael, a math and computer whiz, put it—it's at the core of all modern networks. We live in a digital world that's defined by networks. And all of those networks—telephones, global positioning systems, airline reservation systems, traffic lights, the list is endless—depend on electricity. In short, the network is the electric grid and the grid is the network. If you are lucky enough to be connected to an electric grid, you can connect to the digital information network.

Thanks to electricity, Samuel Morse's telegraph enabled near-instant communication between distant locations. In 1866, a telegraph cable was laid under the Atlantic, thereby creating the first continuous line of communication between the United States and Europe. A decade later, Alexander Graham Bell snared a patent for the telephone. Unlike the telegraph, which required an operator to send and receive messages, anyone could operate a telephone, and those phones began connecting businesses and homes to each other. In his 2011 book, *The Information*, James Gleick writes that the telephone and telegraph "ripped the social fabric and reconnected it, added gateways and junctions where there had only been blank distance." The telegraph and telephone, Gleick explains, began "to turn human society, for the first time, into something like a coherent organism"—and that coherence was only made possible by electricity.[7]

Electricity is the apex predator of the energy kingdom. We convert lots of primary sources of energy—coal, natural gas, oil, biomass, sunlight, wind, water, and nuclear reactions—into electricity, which is a secondary form of energy. Other forms of secondary energy include gasoline, which must be refined from crude oil, and hydrogen, which is derived from natural gas. (Hydrogen can also be produced from water, but splitting water molecules requires large amounts of energy.)

The reason we convert so many fuels into electricity is that it is the most useful form of energy. Among its many wonderful properties, it has no inertia. That means it doesn't have to warm up. It can provide full power in a split second and be turned off just as quickly. Electricity allows us to harness the motion of electrons. We can generate flows of electrons from both kinetic energy and potential energy, and we can make those forms of energy switch places.[8] That is, we can convert potential energy into kinetic energy and vice versa. An obvious example of that: electric energy can be used to charge a battery that contains chemical energy. We can then use the chemical energy in that battery and convert it back into electric energy whenever we need to make a phone call or order a bag of Osmocote from Amazon.

While electricity can be generated in many different ways, and harnessed in even more, it comes with significant downsides. Electricity is persnickety. It must be consumed at almost the same instant it is generated. That differentiates it from wood, coal, oil, and natural gas, all of which can be stored relatively easily. Of course, we can store modest amounts of electricity. The batteries that energize the phones we carry in our pockets hold a few watt-hours of electricity. And that tiny quantity of energy is enough for us to text our pals, map a route in the car, and chat on the phone. But economically storing large quantities of electricity—enough to power a city for a day or more—remains beyond the reach of current technology. In fact, if you were somehow able to collect all of the world's automobile batteries, charge them up, and link them together, they would be able to hold only enough electricity to power the globe for less than thirty minutes.[9]

Although we can't store electricity in large quantities, our ability to store and manipulate relatively small flows of it has been transformative. Batteries allow us to have something that is truly new: lightning in a bottle. Armed with ever-better chemistries and metallurgies, companies all over the world are producing a staggering array of batteries that vary from pacemaker units that get implanted inside the human body to vanadium flow batteries that require thousands of liters of liquid chemical to be stored in giant tanks.

Now that we have a better understanding of what electricity is, we must differentiate between two terms: energy and power. They are commonly confused. They are *not* the same thing. Energy is the ability to do work. It is measured in joules (J), watt-hours (Wh), or British thermal units (Btu). Power is the rate at which work gets done. It is measured in watts (W) or horsepower.[10] The equation for power is simple: 1 joule per second equals 1 watt. Which looks like this:

$$1\ \mathrm{J/s} = 1\ \mathrm{W}$$

Another way to think of these terms is to remember that energy is a quantity, such as a liter of oil or a ton of coal. Power is a rate—that is, it's a measure of energy flow over a given period of time. A helpful way to understand the difference between energy and power is to recall the generator that Wilfredo Roque and Iris Ortiz relied on to power their home in Barrio Antón Ruíz. Their machine had a power rating of 4,500 watts, meaning at full capacity it could produce 4,500 watts of *power*. If Wilfredo runs it for one hour, it will produce 4,500 watt-hours (4.5 kilowatt-hours) of *energy*.

Finally, a quick primer on the International System of Units, or SI (the acronym for Système International), which specifies symbols for units and for the numbers that represent multiples and submultiples of those units. When looking at the scale of electricity generation and usage, it's helpful to remember a few SI prefixes, including this sequence: KMGT.

That's short for kilo, mega, giga, and tera. Those are the prefixes for units of power and energy in, respectively, thousands, millions, billions, and trillions. Thus, you will see references to power in kilowatts, megawatts, gigawatts, and terawatts. You'll see those same units expressed in energy terms: kilowatt-hours, megawatt-hours, gigawatt-hours, and terawatt-hours. Be not afraid. To put those prefixes into context, remember that we use electricity in our homes at the kilowatt scale: your hair dryer uses about 1,800 watts or 1.8 kilowatts. Electricity demand for a small town will likely be measured in megawatts. At the big-city level, demand is often measured in gigawatts, and at the country level, in terawatts. For instance, the United States is energized by an electric grid that has a total installed generation capacity of about 1 terawatt, or 1 trillion watts.[11]

Okay. That's the end of Electricity 101. Now that we have a better understanding of what electricity is, let's answer the second question: Why has it been so transformative?

2

THE TRANSFORMATIVE POWER OF ELECTRICITY

> We will make electricity so cheap that only the rich will burn candles.
>
> —THOMAS EDISON

There are three reasons why electricity has led to such profound human flourishing: lighting, power, and density.[1] Electricity made lighting cheap, abundant, and reliable, which fundamentally changed how people could spend their days and nights. Electricity provides instant power, which transformed everything from manufacturing to urban transportation. Finally, electricity gives us the ability to concentrate energy flows like never before. Those highly concentrated energy flows have shaped everything from the height of our cities to the productivity of our factories and microprocessors.

Let's look at lighting first. Electricity has allowed us to slay one of our oldest foes: darkness. For millennia, the cost of having well-lit spaces at night was so high that only the very rich could afford it. That meant that the poor were usually left in the dark, where their fears of the dark—and all the evil things that inhabited the darkness—could be preyed upon by mystics, priests, and shamans.

If you wanted to read or work after dark, the choices for illumination were few: fireplaces, torches, lanterns, or candles—all of which involved burning something.

For centuries, as soon as the sun went down, people had no choice but to lock themselves in their homes. In 1380, a decree required that residents of Paris be off the streets, and "at night, all houses are to be locked and the keys deposited with the magistrate. Nobody may then enter or leave a house unless he can give the magistrate a good reason for doing so." About that same time, in England, anyone who was walking on the street after dark was considered a suspect and was subject to immediate arrest. In 1467, a decree in England made clear that "no man walke after IX of the belle streken in the nyght withoute lyght or without cause reasonable in payne of empresonment." Furthermore, as recounted by author Wolfgang Schivelbusch in his book *Disenchanted Night*, "in big cities like Berlin and Vienna, similar regulations remained in force until well into the nineteenth century."[2]

The industrialization of lighting—that is, the key breakthrough that dramatically lowered its cost and improved its availability—began in the early 1800s, when municipalities, factories, and homes began using methane derived from coal, which was first used to fuel streetlights and then indoor lights. Known as "town gas," it would play a major role throughout the nineteenth century and well into the twentieth century. Town gas was produced by heating coal to a high temperature. The gas was then captured, stored, and delivered through often-leaky pipelines to customers. Gas lighting gained customers quickly because it was cheaper than lanterns or candles. By 1822, London was leading the world in gas lighting, with four companies operating a total of 200 miles (322 kilometers) of pipeline. Over the next few decades, municipal gas systems became common in Britain, France, and Germany. But as gas lighting spread, so, too, did complaints about it. According to one account, gas lighting "consumed enormous quantities of oxygen and often raised the temperature in closed rooms to tropical levels."[3] The author Edgar Allen Poe wrote that gas lighting is "totally inadmissible within doors. Its harsh and unsteady light offends." In

1878, a British publication on proper homemaking declared that "few have felt the overpowering and sickening influence of a room liberally lighted by gas, and closely shut up, as frequently rooms are, at the time when gas is most required." It continues, saying gas lighting is "equally injurious to decorations, be they pictures, papers, ceilings, or hangings quickly making them dingy and dirty."[4]

Gas lighting had other drawbacks. Each light had to be lit every night by hand. The glass fixtures that contained the gas lights had to be cleaned regularly, because the smoke produced by the flame often left carbon deposits. The systems that provided the fuel were also notoriously dangerous. Town gas was often stored in aboveground tanks called gasometers. In 1865, a gasometer in London exploded, killing ten workers. The accident led the *Times* newspaper to write that gasometers were a public health danger and that "those who live near them and the buildings in their neighborhood, are exposed to as serious consequences as if they were placed over a powder magazine."[5]

Lamps fueled by kerosene and whale oil were also widely used. But, like gas lights, they were hot and depended on a live flame, which meant an ever-present danger of fire. A dropped lantern or unattended candle could result in a fire that could burn down your house, or much of a city. The Great Chicago Fire of 1871 killed about three hundred people and destroyed more than three square miles (eight square kilometers) of the city. The inferno, which in popular lore was started when Mrs. O'Leary's cow kicked over a lantern, left 100,000 people homeless.[6] Arc lights also enjoyed a period of popularity prior to widespread electrification. Arc lights produce illumination by sending an arc between two carbon electrodes heated to a white-hot temperature. Able to illuminate large areas, arc lights were used for lighting streets and outdoor areas.[7] But due to their intense heat output, they were impractical for indoor use.

Thomas Edison's incandescent lamps offered a safer form of lighting than anything that had come before. If gas lights failed, or leaked, they could fill a room with combustible fumes that could explode. For that reason, some of the earliest businesses to adopt electric lights were flour mills, textile factories, and other facilities that handled flammable materials.[8] Edison's electric lights were also more agreeable to consumers. The first reviews of his lighting system were all-out raves. The *New York Times*, which was one of Edison's first customers, reported that Edison's new lights were "as thoroughly tested last evening as any light could be tested in a single evening, and tested by men who have battered their eyes sufficiently by years of night work to know the good and bad points of a lamp, and the decision was definitely in favor of the Edison electric lamp, as against gas."[9] The *Times* also reported, "You turn the thumbscrew and the light is there, with no nauseous smell, no flicker, no glare . . . more brilliant than gas and a hundred times steadier."[10]

Those positive reviews stoked demand. On the first day of operation, the Pearl Street plant was energizing 1,284 of Edison's incandescent lights. By the end of 1882, that number had nearly tripled. By October 1883, his central station had 508 customers who were using 10,164 lamps.[11] As electricity systems proliferated, the cost of lighting declined. According to statistician Max Roser of *Our World in Data*, in 1880, the year after Thomas Edison devised a workable electric light at his laboratory in Menlo Park, New Jersey, the price of lighting was about 530 British pounds per million lumen-hours. By 1900, the price had dropped to 236 British pounds. By 2000, it had dropped to less than 3 British pounds, or roughly 176 times cheaper than it was in 1880.[12] Today, lighting has become so cheap we scarcely think about it. Every year, Americans use nearly 7 terawatt-hours of electricity solely for Christmas decorations.[13] That's about as much as the entire annual electricity consumption of countries like Albania and Latvia.[14]

We now use many types of lighting, including halogen, incandescent, light-emitting diode (LED), high-pressure sodium, metal halide, and ceramic-metal halide. Each has advantages and disadvantages. The key is that all of them are readily available and affordable. We convert watts into lumens without a second thought. Nevertheless, by conquering darkness, by slashing the cost of lighting, electrification fundamentally changed the course of human history.

The second reason for electricity's transformative effect is that it provides instant power for nearly any purpose: communication, computation, heating, lighting, and motive power. Electric power allows us to attain precision—in both speed and control—that cannot be achieved with other forms of energy, and it is convertible into work at very high efficiency with no smoke or odor.

Throughout all of history, human beings have been trying to harness energy so they can do more work. Whether the harness was an actual leather and rope contraption attached to a pair of oxen pulling a plow, or a steam-engine-driven water pump draining the water from a coal mine, the aim has been the same: to get more energy applied to a given task so that more work gets done faster and cheaper. For millennia, the only sources of power were what could be obtained from human muscle, draft animals, wind energy, biomass (derived from the sun), and waterwheels.

James Watt changed how humans obtain power by improving the steam engine. Watt, a Scottish instrument maker, tinkered with a design pioneered by Thomas Newcomen. Watt estimated that Newcomen's engine wasted about 95 percent of the fuel it used. By improving the efficiency of the steam engine through the addition of an external condenser, Watt liberated industry from the geographical constraints of rivers and streams. Watt's engines revolutionized both industry and transportation.[15] A typical Boulton & Watt steam engine from the early nineteenth century was capable of producing about 24 horsepower (18 kilowatts).[16] But unlike a horse, the steam engine could be worked around the clock, produced no manure, and didn't require grain, meaning it didn't compete with humans for food or available farmland. Watt's steam engine ignited the Industrial Revolution, and he is remembered today because his name is the metric we use for power. By 1905, the steel magnate Andrew Carnegie estimated that the world had about 150 million horsepower (110 gigawatts) of steam capacity at work.[17]

Throughout the nineteenth and twentieth centuries, the steam engine was improved and improved some more. Those improvements played a direct role in the Electric Age, as steam engines (and later, steam turbines) were used to drive generators to produce electricity. Today, the majority of the world's electricity-generation plants are still using high-pressure steam to produce electricity. We also generate electricity with internal-combustion engines, geothermal plants, hydropower dams, wind turbines, and solar panels. Continuing improvements have helped us generate more and more electricity and in ever more efficient ways. By converting increasing amounts of primary energy into electric power, humans have thrived like never before. Humans are flourishing because, with electricity, they have access to nearly infinite amounts of power that can be applied to nearly any kind of work.

Now to the third point: density. Electricity allows us to concentrate energy flows in unprecedented ways. We can concentrate those flows because the electricity we consume is highly ordered energy.[18] We convert primary energy sources into electricity and then distribute that energy over wires in carefully calibrated doses of voltage and amperage. Those highly ordered flows of electrons mean we can, in effect, stack them in ultradense packages. This allows us to concentrate and harness the energy of those moving electrons in far greater quantities—and with far greater precision—than could ever be achieved by using wood, steam, or the crankshaft of an internal-combustion engine.

Concentrating energy flows allows us to do work and therefore create wealth. Regardless of the work to be done—repairing a cornea with a laser, crunching data with a computer, or growing marijuana—we need dense flows of energy. Electricity is the ultimate energy concentrator. It allows us to put massive flows of energy in very small spaces.

In physics lingo, electricity allows us to boost power density, which is the amount of energy flow that can be produced or harnessed in a given volume, area, or mass.[19] Examples of power density include watts per square meter, watts per liter, and watts per kilogram. Power density allows us to have a common denominator that we can use to compare energy-system outputs and inputs across centuries and industries. Vaclav Smil, the author, polymath, and energy analyst, says that power density is a "key analytical variable to evaluate all important biospheric and anthropogenic energy flows."[20] In his 2015 book *Power Density*, Smil explains that, for millennia, humans existed on the ragged edge of starvation and disease because they had to scrape by on the meager energy flows that could be obtained from farming, which relies on turning sunlight into usable biomass (grain, wood, or fodder), which is then consumed by humans or animals to do work. Smil concluded that, no matter whether you are growing corn, which humans have been doing for about 9,000 years, or planting trees for use as fuel for the stove, the power density of farming is limited to roughly 1 watt per square meter.[21] Not only is that a tiny level of power density, the productivity of any given farm is continually under threat due to lack of rain or too much. It's also vulnerable to insects, high winds, thieves, and wildlife.

The Industrial Revolution allowed us to escape the meager energy budgets of farming. By using hydrocarbons (at first coal, then later oil and natural gas) humans were able to harness ever-increasing quantities of power and do so in ever-denser packages. In place of animal power, sun power, and wind power, factories began using advanced waterwheels and coal-fired steam engines. In the 1820s, the Merrimack Manufacturing Company, a major clothing producer, began churning out fabric on the banks of the Merrimack River in Lowell, Massachusetts. The factory, which produced calico and other fabrics, was powered by a waterwheel that gave it a power density of about 20 watts per square meter.[22] That rate was a big improvement over what could be obtained with a draft animal, but it was only a foreshadowing of the densities that would come with steam engines and electrification.

For instance, by the late 1920s, manufacturing at Ford Motor Company's River Rouge plant in Dearborn, Michigan, depended on electricity produced by a 315-megawatt on-site generation plant that burned two hundred tons of coal per hour.[23] Smil calculates that River Rouge had a power density of about 1,000 watts per square meter, fifty times higher than what was obtained by the waterwheel at the Merrimack Manufacturing Company a century earlier.[24] Electric power allowed Ford's factories to operate drills and other precision equipment at speeds that were unimaginable on the old pulley-driven systems, which relied on shafts, belts, or chains that were driven by waterwheels or steam engines. At its peak, River Rouge employed more than 100,000 workers and was turning out a new car every forty-nine seconds.[25] Those enormous production levels were only possible because electricity allowed the components of Ford's assembly line to be arranged for optimum output rather than their proximity to a steam engine or waterwheel. As Ford put it, electricity "emancipated industry from the leather belt and the line shaft."[26]

The power density we get thanks to electricity has fostered our ongoing migration into cities. We live in an urban-majority world today, and all of the people who live in those cities depend on electricity to fuel their everyday lives. The importance of power density can be seen by looking at Rockefeller Center. The iconic cluster of high-rise buildings in Midtown Manhattan covers a surface area of roughly 103,600 square meters. According to a 1999 report from Consolidated Edison, the utility that serves central Manhattan, the electric load from the buildings in Rockefeller Center is about 93 megawatts.[27] Therefore, the average power density across all of the buildings in Rockefeller Center is nearly 900 watts per square meter. That level of power density could never have been achieved with teams of horses or a steam engine. It could only be done with the highly ordered energy we get from electricity.

The importance of power density that is evident in the skyscrapers at Rockefeller Center can also be seen in the chips that run our computers. For instance, one of Advanced Micro Devices' latest microprocessors, the Phenom II X940, has a footprint of 12.25 square centimeters and draws 45 watts of power. If we supersize that microprocessor to a full meter, it would have a power density of 3,672 watts per square meter, nearly four times the power density found in Ford's River Rouge plant.

Lasers are among the best examples of how the power density we get from electricity allows us to do things that could never be achieved with other forms of energy. For instance, scientists like Canadian physicist Paul Corkum routinely use lasers that achieve power densities of 10^{18} watts per square meter.[28] Written out, that looks like this: 1,000,000,000,000,000,000 watts. Corkum, who works in the field of high-harmonic interferometry and high-harmonic spectroscopy, uses that incredibly high power density to create ultrashort light pulses that allow him to photograph electrons. Of course, we now use lasers for all kinds of things, including surgery, tattoo removal, and fiber optics.[29]

In 2000, the National Academy of Engineering chose the twenty greatest engineering achievements of the twentieth century. Electrification ranked first. Not only that, but of the top twenty achievements, thirteen are directly dependent on electrification, including electronics, computers, and air conditioning, as well as health technologies, lasers, and household appliances.[30] It could easily be argued that every item on the list, including automobiles, airplanes, water supplies, and agricultural mechanization, are also dependent on electrification. And here's the key thing: all of those technologies have helped humans live longer, freer, and richer lives.

TABLE 1

THE NATIONAL ACADEMY OF ENGINEERING'S LIST OF THE TOP TWENTY ENGINEERING ACHIEVEMENTS OF THE TWENTIETH CENTURY

Electrification	Spacecraft
Automobile	Internet
Airplane	Imaging
Water Supply and Distribution	Household Appliances
Electronics	Health Technologies
Radio and Television	Petroleum and Petrochemical Technologies
Agricultural Mechanization	Laser and Fiber Optics
Computers	Nuclear Technologies
Telephone	High-Performance Materials
Air Conditioning and Refrigeration	
Highways	

The transformative power of electricity has been the subject of numerous academic papers. In 2014, two Turkish researchers, Yilmaz Bayar and Hasan Alp Özel, performed an analysis of about two dozen published papers on electricity and economic growth. They found "unidirectional causality between electricity consumption and economic growth." That is, electricity use drives economic growth.[31]

While electricity drives economies, it is also clear that greater wealth increases electricity consumption. That makes sense. As people get wealthier, they consume more electricity because they can afford more electrical devices. For instance, a person living in a newly electrified home may first buy a refrigerator and some lights. After that, they will want an air conditioner and perhaps an electric stove. This bidirectional effect of wealth and electrification was discussed by energy writer Roger Andrews in a 2015 study in which he concluded that in developing countries "wealth creates electricity and not the other way round. There is no question, however, that once a country gains wealth it cannot sustain it without electricity. When the electricity disappears the wealth goes with it."[32]

Another paper, published in 2010 by two academics at the University of Karachi, examined the "causal relationship between energy consumption and economic growth" in Bangladesh, India, and Pakistan. The analysis studied data from between 1971 and 2008. While the analysis did not focus specifically on electricity, the conclusion of the authors, Kashif Imran and Masood Mashkoor Siddiqui, was clear: "Energy serves as an engine of economic growth and economic activity will be affected in the result of changes in energy consumption. . . . GDP is basically determined by energy."[33]

The close correlation between electricity use and human health and economic growth has become so obvious that international investment bankers have adopted electricity use as a measure of economic activity. In the late 2000s, when China's economy was faltering, equity analysts at JPMorgan Chase and other investment banks used China's electricity production data as a proxy metric for industrial output.[34] What is true for China is also true for the world at large. The correlation between electricity use and economic growth can be seen in the chart on page 21, which shows that the two move in near unison and they've been dancing cheek-to-cheek for decades.

Another way to look at how electricity use and wealth are intertwined is to look at nighttime luminosity—that is, the amount of light emitted by a region at night. In 2010, William Nordhaus, an economist at Yale University, published a paper that found that nighttime luminosity—which was measured using images captured by satellites orbiting the earth—is closely correlated with personal incomes. Nordhaus, who won the 2018 Nobel Prize in Economics, determined that luminosity isn't particularly effective for analyzing wealthy countries. But it is useful in analyzing wealth in developing countries where traditional statistical information is not readily available.[35] In 2012, a similar technique was used by three researchers from the National Bureau of Economic Research, who published a paper titled "Measuring Economic Growth from Outer Space." That paper concluded that nighttime luminosity provides "a very useful proxy for GDP growth over the long term and also tracks short-term fluctuations in growth."[36]

Of course, you don't have to look at satellite images to see the correlation between electricity use and wealth. It can be seen in the World Bank's consumption numbers. In 2014, Iceland, Norway, Bahrain, Canada, and Qatar had the world's highest per-capita electricity use, and all five were among the world's wealthiest countries. Conversely, places where electricity consumption is extremely low—such as Haiti, Gaza, South Sudan, Niger, Ethiopia, and Tanzania—are among the poorest.[37]

The punch line here is obvious: increased electricity use fosters economic growth, which, in turn, means better living conditions for humans. Electricity use provides a reliable barometer for the health and wealth, or poverty, of individuals and societies. The International Energy Agency (IEA) has called electricity "crucial to human development" and said that electricity use is "one of the most clear and undistorted indications of a country's energy poverty status." Put another way, electricity bolsters economic growth and economic growth bolsters electricity demand. Together, those things help people escape poverty. As Paul Collier, the author of *The Bottom Billion: Why the Poorest Countries Are Failing and What Can Be Done About It*, famously put it, "Growth is not a cure-all, but lack of growth is a kill-all."[38]

Perhaps the easiest way to understand why electricity has been so transformative is to look at one of humanity's greatest inventions: the city.[39]

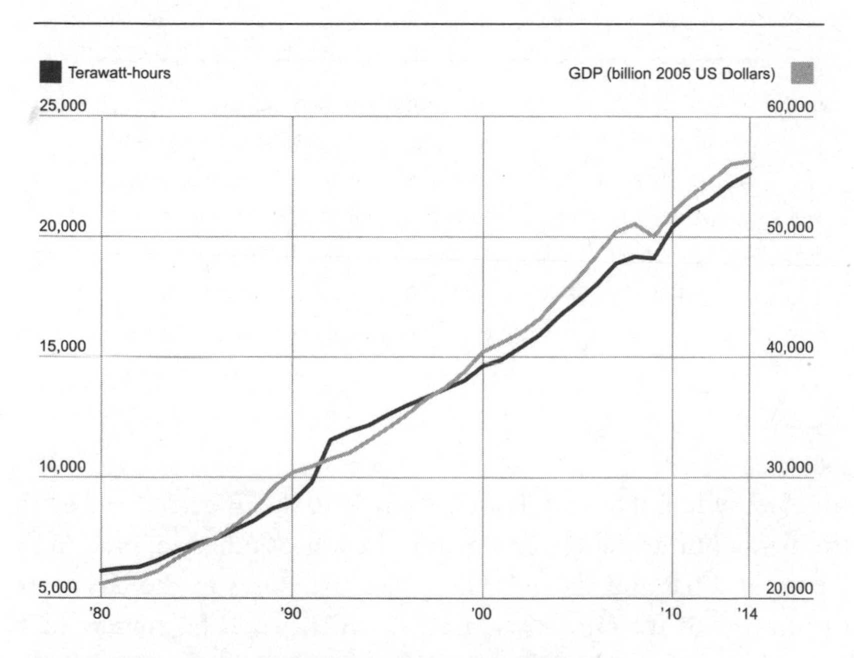

Global GDP and Global Electricity Use: 1980 to 2014

Sources: HumanProgress.org, Energy Information Administration

3

THE VERTICAL CITY

> At last the town was lighted and we had ocular evidence of our success. We made a gala night of it. The streets and stores were crowded with people, the big 150-candle-power lamps were running at about double their candle-power, and my townsmen, though very skeptical as to the dangers to be encountered when going near the lights, rejoiced with me.
>
> —**WILLIAM STANLEY,** electricity industry pioneer[1]

In 1882, when Thomas Edison launched the Electric Age with the first commercial electric grid in Lower Manhattan, New York City was squat and squalid. The tallest structures in the city were the towers of the Brooklyn Bridge, which stood 84 meters (276 feet) high. The great bridge, designed by the engineer John Roebling, had been under construction for thirteen years and was finally nearing completion.[2] Edison and his employees were likely monitoring the work on the landmark bridge, which was being erected just a few blocks east of Edison's generating station at 255-257 Pearl Street.

New York's streets, as usual, reeked of animal feces and urine. Horses and horse-drawn wagons and streetcars dominated the thoroughfares. The air was usually fouled with smoke from the plethora of coal-fired steam engines that were providing power to factories, print shops, and other industries. Inside the city's homes, offices, and factories, gas lighting—and all the heat and noxious fumes that came with it—dominated. Locations that lacked a gas connection relied on kerosene lamps.

The first skyscraper in the city was the Equitable Life Assurance Building at 120 Broadway.[3] Located about half a mile west of Edison's power plant, the Equitable Building stood 40 meters (130 feet) high and was nearly twice the height of all previous business buildings. Completed in 1870, it was billed as a fireproof building. It was constructed with iron framing and had ten floors that were served by five steam-powered elevators. Tourists flocked to the building to ride the elevators and visit the rooftop observation deck, which also served as the site of New York City's weather bureau.[4] New Yorkers marveled at the Equitable Building because nearly all of them lived and worked in buildings that were only half as tall, if that.

For nearly all of human civilization, the height of buildings was limited by people's willingness to climb stairs. Any building taller than four, five, or six stories was impractical. Walking up two or three flights of stairs isn't terrible. Carrying a load of groceries and a screaming infant up four or five flights of steep, dark stairs, is, pardon the pun, another story.

By electrifying part of Lower Manhattan, Edison ignited the rise of the vertical city. Indeed, the Electric Age and the skyscraper were birthed at about the same time, in about the same place: an area of Lower Manhattan that covers about one square mile.[5] There, Edison, Nikola Tesla, and George Westinghouse converged to pioneer the shape and components of the modern electric grid. But a lesser-known inventor—a man who, like Tesla, briefly worked for Edison—would also leave an indelible mark on the urban environment. His name was Frank Julian Sprague.

Sprague doesn't have the name recognition of Tesla, the Serbian-born genius who invented the alternating-current (AC) motor. Nor is Sprague as famous as Westinghouse, the ingenious industrialist who commercialized the transformer, the device that allows electricity producers to boost (or decrease) the voltage carried on a given electricity line.[6] Tesla, Westinghouse, and Edison get plenty of well-deserved attention. But it was Sprague who developed the first electric motors that ran on Edison's grid. Sprague would go on to electrify urban transportation by putting his motors on streetcars, subways, and railroads. It was at 253 Broadway, about half a mile northwest of Edison's original power plant on Pearl Street, that Sprague deployed the first set of electric elevators, a technology that would radically reshape our cities. By pioneering electric motors, electric railways, and electric elevators, Sprague fundamentally changed how and where humans live.

"Electricity is why we have modern cities," says Jesse Ausubel, the director of the Program for the Human Environment at Rockefeller University in New York City. Ausubel, a lifelong New Yorker, summed up the importance of power by telling me that electrification "completely transformed the structure, the geography, the geometry of cities." He went on, "Rome, 2,000 years ago, had a million people. But essentially cities remained the same for hundreds, thousands of years." The limit of all the old great cities, "whether it was Beijing or Baghdad or Rome or Cairo," was about one million people. After electrification, he said, "suddenly we were able to move into three dimensions." With electricity, we can "have these cities not just of one million, but of two, and three, and five, and ten million people."

Ausubel continued, "Basically, height is electrical." In other words, the taller the building, the more restricted we are in the types of energy that can be used. In a one-story building, Ausubel said, "it's not much of a problem to bring some hay, or a cord of wood, or whatever, to create heat or other forms of energy. However, if you have a ten-story building, wood and hay don't work very well."

Numerous factors fostered the rise of the vertical city. But it was Sprague's daring, drive, and ingenuity that turned the promise of electricity-fueled transportation into a reality.

•

Born on July 25, 1857, in Milford, Connecticut, Frank Julian Sprague knew hardship from an early age. His mother died when he was eight years old. A few months after her death, Frank and his brother, Charles, were abandoned by their father. They were taken in by their aunt, Elvira Betsy Ann Sprague, who lived in North Adams, Massachusetts. In 1874, Frank was admitted to the Naval Academy, which was then considered one of the best universities in the country. Midshipmen were schooled in a wide variety of topics, including navigation and geometry. As one Sprague biographer put it, at the Naval Academy students were "taught to think concretely, abstractly, and above all, systematically." Those skills would be critical to Sprague's success. He graduated in 1878, ranking seventh out of his class of fifty, earning honors in math, physics, and chemistry. He later recalled that, while in Annapolis, he "developed something of a flair for mathematics, and particularly for naval architecture and physics."[7]

Sprague took an early discharge from the navy to work for Edison. His first day working for the great inventor was the same day the Brooklyn Bridge opened: May 24, 1883.[8] Edison immediately put Sprague to work overseeing the installation of new electric grids in Sunbury, Pennsylvania, and Brockton, Massachusetts.[9] Sprague proved his worth almost from the outset, showing Edison's men how to use mathematical calculations to determine the size of wiring needed to serve the demand on a given electric grid.

Sprague's technical training and mathematical skills stood in sharp contrast to Edison's method of endless experimentation. Sprague was impressed by Edison's operation, but he wasn't content with his job. His ambition was to develop and deploy electric motors for industry and transportation. Joseph J. Cunningham, the preeminent historian on the electrification of New York City and the author of the excellent 2013 book *New York Power*, told me that "the value of a practical motor had become obvious to Sprague" while he was working for Edison. But Edison showed no interest in developing a motor and instead focused on lighting.

In April 1884, after just eleven months with Edison, Sprague set out on his own, a move that, he admitted later, came with

Frank Sprague in 1892, the year he won the elevator contract for the Postal Telegraph Building.
Source: Cassier's Magazine II (May–October 1892), available from Wikimedia Commons.

"considerable risk." That's an understatement. When he quit Edison, Sprague was twenty-six years old, new to New York City, and had little money. His salary with Edison had been modest: $2,500 per year (about $68,000 in 2018 dollars). Quitting the most famous and successful inventor of the century was one of many risky bets that Sprague would make over the next two decades.

Sprague may have frequently lacked the money needed to commercialize his inventions, but he never lacked for confidence in his ability and intellect. Prior to Sprague, electric motors were largely theoretical. Edison and others had fashioned motors out of generators by running them backward; that is, rather than spinning the rotor on the generator to produce outgoing current, they would feed electric current into the generator. Edison thrilled visitors to his laboratory in Menlo Park by giving them rides on a miniature train that he had built on the grounds. The track was about a mile long and provided power to the train through electrified rails. The train could reach speeds of forty miles (sixty-four kilometers) per hour, but it didn't have a proper motor; instead, it used a generator that was running in reverse. Although such an arrangement could work, it forced the generator to perform a task for which it hadn't been designed.

Sprague saw an opportunity. His confidence attracted investors, the most important of whom was Edward Hibberd Johnson, who was also one of Edison's early backers. With Johnson's backing, Sprague set up his own shop and began designing and building direct-current (DC) electric motors. Within a few months, he had several working prototypes that he showed at the Philadelphia Electrical Exposition in September 1884.[10]

As Cunningham explains, Sprague's motors "took industry by storm." They were powerful, compact, and operated at constant speed with little or no sparking. Edison himself endorsed Sprague's invention, calling it "the only true motor; the others are but dynamos turned into motors. His machine keeps the same rate of speed all the time, and does not vary with the amount of work done."[11] For a time, Sprague's motors were the only ones allowed to run on Edison's New York grid. Textile producers and other manufacturers were among the eager customers for Sprague's products. Within a year of launching his new venture, Sprague had sold 3,000 motors. Within a decade, forty-seven different types of electric motors produced by Sprague and other companies had "found application on cranes, lifts, machine tools, and other equipment."[12]

While his motors were an immediate success, Sprague wasn't satisfied with putting them into cranes, lifts, and looms. He wanted to electrify transportation. There were plenty of reasons for that. By the 1880s, New York and other bustling cities were desperate to find an alternative to horse-drawn trolleys. Some forty horses per day were dying on the streets of New York City, victims of accidents, overwork, or broken legs from slipping on slick cobblestone streets.[13] The average life expectancy for a horse pulling a streetcar was less than two years.[14] Those herds of horses were also dropping hundreds of tons of manure and tens of thousands of gallons of urine per day on the city's streets.[15] Vacant lots were piled high with enormous mounds of stinking dung. The manure provided a breeding ground for swarms of disease-carrying flies, which were blamed for outbreaks of typhoid and other health problems. During dry periods, the manure was a stinky nuisance that mixed with the city's airborne dust. During heavy rains, the shit merged with mud to form a slimy, pungent pollutant that stuck to everything it touched.

Sprague knew that his electric motor could provide a faster, cheaper, cleaner method of urban transport. In 1884, shortly after showing his motors in Philadelphia, Sprague and Johnson formed the Sprague Electric Railway and Motor Company. Over the next few months, he had, in his words, "schemed out" plans for an electric railway and began testing a prototype in New York City. Sprague hoped his first customer would be the Manhattan Elevated Railway, which was controlled by the financier Jay Gould, who was one of the richest people in America. He was also one of the most reviled. (In 1869, Gould and another financier, Jim Fisk, tried to corner the gold market. They failed to do so, but their attempt caused the prices of stocks and agricultural products to plummet.)[16] Gould owned a controlling interest in the Manhattan Elevated Railway Company, which ran the Second, Third, Sixth, and Ninth Avenue elevated lines, all of which used steam-powered locomotives.[17] The steam locomotives provided transportation, but they were hated by New Yorkers because they were dirty and noisy. And because they were elevated, they often showered sparks and coal dust onto pedestrians below.

Sprague got an opportunity to demonstrate his technology for Gould. But the demonstration ended badly after Sprague—who was operating one of his prototype rail cars—suddenly reversed the current in the controller, which caused a loud explosion and a load of flying sparks. Gould, startled by the explosion, tried to jump off the train and left in a huff. Unable to prove his technology in New York, Sprague signed a contract in May 1887 to install an electric rail system in Richmond, Virginia. The risk of the deal was obvious. The contract called for the Richmond Union Passenger Railway to pay Sprague $110,000 (about $3.1 million in 2018 dollars), but only once the entire system—forty cars, eighty motors, and a central station generating 280,000 watts (375 horsepower)—was operating satisfactorily. Thus, Sprague was going to have to construct a power plant that was about half the size of Edison's Pearl Street station (which was rated at 600,000 watts), and he was going to have to do something far more complicated than what Edison had achieved. Sprague's grid would have to do more than merely provide illumination. Sprague needed to build a grid that could power heavy, moving cars loaded with dozens of people. He would have to engineer a motor strong enough and durable enough to handle highly variable loads. He would have to figure out how to mount the motors on the carriages and design a system that would supply continuous power to the moving cars. Moreover, he would have to prove that it could all be done safely, cheaply, and under a crushing deadline. Sprague later admitted that the contract for the Richmond system was one that a "prudent businessman would not ordinarily assume." If he failed to make the rail system work, it would mean "blasted hopes and financial ruin."[18]

When he started, Sprague had only a blueprint of the Richmond system. He had to design, test, and deploy everything that would be needed, from the overhead wire system to how to attach his motors to the axles of the cars. Making matters worse, Richmond was a terrible place to test his system. Shortly after the project got underway, "serious problems emerged. The steep grades and sharp turns of Richmond's topography, its unpaved streets, and its clay soil rendered the route a 'horse killer' that was ruinously expensive to operate by traditional technologies."[19]

As author Frank Rowsome Jr. explains in his excellent biography of Sprague, *The Birth of Electric Traction*, the streets of Richmond would challenge the great inventor at every step.[20] But Sprague had a genius for machines and the tools needed to build and repair them. As Rowsome put it, Sprague "was a man for whom machinery wanted to work. . . . He had an instinct for elegant simplicity in design and for arrangements of mechanism that could be persuaded to work without endless finagling."[21] Despite his genius and aptitude for hard work, Sprague and his team endured myriad setbacks, including derailments and burned out motors. The setbacks were costly and forced Sprague to miss a deadline. The owners of the railway demanded, and got, a reduction in the final price, to $92,000. But Sprague and his team of engineers and mechanics pressed on, and in May 1888, just twelve months after he signed the contract, the Richmond Union Passenger Railway was declared a success. Sprague's rail cars were traveling a total of 11,000 miles (17,700 kilometers) and carrying some 40,000 passengers per week. The owners of the railway were ecstatic. By the time they paid Sprague, the new railway was earning its investors about $6,000 per month.[22]

By completing the system in Richmond, Sprague had designed, built, and successfully deployed the world's first full-scale commercial electric railway. His success in Richmond helped fuel the deployment of similar systems all over the world. Not only that, the designs that Sprague developed for the Richmond system—including the way he attached his motors to the axles and truck frames of the railcars—would become the standard for electric trolley and rail systems around the world.

After figuring out how to use his electric motors to move passengers across the horizon, Sprague focused on moving them skyward. In 1892, he formed the Sprague Electric Elevator Company. The new venture got its first opportunity when the Postal Telegraph Cable Company decided to build a skyscraper in lower Manhattan at 253 Broadway. Formed in 1883, Postal Telegraph was among the fastest-growing companies of that era and quickly became a key competitor of Western Union, which was controlled by Jay Gould. The new building was to be among the tallest in the city and thereby show that the young Postal Telegraph company could compete with Western Union. Due to the prestige that would inevitably come with such a project, the competition for the elevator contract was fierce. At that time, Elisha Otis's elevator company dominated the market with its hydraulic elevators. But the contract was nevertheless won by the Sprague Electric Elevator Company.

The deal required Sprague's new company to install six elevators: Two of them were to be express elevators. The remaining four would provide local service. The contract called for Sprague's electric elevators to equal, or exceed, the speed of hydraulic elevators. In addition, they were to require less maintenance and use less space. As with the Richmond railway contract, the deal was dangerously one-sided. If his elevators didn't work as Sprague had promised, the contract required him to rip them out and replace them, at his cost, with hydraulic elevators. Thus, Sprague had to—from scratch—design, test, and install everything he was going to need. If he failed, it meant buying a complete—and, in his view, inferior—elevator system from his rivals. Nevertheless, on October 8, 1892, Sprague signed the contract to install electric elevators in the Postal Telegraph Building.[23]

Over the next two years, Sprague and his team overcame dozens of problems. Among the most difficult to resolve was making the elevator cars run smoothly, without sudden jerks or starts. Passengers on horse-drawn streetcars were used to being aboard a vehicle that stopped or started suddenly, and knew to hold tight to the carriage in case a horse slipped, or started up suddenly, so they wouldn't be tossed overboard. They could anticipate bumps or sudden starts by watching the traffic around the streetcar and the horses pulling it. Elevator passengers, however, have no way to anticipate what might happen. There are no horses, landmarks, or traffic to monitor for visible or audible clues.

Through patient testing and development, Sprague solved the problem of jerky starts and stops. He also made other key refinements to vertical transportation, including an automatic floor-alignment system and a self-centering operator's control that stopped the elevator car if it was released, a mechanism that became known as the "dead-man's control."[24] By 1894, when the Postal Telegraph Building was finished, Sprague's elevators worked perfectly. Sprague received a letter of praise from the architects saying that, even though the electric-drive system was "a radical departure in elevator practice," the new elevators, in "economy of space occupied, speed, ease of motion, safety, and cost of running,

Left: Architect rendering of the Postal Telegraph Building in 1893. Note the height of the buildings next to the skyscraper: most are four or five stories high. Right: the Postal Telegraph Building in 2017.

Sources: Moses King, King's Handbook of New York City: An Outline History and Description of the American Metropolis *(Boston: M. King, 1893), available from Wikimedia Commons; photo by author.*

the building committee as well as ourselves, are decided of the opinion that the results have far exceeded our expectations."[25]

Another remarkable fact about Sprague's first set of elevators at the Postal Telegraph Building is this: they operated at speeds comparable to those of modern-day elevators.[26] Four of the elevators Sprague installed at the Postal Telegraph Building operated at a speed of 325 feet per minute (5.9 kilometers per hour). That's the speed of a brisk walk.[27] The two express elevators whisked passengers skyward even faster: at 400 feet per minute, or 7.3 kilometers per hour.[28]

That speed was key to their success. Elevators have to be fast. In cities, we want to travel as fast—or faster—when we travel vertically as we do when traveling horizontally. People in cities are in a hurry. A New York minute doesn't last very long, because residents of Manhattan, Brooklyn, and the other three boroughs have places to go and people to see.

Our travel habits are determined not necessarily by distance, but by time. That is, we don't worry so much about how far we need to travel as about how long it will take us to get where we're going. This time budget has become known as Marchetti's constant. Named for the Italian physicist Cesare Marchetti, it says that people spend about an hour per day getting around—going to work, shopping, and going to school. Further, Marchetti found that this hour-per-day rule also applied to ancient cities such as Rome and Marrakesh: those cities were approximately five kilometers across, which was about the distance an average person can walk in an hour. To prove his point, Marchetti compared the geographic growth of Berlin over time and found that it expanded concentrically over the years as advances in transportation increased the distances people could travel during that one-hour time period.[29]

Our travel-time constraints also apply to vertical transportation. With the electric elevators at 253 Broadway, Sprague proved that vertical transportation could be just as fast—and far safer—than traveling by foot on level ground.[30] That was a crucial turning point for New York's real estate market. Sprague's elevators helped make the vertical spaces above the streets as valuable—or more valuable—than what lay along the bustling boulevards. Today, the world's most prosperous cities are routinely populated by office and residential buildings that are fifty stories tall and taller, because occupants and visitors can get from the street level to the penthouse in a minute or two.

The impact of electrification on New York City and its population can be seen in the numbers. In 1880, two years before Edison launched the Pearl Street plant, the city held 1.2 million people. Two decades later, the population had nearly tripled to 3.4 million. By 1930, the population had doubled again to 6.9 million.[31] Electricity unleashed the beast of what New York City is today.

It allows 8.6 million people to thrive in one of the world's tallest, densest, and most vibrant cities.

The electrification that began in New York City not only shaped cities; it also shaped the natural environment. Tall buildings fueled by electricity allow cities to have greater population density, which helps reduce the human footprint on the natural world. "Cities are the salvation of nature," Ausubel told me. "It's only by concentrating a significant portion of humanity in livable, attractive cities, that we have the chance to spare the rest of nature for the lions, and the tigers, and the eagles."

Today, the world's cities contain more than half of the world's population but cover less than 3 percent of its land.[32] Those small footprints are obvious in New York City. People who live in the Big Apple use less energy and less stuff than their counterparts in suburban America because they live in smaller spaces, many of which are stacked on top of each other. It takes far less cement, wood, and copper to provide living space for an apartment dweller in Midtown Manhattan than it does to house a suburbanite living in Dripping Springs, Texas.

By pioneering electric elevators and electric transportation, Sprague not only helped spare nature; he also had a hand in creating an urban real estate boom. Some of the world's most expensive real estate can be found atop the tallest buildings. In 2014, the penthouse at the Odeon Tower in Monaco, a 170-meter (560 feet) skyscraper, was put on the market for $400 million. In Singapore, the "super penthouse" in Clermont Residence, the tallest residential structure in the city, was available for a mere $47 million.[33] About that same time, a penthouse pad at the Pierre Hotel in New York was selling for $125 million, and three other high-rise apartments in Manhattan were selling for about $100 million.

Those swanky apartments are only part of the real estate story. In 2017, a group of economists estimated that the land in New York City—just the land, not the buildings—was worth about $2.5 trillion.[34] That land is worth trillions because of what can be built on it, or rather, what can be built above it. Electrification allowed New York to become one of the world's most vertical cities. Only

Hong Kong has more skyscrapers—a term that currently refers to buildings that are at least 150 meters (which is roughly fifty stories, or 492 feet)—than New York City.[35] Today's New Yorkers navigate the vertical city thanks to the city's 71,000 elevators, many of which are direct descendants of the ones that Sprague installed at the Postal Telegraph Building.[36]

The pioneering work done by Edison, Sprague, Tesla, Westinghouse, and others provided the blueprint for nearly every electric grid that followed. Edison's use of direct current, which required power plants to be located close to customers, was supplanted by alternating-current systems that could economically serve customers over much larger areas. Westinghouse's transformers allowed electric utilities to dramatically increase the voltage on their transmission lines, which allowed generation stations to be located tens, or even hundreds, of miles away from customers. Those innovations set the stage for a wave of electrification that swept the globe and continues to this day. But the electrification that began in Lower Manhattan in the 1880s didn't reach everyone. By the early 1930s, half a century after Edison's pioneering work on Pearl Street, millions of rural Americans were still stuck in the dark.

4

THE NEW (ELECTRIC) DEAL

> Communism is Soviet power plus the electrification of the whole country. Otherwise the country will remain a small-peasant country. . . . Only when the country has been electrified, and industry, agriculture, and transport have been placed on the technical basis of modern large-scale industry, only then shall we be fully victorious.
>
> —**VLADIMIR LENIN**[1]

The success of the Pearl Street plant in Lower Manhattan ignited a frenzy of electrification. By 1890, just eight years after Pearl Street, the United States had about one thousand central power stations.[2] That rapid growth would continue for the next several decades. Between 1900 and 1930, US electricity production grew nearly twentyfold.[3] While that expansion allowed tens of millions of people to be plugged in for the first time, the rapid expansion of electricity service also led to a concentration of political and economic power that ignored the needs of millions of rural Americans. This chapter looks at a trio of New Dealers who worked together to make sure those rural Americans weren't left in the dark.

•

AMONG THE TITANS OF INDUSTRY who dominated American business in the late 1800s and early 1900s, a few names stand out: John D. Rockefeller, Andrew Carnegie, and J. P. Morgan are among the most obvious. They dominated in oil, steel, and finance. Samuel Insull was the titan of the electricity sector. And while he brought modern business practices to the industry, he also became a symbol of the evils of big business.

Born in London in 1859, Insull emigrated to the United States in 1881 to take a job working for Edison. Although he had read everything he could find on the great inventor, his first encounter with Edison left him unimpressed. "With my strict English idea as to the class of clothes to be worn by a prominent man, there was nothing in Mr. Edison's dress to impress me," Insull later recalled.[4] It's not clear what Edison thought of Insull's appearance, but shortly after he hired Insull, who had been working for an American banker in London, the storied inventor put his new employee in charge of nearly all of his financial affairs. Insull, who had a gift for mathematics and business, was later put in charge of Edison's manufacturing operations. But Insull grew tired of Edison's demands, and in 1892 he left New York to become the president of a small electric utility in Chicago.

Over the next three decades, Insull would accomplish one of the most breathtaking business expansions of the twentieth century: he started with a single electricity-generation station that had about 5,000 customers. By 1920, he had more than 500,000 customers.

From the outset of his stint in Chicago, Insull understood that if he was going to expand, he had to make electricity cheap. "If we ever expected to offer energy at low prices to our customers," Insull said, "we must produce that energy at the lowest possible price."[5] To make that happen, Insull invested in larger and more efficient generators. He also sought, and was regularly granted, monopoly franchises for electricity service in given territories. He did so by befriending local regulators and by claiming that electric power companies were natural monopolies, and thus were the most efficient way of providing electric service.[6]

Insull maintained control over his sprawling empire through a web of what became known as public utility holding companies. Those holding companies allowed a small group of investors to maintain control over a large number of companies. Here's how it worked: The investors who controlled the holding company raised capital by selling stocks and bonds to other investors. That capital was then used to acquire electric utilities, which were controlled by subholding companies that were also controlled by the top investors. Then, through financial engineering—including selling stocks of untradeable securities between the various companies—the top investors could inflate the value of their holdings and maintain control of the entire operation while remaining insulated from state regulations. In addition to their stranglehold on the electricity market, the holding companies made money by extracting large service payments from the utilities they controlled. Thus, if a small utility needed engineering or accounting services, the holding company would inflate the price of those services. An early 1930s report by the Federal Trade Commission found that in some instances the holding companies were "exacting profits ranging from 50 percent to over 300 percent of the actual cost of the services."[7]

As the holding companies grew, so did their geographic reach. By the early 1930s, just sixteen electric holding companies were producing about 77 percent of all the electricity produced by privately owned power plants in America. In addition, over 80 percent of the country's natural-gas pipelines were controlled by just fifteen holding companies.[8]

At its peak, Insull's primary holding company, Middle West Utilities Company, had subsidiaries in thirty states and was supplying electricity to about 5,300 communities.[9] While Insull's empire was huge, the biggest of the holding companies was Electric Bond and Share. It was formed in 1905 by General Electric to market GE's equipment and engineering capabilities.[10] By the 1930s, Electric Bond and Share controlled about 10 percent of all the electricity produced in the country.[11] Its corporate structure included five subholding companies, which in turn managed 121 subsidiaries. Electric Bond and Share was particularly powerful in Texas, where it provided electricity to more than five hundred communities in sixty-three counties.[12]

Electric Bond and Share wanted to keep competition at bay so that it could continue charging exorbitant rates for its electricity. For example, in 1925, Texas Power & Light, which was part of Electric Bond and Share, was providing electricity to the central Texas towns of Kerrville and San Marcos at rates as high as 15 cents per kilowatt-hour.[13] That would be about $2.15 in 2018 dollars.[14] That's astonishingly expensive given that the average price of residential electricity in the United States in 2018 was about 12.9 cents per kilowatt-hour.[15] When measured in constant-dollar terms, in the mid-1920s, residents of some rural Texas towns were paying about seventeen times as much for their electricity as residents of those towns are paying today.

Not only were the holding companies extracting big profits from consumers in small towns, they were also refusing to extend electric service to rural villages, farms, and ranches. The holding companies wanted big profits. They were happy to provide electricity service in cities and towns where population density was high. Lots of customers in small geographic areas meant that the electric companies could serve dozens, or even hundreds, of customers on a single distribution line. More customers on few lines meant lower costs per customer and, therefore, higher profits. That wasn't the case in rural areas, where the utility might have to put up several miles of line to serve a handful of customers. The result of the holding companies' refusal to serve rural customers was obvious: by the early 1930s, nine out of ten US farms lacked electricity.[16]

The disparity in electrification between urban and rural areas, along with the tremendous concentration of wealth and power in the holding companies, led to increasing resentment from consumers and politicians. By the early 1930s, as the Great Depression worsened, the utility holding companies' dominance of the electricity business had become a major political issue. Breaking their stranglehold would become a signature achievement of the New Deal.

•

THE FIERCEST SPEECH BY A presidential candidate on the topic of electricity—in fact, one of the few presidential speeches to ever focus on electricity—was delivered on September 21, 1932, in Portland, Oregon. "Electricity is no longer a luxury. It is a definite necessity," Franklin Roosevelt told the crowd at the Portland Municipal Auditorium. He explained that we are "most certainly backward in the use of electricity in our American homes and on our farms. In Canada, the average home uses twice as much electric power per family as we do in the United States." The reason for that, Roosevelt thundered, was that "many selfish interests in control of light and power industries have not been sufficiently far-sighted to establish rates low enough to encourage widespread public use."[17]

In 1932, more than twelve million workers—a full 23 percent of the US labor force—were unemployed.[18] As a percentage of workers, more people were unemployed in 1932 than at any other time in the Great Depression.[19] Roosevelt criticized the incumbent, Herbert Hoover, not only for his mishandling of the economy in general, but specifically on the issue of electrification. As president, Hoover had refused to use federal authority to restrain the public utility holding companies. Roosevelt, on the other hand, campaigned hard on promises to bring power to the people by busting the holding companies. He told the Portlanders that, while serving as the governor of New York, he had been attacked "by the propaganda of certain utility holding companies as a dangerous man. I have been attacked for pointing out the same plain economic facts that I state here tonight." He also told them that he had instructed the New York Public Service Commission to make sure that electric utilities were charging reasonable rates. While talking to the Portlanders, Roosevelt made sure to include a slap at Samuel Insull, declaring that, by taking on the holding companies, he had "created horror and havoc among the Insulls and other magnates of that type."

Roosevelt promised his listeners that if he won the White House, he would make sure that the public gets a "fair deal, in other words to insure adequate service and reasonable rates." He said that communities that were not satisfied with the service being provided to them by a private utility have "the undeniable basic right, as one of its functions of government," to have their own "governmentally owned and operated service."[20] Then, appealing directly to Oregon voters, Roosevelt also came out squarely in favor of putting dams on the Columbia River, dams that would help the Pacific Northwest develop its economy. He said that federally funded electricity projects would "forever be a national yardstick to prevent extortion against the public and to encourage the wider use of that servant of the people—electric power."

Roosevelt's tough attitude toward the holding companies was one of many reasons why he won the White House. And he won it in a big way. Less than two months after the Portland speech, Roosevelt crushed Hoover at the polls. When the final tallies were done, Roosevelt had defeated Hoover in the popular vote by nearly 18 percentage points.[21] Hoover managed to win only six states and just fifty-nine electoral college votes. He didn't even win his home state of California or the state where he was born, Iowa.[22] Meanwhile, Roosevelt won forty-two states and 472 electoral college votes. It was one of the worst defeats of an incumbent president in US history. In the twentieth century, only Woodrow Wilson's 1912 drubbing of the incumbent William Howard Taft, an election in which Taft won just two states, was more lopsided.[23]

By the time Roosevelt took office, the federal government was already playing a significant role in the electricity sector. In 1918, the government began construction of the Wilson Dam at Muscle Shoals, Alabama. The project, which dammed the Tennessee River, was designed, in part, to provide electricity to a nitrate plant that could help supply the US military with explosives.[24] In 1931, federal workers began construction of the Hoover Dam, on the border of Nevada and Arizona. That project would become a critical part of water management and electrification in the region. The dam would also become an integral part of economic development efforts. As historian David Kennedy has noted, "In the success of the western expansion of the United States, the single most important factor was federal investment in water management" in the western states.[25] In 1933, the federal government created the Tennessee Valley Authority, which had broad authority to provide flood control, electricity production, and economic development to the people living in the southeastern United States, a region that was plagued by poverty.

During the first two years of his administration, Roosevelt implemented several key elements of the New Deal, including breaking up the banks with the Glass-Steagall Act. He also established the Public Works Administration, which would spend billions of dollars to build new schools, roads, and government buildings. But by 1935, Roosevelt was ready for more aggressive action to stimulate the still-languishing economy, including a full-scale push for federal intervention in the electricity sector. In his 1935 State of the Union address, Roosevelt made clear that he planned to make good on his campaign promise to bust the holding companies, saying it was time for the "restoration of sound conditions in the public utilities field through abolition of the evil features of holding companies."[26] To beat the utilities, Roosevelt needed powerful allies on Capitol Hill who would get the legislation passed. He found three, all of whom were as committed as he was. Two—Sam Rayburn from Texas and Burton Wheeler from Montana—were Democrats. The third, George Norris, was a Nebraska Republican. Their names are on the two bills—the Rayburn-Wheeler Act of 1935 and the Norris-Rayburn Act of 1936—that busted the monopolists and assured lighting and power for rural Americans.

Rayburn and Wheeler were both born in 1882, the same year Edison started producing power on Pearl Street. Rayburn knew from personal experience the drudgery that came from living in the dark. He'd grown up working on his family's forty-acre cotton farm near the north Texas town of Bonham, a few miles south of the Red River. Rayburn began his stint in the US House of Representatives in 1913 at the age of thirty. He never forgot where he came from or how hard life was for farmers.[27] Without electricity, Rayburn said, rural farmers and ranchers were merely "unwilling servants of the washtub and water pump."[28] He famously said, "I want my people out of the mud and I want my people out of the dark." Rayburn knew what electricity could mean to farm families. "It will take some of the harsh labor off the backs of the farm men and women. Can you imagine what it will mean to a farm wife to have a pump in the well and lights in the house?"[29]

Wheeler was—as he later titled his autobiography—a *Yankee from the West*. Although he represented Montana in the US Senate, he was born and raised in Massachusetts and attended law school at the University of Michigan. In 1905, he settled in Butte, Montana, after losing his traveling money in a poker game. Montana turned out to suit him just fine. In 1922, his constituents elected him to the Senate.[30] By the mid-1930s, Wheeler had been involved in many tough fights on Capitol Hill. But the clash over the holding companies, he would later write, was "the biggest, bitterest, and most extravagant" of his career.[31]

Wheeler was among Roosevelt's earliest and most enthusiastic supporters. In 1930, he became the first nationally prominent Democrat to back Roosevelt for the 1932 nomination for president. In a speech in New York, Wheeler declared that Roosevelt was the one "general" who could lead the Democrats when it came to the "control of power and public utilities."[32] In *Yankee from the West*, Wheeler made clear his distaste for, or perhaps outright hatred of, the holding companies, calling them "unsound scalping operations." He added that "this kind of bloodsucker not only drained the investor, but through fraudulent overcapitalization of public utilities also fastened outrageous prices on the light, gas, water, and power consumers."[33]

While Rayburn and Wheeler were essential players in the fight for rural electrification, they might not have succeeded without Norris. Born in 1861, Norris was instrumental in creating the structure of Nebraska's state government, which is the only one in the country with a unicameral legislature.[34] In 1903, he began the first of five terms in the House of Representatives representing Nebraska's Fifth District. From there, he moved on to the Senate, where he began his three-decade stint on March 4, 1913, the same day that Rayburn began his first term in the House of Representatives.[35] A flinty lawyer who abstained from drinking alcohol his

US Senator Burton Wheeler of Montana in an undated photo. *Source: Library of Congress.*

entire life, Norris was an ardent believer in the power of government to help people. He was deeply suspicious of big business, was an enemy of Henry Ford, and was convinced that the federal government should be using hydropower to benefit citizens and farmers. As one account explains it, Norris saw government-funded dams as a "model for governments all over the world to control river flooding, provide cheap electricity, and eliminate food shortages and poverty."[36] In 1933, Norris led the fight for the successful passage of the Tennessee Valley Authority Act.[37]

By 1935, Rayburn, Wheeler, and Norris were all formidable players on Capitol Hill. That year, they introduced the Rayburn-Wheeler Act—better known as the Public Utility Holding Company Act—the key provision of which was known as the "death sentence." It required all of the holding companies that "were not parts of geographically or economically integrated systems to dissolve or reorganize themselves" by 1938. That meant that, for instance, Electric Bond and Share had to divest itself of small utilities that it owned in places like Kerrville and San Marcos.

US Senator George Norris of Nebraska in an undated photo.
Source: Library of Congress.

In addition, Rayburn-Wheeler outlawed the pyramidal structure of interstate utility holding companies and required holding companies that owned 10 percent or more of a public utility to register with the Securities and Exchange Commission and give detailed accounts of their holdings. The purpose of the legislation, Wheeler explained, was to "bring reduced rates to consumers by eliminating padded valuations, various schemes for milking subsidiaries by holding companies, and irregularities in securities corporations."[38]

The holding companies reacted to the prospect of Rayburn-Wheeler with fury. They spent some $1.5 million (about $27.5 million in 2018 dollars) trying to defeat the legislation, hired some six hundred lobbyists to push their case with members of Congress, and convinced supporters to send some 250,000 telegrams and five million letters to members of the House and Senate.[39] As the fight heated up, Will Rogers weighed in on the debate.[40] In the mid-1930s, Rogers, a Cherokee who was born in the tiny town of Oologah, Oklahoma, was among the most famous people in

America. He was a movie actor, humorist, and trick roper, as well as one of the most astute political critics of the day. He had a weekly column in the *New York Times*, and his syndicated columns were carried in newspapers read by forty million Americans.[41] Rogers wisecracked that the holding company is "something where you hand an accomplice the goods while the policeman searches you."[42]

In 1935, despite the holding companies' campaign to stop the legislation, the Public Utility Holding Company Act became law.[43] That same year, Congress also passed another law, the Federal Power Act, which empowered the federal government to oversee interstate electricity markets and gave the new Federal Power Commission a mandate to insure that electricity prices were "reasonable, nondiscriminatory and just to the consumer."[44] (That agency is now known as the Federal Energy Regulatory Commission.) In 1936, Congress passed the other piece of critical legislation, the Norris-Rayburn Act—also known as the Rural Electrification Act. It established the Rural Electrification Administration (REA) as an independent federal agency and charged it with providing loans for the construction and operation of generating plants, as well as distribution and transmission lines, for the provision of electricity "to persons in rural areas."[45]

In 1945, in his memoir, *Fighting Liberal*, Norris wrote that federal backing for rural electrification "constitutes one of the largest organizations of a governmental nature ever undertaken in the United States. Its benefits to the rural population have been of mammoth proportions and will grow constantly as electricity is carried to thousands more farms in all areas of the country."[46]

Norris, Wheeler, and Rayburn deserve the lion's share of the credit for assuring rural electrification in America. But another New Dealer, Lyndon Johnson, also deserves credit. During his long political career, Johnson would become one of America's strongest proponents of rural electrification. For him, the grinding poverty of farm living without electricity wasn't an abstract notion. He was "raised by the light of lanterns and cooked for on a wood-burning stove. He had seen his mother scrubbing clothes in a washtub. He knew the insides of outhouses," explains Ronnie Dugger in his biography of Johnson, *The Politician*. By the time Johnson arrived on Capitol Hill in 1937 as the newly elected representative from Texas's Tenth Congressional District, rural electrification was still largely bypassing his constituents in the Texas Hill Country.

5

WIRING THE SUPERPOWER

> It is not the damming of the stream or the harnessing of the floods in which I take pride, but rather in the ending of the waste of the region. . . . New horizons have been opened to young minds, if by nothing more than the advent of electricity into rural homes. Men and women have been released from the waste of drudgery and toil against the unyielding rock of the Texas hills. This is the true fulfillment of the true responsibility of government.
>
> —LYNDON JOHNSON[1]

Lyndon Johnson's first meeting as a US congressman with President Franklin Roosevelt didn't go well. It was June 1938. Johnson had joined the House of Representatives fourteen months earlier and he was focused on one task: he needed the Rural Electrification Administration to extend a loan to the Pedernales Electric Cooperative back in Johnson City, and he needed Roosevelt's help to get that loan.[2]

Johnson had met Roosevelt for the first time about a year earlier when the two had shared a train ride from Galveston. Roosevelt took a liking to the six-foot, four-inch Texan, who at the time wasn't even thirty years old. Johnson had won his seat in the House of Representatives in large part because he promised voters in the Tenth District that he would be an unwavering supporter of Roosevelt and his policies. But when Johnson met Roosevelt in Washington to talk business, the president was distracted. Rather than discuss Texas hydropower projects and electrification, the president was reportedly more interested in knowing whether Johnson had ever seen a Russian woman naked.[3] Probably nervous, and no doubt flummoxed by Roosevelt's question, Johnson didn't even get to ask the president about the loan before he was ushered out of his office. Though he failed at his initial meeting with Roosevelt, Johnson didn't give up. He couldn't. He had campaigned on a pledge to bring electricity to the Texas Hill Country, and he was going to make good on that promise.

By the time of his meeting with Roosevelt, the Rural Electrification Act had been in effect for two years, but the hard-luck farmers and ranchers in the Texas Hill Country were still largely stuck in the dark, and it looked like that was where they were going to stay. The problem was too few people. The Rural Electrification Administration was authorized to make loans for up to thirty-five years at 2 percent interest.[4] Under the rules set by the REA, rural electric cooperatives had to be able to show that they could connect at least three customers for every mile of new distribution wire. That was a problem. The population density of the rocky, Ashe-juniper-infested hills of the Edwards Plateau west of Austin was simply too low. Even if the region had had more people, there wasn't enough money to make rural electrification happen. The farmers and ranchers who worked the region's thin, rocky soil were barely eking out a living as it was.

Nevertheless, Johnson—who was born in 1908 near the tiny town of Stonewall—knew that if he was going to make his mark in Washington, he had to bring electricity to the Texas Hill Country. When he arrived in Washington in 1937, two major hydro projects were being built on the Colorado River in central Texas: the Buchanan Dam, located about seventy miles northwest of Austin, and the Mansfield Dam (then known as the Marshall Ford Dam), which was located about twenty miles northwest of the Texas capitol building.[5] The dams were enormously expensive—and enormously

Speaker of the House of Representatives Sam Rayburn gets a kiss from Senator Lyndon Johnson on Rayburn's seventy-fourth birthday, January 8, 1956. Rayburn was the longest-serving Speaker of the House in US history.

Source: Dolph Briscoe Center for American History, University of Texas at Austin, Sam Rayburn Papers, di_04781.

valuable—assets that would provide flood control, recreational opportunities, and, of course, water for irrigation and municipal use. But little had been resolved about the electricity that was going to be produced by the dams' powerhouses. Together, the two dams were to have more than 100 megawatts of electric generation capacity.[6] But the construction budgets didn't include any cash for poles and wires to carry that hydropower to local residents living in tiny places like Wimberley, Oatmeal, and Comfort.

Johnson knew that if the federal government didn't step in, the electricity from the dams would likely be controlled by the big utilities. Like the man who would become his mentor, Sam Rayburn, Johnson well understood that, by the mid-1930s, the Texas economy was largely controlled by eastern financial interests. The public utility holding companies had a stranglehold on the state's electricity production. Out-of-state companies owned 95 percent of the electric power sector, 83 percent of the oil refineries, and 99 percent of the railroads. Correcting that imbalance, and giving local interests more say—or, rather, an ownership stake—in the production of electricity was critical if Texas was to thrive. As Ronnie Dugger explains, "The case for the limited socialism called 'public power' was plain, and Johnson knew and felt the sting of his position." Dugger then quotes Johnson, who told him: "They hated me for these dams. . . . The power companies gave me hell. They called me a Communist."[7]

Johnson didn't care. In a speech on the subject, he told his constituents, "We should use that power. I believe that river is yours, and the power it can generate belongs to you."[8] In other words, if having electricity in the Hill Country required socializing the cost of electrification, or being called a Communist, Johnson was just fine with that.

In 1938, a group of ranchers and farmers formed the Pedernales Electric Cooperative with the hope that they could qualify for a federal loan from the REA. But the early efforts didn't yield much success. The population-density problem seemed insurmountable. As another Johnson biographer, Robert Caro, recounted in his 1982 book *The Path to Power*, despite months of effort to sign up local residents for the co-op, "the county agents had collected nowhere near the number required—not three per mile, but fewer than two. The Pedernales Electric Coop appeared stillborn."

In his pursuit of federal money for the new co-op, Johnson made a direct appeal to John Carmody, the head of the REA, asking for an exemption to the three-customers-per-mile rule. Carmody refused.[9] With few other options, Johnson turned to a friend, Tommy Corcoran, who was one of Roosevelt's top advisers, and asked him to arrange another meeting with the president. Corcoran agreed to do so and offered some advice: "Show him what Austin will look like. . . . Don't argue with him, Lyndon, show him."[10]

At his next meeting with Roosevelt, Johnson strode in with poster-size photos that showed the progress being made on the Buchanan Dam, as well as images of long-distance transmission lines and an electrified rural home at night. Roosevelt agreed on the spot. According to Johnson, the president immediately called Carmody at the REA. Roosevelt instructed Carmody to make the loan to the Pedernales Electric Cooperative and to "charge it to my account." Yes, Roosevelt understood the REA's three-customers-per-mile rule for new loans, but he assured Carmody that the people in the region would "catch up to that density problem because they breed pretty fast."[11]

On September 27, 1938, Pedernales Electric Cooperative got a telegram saying that it had been granted a federal loan for $1.3 million (about $23.5 million in 2018 dollars). That was enough to build about 1,800 miles (2,900 kilometers) of distribution lines, which provided electricity to about 2,900 families in the rocky scrubland around Johnson City. The co-op's first customers were required to put down a $5 deposit and pay the minimum charge for electric service: $2.45 per month, which is about $43 in 2018 dollars.[12] Looking back, it's hard to imagine that central Texas was once so sparsely populated that the Pedernales Electric Cooperative couldn't qualify for a loan. But as it turns out, Franklin Roosevelt was right: they do breed pretty fast in Texas. Today, the little co-op that couldn't get a federal loan is the largest electric cooperative in America, and therefore the largest electric cooperative in the world.

In the mid-1930s, the city of Austin had just 60,000 residents and Texas had about 6.1 million people. But the dams and electrification that LBJ pushed for helped attract people to central Texas. Lots and lots of people. By 2016, Austin's population exceeded 900,000 and that of the Lone Star State had grown to nearly twenty-eight million.[13] The growth is particularly evident at the Pedernales Electric Cooperative. Between 1938 and 2017, the co-op grew a hundredfold.[14] Today, it serves more than 300,000 customers in a service territory that covers twenty-four counties and is nearly the size of New Jersey.[15] By 2017, it was collecting nearly $600 million in revenue and was adding about 10,000 new customers every year.[16]

The Pedernales Electric Co-op and roughly nine hundred other electricity co-ops in the United States are living remnants of the New Deal. They are an overlooked, but critical, element of the American success story, providing electric service in nearly 80 percent of America's counties. Co-ops own and maintain about 42 percent, or some 2.6 million miles (4.2 million kilometers), of distribution lines and serve some forty-two million people in forty-seven states.[17]

By providing loans for rural electrification and breaking up the holding companies, the federal government fostered a surge of investment in rural America. Tens of billions of dollars were invested in generators, poles, wires, and meters that, in a matter of decades, allowed the United States to achieve almost universal electrification. Between 1940 and 1950, the amount of electricity sold by cooperatives jumped twenty-two-fold, going from about 300 gigawatt-hours to nearly 6,900 gigawatt-hours. By 1970, those figures had grown another tenfold to more than 76,000 gigawatt-hours.[18] Over that same time period, the cost of residential electricity in the United States fell dramatically, going from 3.8 cents per kilowatt-hour in 1940 to 2.1 cents in 1970.[19]

The cheap electricity provided by rural cooperatives nurtured generations of Americans who, thanks to lighting and electrical conveniences, could get an education and join the middle class, or at least scratch out a living. Thanks to rural electrification, manufacturing operations didn't have to be located in, or near, big cities. Instead, they could set up shop in small towns and rural areas. Electrification meant American farmers had access to refrigeration, thereby assuring that their products wouldn't spoil before they could be transported to the market. Electricity also allowed the use of electric water pumps, milking machines, heaters, and food-processing equipment. The combination of all of those things helped American farmers dramatically increase their productivity; and those productivity gains—combined with mechanization, better fertilizers, irrigation, and better seeds—meant fewer workers were needed and more people could leave the farm

Rural electrification, San Joaquin Valley, California, 1938.

Source: Dorothea Lange, Rural Electrification, 1938, Library of Congress.

for opportunities in towns and cities. In 1930, about 21 percent of the US workforce was employed in agriculture. By 1970, that figure had fallen to 4 percent, and by 2000 it was just 2 percent.[20]

Rural electrification meant air conditioning, which in turn meant that rural farmers and ranchers in southern states could get a respite from the searing heat and wilting humidity that defines summer in places like Florida, Louisiana, Texas, and Mississippi. The cheap, abundant, and reliable electricity provided by rural cooperatives not only saved consumers money, it also helped decentralize economic and political power. Rather than allow big-city financial interests to control the vast majority of America's electricity infrastructure, the New Deal assured that power was more equitably distributed. That diffusion of economic and political power assured that the benefits of electrification—jobs, patronage, and ownership—were spread among a wide variety of local people, all of whom shared a common interest in keeping electricity prices as low as possible. The cooperatives assured that the revenues from electricity sales didn't flow solely into the pockets of rich investors in Chicago or New York. Instead, that money stayed in rural areas, where it was reinvested and used to build more power plants and extend transmission and distribution lines to farms and ranches that had once been beyond the profitable reach of the utilities.

The turnabout in rural electrification can be seen in the numbers. By 1950, nine out of ten farms in America were connected to the electric grid, a reversal of the situation that existed just twenty years earlier.[21]

To be sure, the New Deal–era reforms were not perfect. The Federal Power Commission was not good at setting prices, and some of the other federal entities that were created, including the Tennessee Valley Authority, became bloated bureaucracies that were prone to mismanagement. In his excellent 2013 book, *U.S. Energy Policy and the Pursuit of Failure*, Peter Z. Grossman, an economics professor at Butler University, points out that the more the Federal Power Commission intervened in energy markets, the more distorted those markets became. For instance, the Federal Power Commission controlled the price of the natural gas that was shipped on interstate pipelines through the 1970s. But the federal "price controls only made it impossible for market forces to adjust" to events like the Arab oil embargo of 1973, Grossman writes. And, "far from helping consumers by keeping prices down, controls made the disruption of the oil market in 1973-4 worse than it would have been otherwise. In fact, it was U.S. policy that turned the embargo into a major national emergency."[22]

Although Grossman's critique of federal intervention deserves attention, it's also true that the New Deal reforms on electricity were a seminal event in American history. Roosevelt, Norris, Rayburn, Wheeler, and Johnson saw electricity as a necessary public service that was essential for economic growth and development. Electrification was simply too important to be left solely to for-profit companies.

This tension between electricity as an essential public service and private profit was an issue in electricity tycoon Samuel Insull's day, and it remains a problem today. Regulators are continually being lobbied about which forms of energy generation should receive favored status and subsidies (such as solar and wind) and which ones should, at least in theory anyway, be eliminated (coal and natural gas). Those arguments are occurring at multiple levels of government and they are happening with a US electric grid that is a crazy quilt of large and small electricity providers, including cooperatives, city-owned utilities, privately owned utilities, and government-sponsored entities like the Tennessee Valley Authority, Lower Colorado River Authority, and Bonneville Power Administration. Thanks in large part to the New Deal, the US grid has one of the world's most diffuse ownership structures, with some 3,300 electricity providers.

TABLE 2

AMERICA'S CRAZY QUILT ELECTRIC GRID

In 2016, the United States had:

—189 investor-owned utilities (88 million customers)

—2,013 publicly owned utilities (21 million customers)

—877 cooperatives (19 million customers)

—218 power marketers (6 million customers)

—9 federal power agencies (39,000 customers)

Total: 3,306 electricity providers

Source: American Public Power Association, 2015–2016 Annual Directory & Statistical Report, 26–27.

While co-ops, municipal utilities, and other entities play important roles on the electric grid, the US electricity business continues to be dominated by investor-owned utilities, which as a group serve about eighty-eight million customers.[23] But even the biggest electric utilities in the United States are puny when compared to their global counterparts. For instance, Exelon Corporation, one the biggest investor-owned utilities in the United States, serves about ten million customers.[24] By contrast, Électricité de France (EDF), the utility that dominates the French grid, has thirty-nine million customers.[25] But both Exelon and EDF are dwarfed by State Grid Corporation of China, a Chinese state-owned company that supplies electricity to some 1.1 billion people.[26] State Grid Corporation was a central player in the staggeringly fast electrification of China. Between 1990 and 2017, electricity production in China grew tenfold.[27] Over that same time frame, per-capita GDP in the world's most populous country also soared, going from about $317 to more than $8,800.[28]

Obviously, there are differences between gigantic state-controlled grids like the one in China and the US electric grid. Over the past few decades, China achieved nearly 100 percent electrification thanks to the strength of its export-focused economy and heavy state control of nearly every facet of the economy. By contrast, the United States achieved nearly 100 percent electrification thanks to reforms enacted during the New Deal—reforms that assured economic and political power was shared by local communities.

Put short, the New Deal reforms in the electricity sector helped assure America's superpower status after World War II. Of course, numerous factors contributed to American dominance during the second half of the twentieth century. Chief among them: nearly all of the war making and bomb dropping happened far from American soil. That said, the electrification that swept across rural America in the pre- and postwar years assured that soldiers returning from the war could afford home appliances and the electricity to run them. They could afford the televisions and radios that plugged them into popular culture. The electric grid became the network that connected nearly everyone in America to a common beat, one that cycled back and forth at sixty times per second.

Cheap electricity from rural co-ops and federally funded hydropower projects helped fuel a rapid increase in living standards.

Between 1940 and 1970, electricity production in the United States grew ninefold, to more than 1,600 terawatt-hours.[29] Over that same three-decade period, US gross national product increased nearly tenfold, going from $100 billion to $977 billion.[30] Personal incomes soared, going from less than $600 per year to more than $3,900 (in 1970 dollars).[31] That postwar boom was fueled by electricity. By 1970, the average American was consuming about 7,200 kilowatt-hours of electricity per year. To put that in perspective, that's more than twice today's global average of about 3,100 kilowatt-hours per capita per year.[32] Put another way, when looking around the world, the average person on the planet today uses less than half as much electricity as the average American was using fifty years ago.

The wiring of rural America after the New Deal not only set the stage for America's emergence as an economic superpower, it also helped assure that women and girls were not stuck on rural farms and ranches doing menial labor. In fact, the desire to help farm women was one of the reasons George Norris pushed so hard for rural electrification. "From boyhood, I had seen first-hand the grim drudgery and grind which had been the common lot of eight generations of American farm women," Norris wrote in his 1945 memoir, *Fighting Liberal*. "I knew what it was to take care of the farm chores by the flickering, undependable light of the lantern in the mud and cold rains of the fall, and the snow and icy winds of winter. I had seen the cities gradually acquire a night as light as day." The passage continues:

> I knew the heat of those summer days in a farm kitchen . . . where humidity and blazing sun combined with the stove to create unbearable temperatures. I had seen the drudgery of washing and ironing and sewing without any of the labor-saving electrical devices. I could close my eyes and recall the . . . unending punishing tasks performed by hundreds of thousands of women . . . growing old prematurely; dying before their time; conscious of the great gap between their lives and the lives of those whom accident of birth or choice placed in the towns and

cities. Why shouldn't I have been interested in the emancipation of hundreds of thousands of farm women?[33]

Norris wasn't the only one who knew how electricity could help farm women. During his push to get customers for the Pedernales Electric Co-op, Lyndon Johnson made appeals directly to women. As Caro explains in *The Path to Power*, as the young politician traveled from town to town promoting electrification, he would tell listeners about his mother, how she had hauled water in buckets from the river and scraped her knuckles on the washboard. Electricity would mean water pumps and washing machines. With refrigerators, he told the women, they wouldn't have to "start fresh every morning" at the cookstove. Relieved of that drudgery, he told them, "you'll look younger at 40 than your mother."[34]

Looking back, there's no doubt that rural electrification helped emancipate untold numbers of American women from the drudgery of farmwork. But rural electrification remains a challenge in dozens of countries around the world. Indeed, hundreds of millions of women and girls in rural parts of India and other countries are still enduring the "unending punishing tasks" that Norris wrote about back in 1945.

6

WOMEN UNPLUGGED

Remember: Just plug in. I'm ready!

—Jingle sung by cartoon character **REDDY KILOWATT**[1]

When I met Rehena Jamadar, she was forty-four years old. A soft-spoken, elegant woman, she had her first child, a girl, when she was sixteen. Two other children, a boy and a girl, came shortly afterward.

Rehena lives in the village of Majlishpukur, a tiny agricultural settlement located southeast of central Kolkata in a region known as South Twenty-Four Parganas. The road to the village is intermittently paved with bricks and is barely wide enough for two full-size vehicles to pass safely. Bicycles are the most common conveyance. When I arrived in the village along with my wife, Lorin, and film director Tyson Culver, the cars we arrived in were the only four-wheeled vehicles in sight. Chickens, pigs, and dogs roamed freely. Many of the children—all of them smiling and curious about the strangers who'd suddenly appeared in their village—were barefoot. Rainwater collected in trash-strewn drainage ditches on both sides of the road. Smoke from dozens of small cooking fires had left a blue-gray haze in the air that softened the December morning sun.

My friend Joyashree Roy, an economics professor at Jadavpur University in Kolkata, and a lead author of several reports issued

Rehena Jamadar, left, outside her home in Majlishpukur, West Bengal, India, talking with Joyashree Roy, 2016.
Source: Photo by author.

by the Intergovernmental Panel on Climate Change, was kindly acting as our guide and interpreter.[2] Rehena listened quietly as Joyashree explained, in Bengali, the reason for our visit. Joyashree and her team of graduate students, including Mriduchhanda Chattopadhyay, who was leading the research effort, had been studying energy use in the area for several months. Rehena told us that many of the women rely solely on biomass—principally wood and straw—for their cooking fuel. She said that she often uses liquefied petroleum gas (LPG) in her kitchen. But she added that the gas wasn't always available. Further, even though it was being subsidized by the federal government, LPG was still too expensive for some of the villagers to purchase. That left them with no other option except straw and wood.

When Joyashree told her that we wanted to talk about electricity, Rehena led us to the side porch of her brightly painted home to show us her electric meter. Her modest house had been connected to the electric grid fourteen years earlier. She explained that she pays 100 rupees per month (about $1.50) for her electricity. She was getting cut-rate electricity thanks to the federal government's electrification program. She paid the bill every three months. Inside her home were a couple of light bulbs, a fan, and an outlet in the kitchen. One of the things she liked best about having electricity was that kitchen work gets done much faster. She was using an electric grinder, which made preparing spices and other food much faster and easier. Before electrification, she had to do that grinding by hand, which consumed much of her time in the kitchen.

What was it like before her home got electricity? I asked. And what was it like now? She immediately began talking about the difference that electricity had made to her children and their education. Thanks to electricity, her children were able to read books, practice their writing, and manage their schoolwork at night. That had had a clear and positive result: one of her daughters was attending college in Kolkata, a fact of which Rehena was clearly proud. After we'd talked for a while longer, I asked Rehena: "If you had lived in a house that had electricity when you were growing up, would you have gone to university too?" A brief smile flashed across her face, and without a nanosecond of hesitation she nodded her head to the right, in the way typical of many residents of West Bengal, and said, "Yes. I would have."

There was no remorse. No bragging. No what-could-have-beens in her reply. Only a direct, matter-of-fact response—almost as if I'd asked her if the sun was going to come up in the east the next morning.

Darkness had kept this gracious and intelligent woman from achieving something that she knew was within her grasp. Here was a person who—had she been born in one of India's cities instead of a rural agricultural area—would have gone to college. With a college education, she might have become a doctor, a lawyer, a nurse, or an engineer. By the time I met Rehena, I knew plenty of facts and statistics: the average resident of India uses about 800 kilowatt-hours of electricity per year, which is about a quarter of the global average.[3] I understood the myriad correlations between

Women in Majlishpukur, 2016. *Source: Photo by Lorin Bryce.*

electricity availability and health and wealth. But that fifteen-minute conversation that I had with Rehena and Joyashree made me see the light: Darkness kills human potential. Electricity nourishes it. It is particularly nourishing for women and girls.

Electricity emancipates women and girls from the pump, the stove, and the washtub. And the largest number of women and girls around the world who need to be rescued from electricity poverty are people like Rehena Jamadar. They live in places like Majlishpukur—small, rural, predominantly Muslim villages far from urban centers.[4] Over the next few decades, bringing abundant, reliable electricity to women like Rehena will be key to global poverty-eradication efforts. Those electrification efforts will be particularly challenging in Islamic countries. Why? Over the next few decades the world's Muslim population is expected to surge. By 2060, according to projections from the Pew Research Center, the world's Islamic population will grow by more than 1.2 billion people, and will be as large as the world's Christian population.[5] Hundreds of millions of those new Muslims will be born in places like Majlishpukur, rural agricultural villages where people scrape by on subsistence farming and day labor.

The Islamic population will surge because of simple demographics. There are more young Muslims than there are young Christians and those young Muslims are having more babies than their Christian counterparts. For instance, in sub-Saharan Africa, the median age for Muslims is seventeen. In North America, the median age for Christians is forty.[6] The Pew report estimates that by 2060, 27 percent of the world's Muslim population will be living in sub-Saharan Africa. In 2015, that figure was 16 percent. Another Pew study, this one published in 2016, about the education gap among the various religions, provides yet more reason to be concerned.

The 2016 study found that Muslims, on average, have just 5.6 years of formal schooling. By contrast, Christians obtain an average of 9.3 years of formal schooling and Jews get 13.4 years (the most of any religion). The disparity is particularly obvious among women. Muslim women trail men in formal schooling by 1.5 years. That is, the average male Muslim gets 6.4 years of formal schooling while the average female Muslim gets just 4.9 years.[7] Thus, the future of the Christian and Muslim worlds appears to be one where there will be an ever-widening gap, where Jews and Christians will

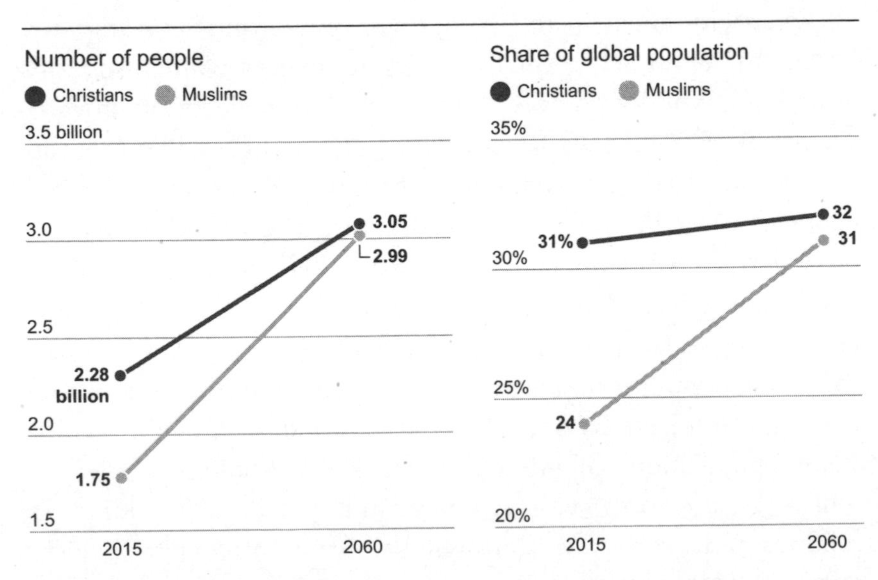

Expected Global Growth in Christian and Muslim Populations, 2015 to 2060

Source: "The Changing Global Religious Landscape," April 5, 2017, Pew Research Center.

continue to be electrified—and therefore educated—but hundreds of millions of Muslims will be undereducated, impoverished, and living without electricity.

Of course, electricity matters to all women, not just Muslim women. According to estimates by Hans Rosling, the late Swedish academic and statistician, about five billion people on the planet are walking around today wearing clothes that have been washed by hand.[8] That means that roughly 2.5 billion women and girls—as part of their daily, or weekly, routine—are washing those clothes. By hand. In buckets or washtubs. And every hour that women and girls spend at the washtub is one missed in the classroom, bookstore, or library. The women and girls who live in electricity poverty have never gotten used to the convenience of a washing machine or refrigerator, to say nothing of a Cuisinart or Instant Pot.

There's an old saying: educate the mother and you educate the child. But mothers can't get much education if they don't have washing machines. And washing machines may be the single most important device that can help elevate the status of women and girls. During the New Deal in the 1930s, increasing the use of washing machines was a major motivation behind the establishment of the Electric Home and Farm Authority. The agency was created to develop and foster "increased use of electric power through the double reduction of cost of electricity to the consumer and the cost of electrical appliances."[9]

The Electric Home and Farm Authority helped rural residents borrow money at low interest rates so they could purchase washing machines, ranges, and refrigerators. The appliances were sold through local utilities and electric co-ops. By 1938, more than 74,000 contracts had been signed, representing more than $11 million in appliance sales. (That would be about $96 million in 2018 dollars.)[10] Utility companies in the United States continued to target women in their advertising throughout the 1940s, '50s, and '60s as a way to increase appliance sales and increase electricity use. The most famous of those marketing efforts was the cartoon character Reddy Kilowatt. Created in 1926 by Ash Collins, the communications manager for Alabama Power, Reddy Kilowatt would go on to star in numerous newsreel commercials and print advertisements. By 1957, more than two hundred utilities around the world had licensed Reddy Kilowatt in one form or another.[11] Dubbed "your tireless household servant," Reddy aimed his message directly at women. In animated thirty-second spots, he could be seen dancing

During the 1940s, '50s, and '60s, American utilities promoted electricity use with a cartoon character named Reddy Kilowatt.

Source: Reddy Kilowatt Records, Archives Center, National Museum of American History, Smithsonian Institution.

around the house, showcasing various electric appliances, while singing the ditty: "I wash and dry our clothes, play your radios. I can heat your coffee pot."

•

NUMEROUS ACADEMIC STUDIES HAVE SHOWN the positive effect electrification has on women and girls. A 2002 study in Bangladesh by Abul Barkat, an economist at the University of Dhaka, found that the literacy rate for females in villages with electricity was 31 percent higher than it was in villages that lacked electricity. The study concluded that the availability of electricity has a "significant influence on education, especially on the quality of education. This influence is much more pronounced among the poor and girls in the electrified households than the poor and girls in non-electrified households."[12]

A 2010 study on post-apartheid electrification in South Africa found that "employment grows in places that get new access to electricity." This was particularly true for women. The study found that electrification led to "large increases in the use of electric lighting and cooking, and reductions in wood-fueled cooking over a five-year period, as well as a 9.5 percentage point increase in female employment."[13]

A 2012 study of rural electrification in India concluded that the availability of electricity had a significant impact on schooling for girls, finding that

> electrification access increases school enrollment by about 6 percent for boys and 7.4 percent for girls. It also increases weekly study time by more than an hour, and the increase is slightly more for girls than boys. As a result of more study hours, children from households with electricity can be expected to perform better than their peers living in households without electricity.

The same study found that "the impact of electrification on labor supply is positive for both men and women; that is, household access to electricity increases employment hours by more than 17 percent for women and only 1.5 percent for men." Further, the study found that electrification reduces the poverty rate by 13.3 percent, and it concluded that "these findings indicate electrification's substantial positive effect on overall household welfare."[14]

Complex studies aren't needed to show that extreme shortages of electricity are the common factor in nearly every country where women and girls are cursed with illiteracy and child marriage.[15] World Bank data shows that the countries with the highest rates of female illiteracy are all in the unplugged world.[16] If you are a female in an impoverished country and you don't have access to electricity, you are effectively a slave to the physical chores of the household: hauling water, making fires, grinding grain, and washing clothes. The United Nations Children's Fund (UNICEF) 2014 *State of the World's Children* report is a sobering document that details the plight of children around the world, and in particular, the plight of girls. Among the aspects that UNICEF examined was the issue of child marriage, that is, cases in which girls are married before the age of eighteen. The report lists the countries with the highest rates of child marriage.

TABLE 3

TOP TEN COUNTRIES WITH THE HIGHEST RATES OF CHILD MARRIAGE ARE ALL ELECTRICITY POOR

Niger: 76 percent

Central African Republic: 68 percent

Chad: 68 percent

Bangladesh: 65 percent

Mali: 55 percent

Guinea: 52 percent

South Sudan: 52 percent

Burkina Faso: 52 percent

Malawi: 50 percent

Mozambique: 48 percent

In eight out of those ten countries, per-capita electricity use is so low that the World Bank doesn't provide data for them. Only two of the ten—Bangladesh (279 kilowatt-hours per capita per year) and Mozambique (440 kilowatt-hours per capita per year)—are using large enough amounts of electricity to be listed in World Bank data sets.[17]

Whether the issue is voting rights, education, or work opportunities, the facts show that electricity helps women and girls. After we visited Majlishpukur, I did a formal interview with Joyashree Roy. I asked her how policymakers should think about electrification in poverty-stricken places like India, particularly given electricity's importance to women and its role in climate change. She said, "You cannot deny them this access to modernity." She argued that for "the rest of the world, who is enjoying modernity, and we are not allowing others to be modern, is a crime. It is a major crime." Electricity, she said, is essential to women. "If your mother is educated then she understands the beauty of knowledge," Joyashree told me. If women don't have electricity, and therefore miss their opportunity to be educated, then, "we are missing two generations, mother and the girls."

As electricity frees women from the washtub, it allows them to become active in civil society. Given that fact, it's not surprising that key advances in electrification occurred at about the same time women were gaining more political power. A look at the timeline of the late nineteenth and early twentieth centuries shows that women's suffrage gained critical momentum during the same decades that the world's major cities were being electrified. For example, 1893 was a turning point for the suffrage movement in America, when Colorado became the first state to adopt an amendment granting women the right to vote. That same year, at the Columbian Exhibition in Chicago, George Westinghouse proved that alternating current would become the standard method of distributing electricity to customers. Also in 1893—eleven years after Edison began producing electricity in Lower Manhattan—women in New Zealand were granted the right to vote. In 1902, Australia gave women the right to vote. In 1906, Finland did the same.[18]

Of course, numerous factors contributed to the ultimate success of the suffrage movement. It's also clear that electrification paralleled the rise of women in society. As electrification swept around the globe, women began leaving rural darkness for brightly lit cities. Electrification increased opportunities for schooling, as well as job opportunities in new manufacturing plants and offices. As women got more jobs and independence, so, too, did they desire more power in the workplace and at the ballot box. That desire gained traction after the deadly fire at the Triangle Shirtwaist Factory in New York City. That blaze, which occurred on March 25, 1911, left 123 women dead and helped spur the growth of the International Ladies' Garment Workers' Union. By 1920, for the first time in US history, more than half of all Americans were living in cities and the pace of electrification was growing at a rapid rate.[19] That same year, as the United States became a country dominated by city dwellers, the Nineteenth Amendment to the US Constitution was ratified. Women in America finally had the right to vote.[20]

Let me be clear that I am not claiming that electrification was the sole reason why women in the United States gained the right to vote. Nevertheless, the evidence shows that electrification played a key role in giving American women and girls in rural areas better economic and educational opportunities.

While there's no doubt that electrification helps women and girls, the enormous gap in electricity use around the world today presents an obvious question: Why have the United States and other wealthy countries been successful at electrification? What do those countries have that the poor, undeveloped countries lack? In the next section, I will use an easily understandable metric to illustrate the enormous disparity in electricity use around the world. I'll then discuss why so many people are still struggling to get the electricity they need and show how the human thirst for electricity is overwhelming other concerns, including those about air pollution and climate change.

PART TWO

Why Are Billions Still Stuck in the Dark? And What Are They Doing About It?

7

MY REFRIGERATOR VERSUS THE WORLD

> To the electron: May it never be of any use to anybody.
>
> —**JOSEPH JOHN THOMSON**,
> Nobel Prize winner and discoverer of the electron[1]

My kitchen refrigerator is not particularly fancy. It's a typical American fridge: it has two doors (one for the freezer, the other for the refrigerator), an in-door ice dispenser, and a cold water tap. My wife, Lorin, and I bought it from Home Depot in about 2007. Aside from a few minor repairs, it has been pretty reliable. That said, it uses a lot more electricity than what Whirlpool said it would. The specification sheet claimed the machine would use about 616 kilowatt-hours per year.

After plugging my refrigerator into my watt meter, it was immediately apparent that the real-world consumption figures were far higher: about 949 kilowatt-hours per year. I wanted to make sure my numbers were correct, so I kept the refrigerator plugged into the watt meter for more than forty-eight hours. Once I was certain of the number, I rounded up to an even 1,000 kilowatt-hours per year so I could have an easy-to-remember benchmark. Other analysts have used their refrigerators as a benchmark for electricity use.[2]

Doing so makes sense for several reasons. First among them: refrigerators are among the biggest electricity users in American homes.[3] Our house in Austin has plenty of other electricity-hungry devices, including two air-conditioning systems, dozens of lights, a washer and dryer, a toaster, an oven, a hair dryer, and numerous electronic gadgets. Those things use plenty of energy, but refrigerators are an essential appliance. They assure food doesn't spoil, medicines stay fresh, and, most important, the beer stays frosty cold.

After settling on 1,000 kilowatt-hours as a benchmark, I worked with Seth Myers, an Austin-based information-graphics expert, to create a database so we could compare per-capita electricity use among the world's countries with the amount used by my refrigerator. In addition to electricity use, we included population data, GDP, mortality rate, life expectancy, and religious affiliation. For consistency, we used numbers from 2012 as much as possible. We then separated those two hundred countries and territories into three groups:

- The Unplugged countries, where per-capita electricity use is less than 1,000 kilowatt-hours per year
- The Low-Watt countries, where per-capita electricity use is between 1,000 and 4,000 kilowatt-hours per year
- The High-Watt countries, where per capita electricity use exceeds 4,000 kilowatt-hours per year

The most surprising number in our database was the number of people who are Unplugged. Roughly 3.3 billion people—about 45 percent of all the people on the planet—live in places where per-capita electricity consumption is less than 1,000 kilowatt-hours per year, or less than the amount used by my refrigerator. According to the International Energy Agency, of those 3.3 billion Unplugged people, about one billion have no access to electricity at all.[4]

The electricity consumption of the Unplugged population also has historical significance: at 1,000 kilowatt-hours per year, the people living in Unplugged countries are using about the same amount of electricity as an average resident of Chicago did back in 1925.[5] Thus, when it comes to electricity use and all the things we consider modern, there are about three billion people on the planet today who are roughly a century behind the people living in High-Watt locales.

The countries in the Unplugged segment include places like El Salvador, the Philippines, Bolivia, Pakistan, and India. On nearly every metric of human well-being, whether it's child mortality or life expectancy, the residents of the Unplugged countries trail far behind the people living in Low-Watt and High-Watt places. For instance, people living in High-Watt countries live, on average, about sixteen years longer than those in Unplugged countries. The child mortality rate in Unplugged countries is nearly ten times higher than the rate in High-Watt places. The average per-capita GDP in Unplugged countries is about $1,973 per year, or about twenty times smaller than in High-Watt countries, where the average is about $38,844.

Trifurcating the world by electricity use also revealed data about religious affiliation: while Christians are spread fairly evenly among the three segments, about two-thirds of the world's Muslims live in Unplugged countries and about 99 percent of the world's Hindus live in the Unplugged world.

After looking at the Unplugged population, we looked at the second-largest group: the Low-Watt countries. We found that roughly 2.7 billion people—about 37 percent of the world's population—are living in Low-Watt countries. Poland (where per-capita electric use is about 3,900 kilowatt-hours per year) as well as Chile, Ukraine, China, and Turkey are all at the top end of the Low-Watt world. Thus, Low-Watt countries are not necessarily poor. But their per-capita GDP lags behind the High-Watt segment.

About 19 percent of the world's population—some 1.4 billion people—are now living in High-Watt countries, places where electricity use is greater than 4,000 kilowatt-hours per capita per year. The High-Watt countries include the United States, Britain, Sweden, Belgium, Germany, and Israel.

Seth and I used 4,000 kilowatt-hours as the minimum level for the High-Watt rating because that level of electricity use is

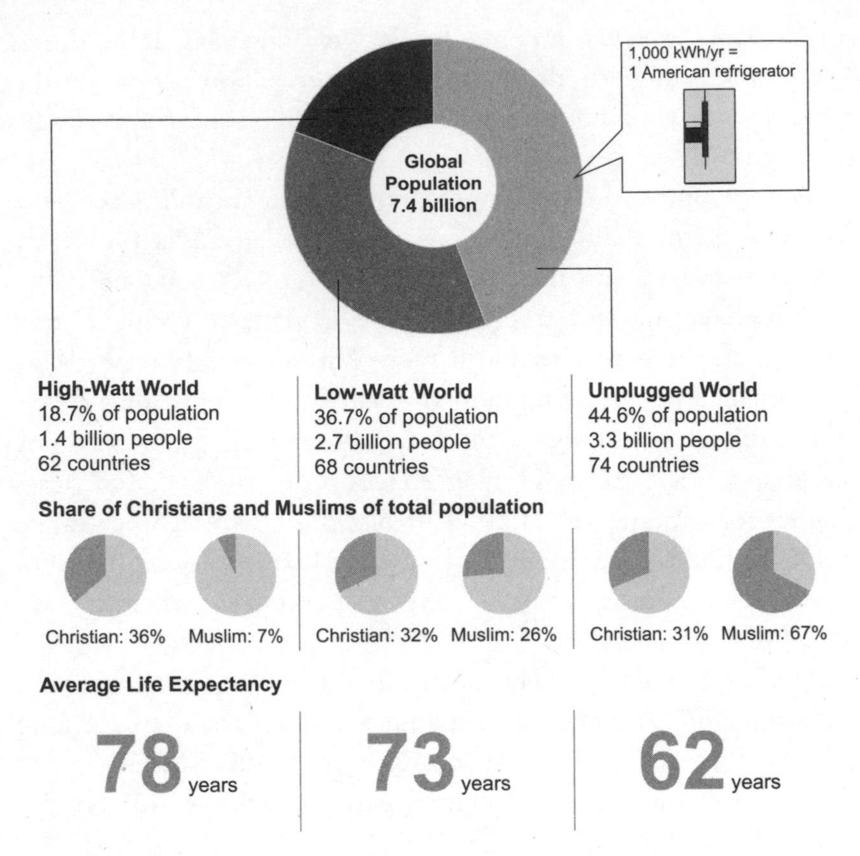

The High-Watt, Low-Watt, and Unplugged Worlds, 2012

Sources: World Bank, author calculations.

considered the minimum for living a long, high-quality life. In 2000, Dr. Alan D. Pasternak, a chemical engineer who worked at Lawrence Livermore National Laboratory, published a well-known paper that analyzed per-capita electricity consumption in sixty countries and then correlated that use with indicators of human health and welfare.[6]

Pasternak used data from the United Nations' Human Development Index (HDI), which ranks countries based on measures like life expectancy, nutrition, health, mortality, poverty, education, and access to safe water and sanitation and calculates a score for each country, with 1.0 being the maximum.[7] The UN estimates that less than 15 percent of the world's population was enjoying an

HDI of 0.9 or greater.[8] In Pasternak's analysis of the connection between HDI and electricity, he wrote that "there is a threshold at about 4,000 kilowatt-hours per capita, corresponding to an HDI of 0.9 or greater, in the relationship between HDI and electricity consumption."

In other words, Pasternak (who died in 2010) found that the 4,000 kilowatt-hours mark was the key dividing line.[9] "As electricity consumption increases above 4,000 kilowatt-hours, no significant increase in HDI is observed," he wrote. Conversely, he found that the lower the electricity consumption, the lower the HDI. In the summary of his findings, Pasternak was blunt, saying that neither the HDI nor the economic output "of developing countries will increase without an increase in electricity use." Pasternak continued, "What is of interest is the fact that large populations of the world are significantly below the electricity threshold level associated with a Human Development Index typical of developed countries." Those low rankings "reflect short life expectancy and low educational attainment—measures that are far more compelling than the purely economic metrics usually associated with energy consumption." Therefore, he said, "there is a compelling need for increased energy and electricity supplies in developing countries."

The essential point of Pasternak's paper is this: small amounts of electricity are not effective at alleviating poverty. Sure, a little juice is better than nothing. But Pasternak's paper shows that if the world's poverty-stricken people are going to come out of the dark and into modernity, they are going to need much more electricity than what is now being produced. Furthermore—and this is the hard part—the vast majority of that electricity will have to be generated inside the countries that need it.

Globalization can be seen in nearly every commodity, from coffee to tennis rackets and molybdenum to maple syrup.[10] Not so for electricity. Yes, there's robust trade in the fuels that we use to produce electricity, including coal, natural gas, and uranium. And there's booming global trade in solar panels, wind turbines, gas turbines, reciprocating engines, distribution wire, poles, and transformers. But very little electricity gets shipped across international borders. In 2013, global trade in electricity totaled about 308 terawatt-hours out of some 23,000 terawatt-hours generated that year.[11] Or think of it this way: In 2013, cross-border electricity trade totaled the equivalent of about 500,000 barrels of oil per day. That same year, global oil trade averaged more than 42.5 million barrels per day—an amount eighty-five times larger than the cross-border trade in electricity. Add coal and natural gas to the amount of oil moving across borders and the totals are yet more striking: cross-border trade in hydrocarbons is about 150 times larger than cross-border trade in electricity.[12]

Some electric grids cross international boundaries. For instance, Europe has an interconnected grid that allows Germany, Denmark, France, and other countries to trade electricity. In North America, the US electric grid is connected to grids in Mexico and Canada. Further, there are plenty of long-distance transmission lines, some of them spanning hundreds of miles, that carry electricity from rural dams and power plants to big cities. But the longer the distance, the more electricity gets lost in the process. That fact, combined with the high cost of building transmission lines, has meant that the cost-effective limit for electricity transmission, which can be done with either alternating current or direct current, is roughly 1,200 miles (1,930 kilometers).[13]

Since its earliest days, the business of generating and distributing electricity has been dominated by electric grids that are owned and operated by local or regional companies. That is particularly true in the United States, where about 250 cities and towns own and operate electric and/or gas utilities.[14] This localism reflects the importance of reliable electricity service. Further, it illustrates the fact that political leaders are not willing to allow their electricity to be controlled by foreign countries. Why? The short answer is that no sovereign country is going to turn over the control of its electric grid to another country, no matter how friendly. The risks of doing so are simply too great. European countries may be connected to a continent-wide electric grid, but Europe's electricity cooperation depends upon long-term peaceful relationships among all of those countries. Any upset of those peaceful relationships could upset the shape of the European grid.

Electric grids provide a near-perfect reflection of the people, communities, and countries they are powering. Those grids range in size from the 4,500-watt grid Wilfredo Roque and Iris Ortiz were using in Puerto Rico after Hurricane Maria to ones as large as the State Grid Corporation of China, which provides electricity to over 1.1 billion Chinese.[15] In Puerto Rico, people like Iris and Wilfredo set up their own grids. In China, the Communist government plugs all of its citizens into a single sprawling provider.

The local—or national—nature of electricity production and consumption is just one of its many peculiarities. It's also one of the reasons why solving the scourge of electricity poverty will be so difficult. Unplugged countries can't import the electricity they need from High-Watt or Low-Watt countries. Each country or region has to build, pay for, and manage their own electric grid. That's no simple task.

8

THE POWER IMPERATIVES

Integrity, Capital, and Fuel

> When Edison . . . snatched up the spark of Prometheus in his little pear-shaped glass bulb, it meant that fire had been discovered for the second time, that mankind had been delivered again from the curse of night.
>
> —**EMIL LUDWIG**, German author[1]

After looking at electrification efforts all over the world, I've concluded that all successful electric grids—no matter how big or small, no matter where they are located—rely on three interrelated factors: integrity, capital, and fuel.

The importance of integrity, capital, and fuel can be seen by looking, once again, at Barrio Antón Ruíz. After Hurricane Maria destroyed Puerto Rico's electric grid, Wilfredo Roque and Iris Ortiz created their own. Their grid had integrity because the couple controlled it entirely: the generator, the extension cord that carried the electricity over the driveway to the house, and all of the devices in the house that were being powered by it. Anyone who attempted to steal electricity from their generator, or who attempted to cart off the generator itself, could easily be detected and stopped.

Wilfredo and Iris's grid also demonstrates the importance of capital. After my visit, I corresponded with Iris by email. She explained that she and Wilfredo would have dearly loved to have a solar system so they wouldn't have to depend on PREPA. But getting enough solar panels to fuel their home would have cost $8,000, a sum far beyond their budget. In addition, she told me, even if they'd had $8,000, that solar system didn't include batteries, so they would have had to continue paying for a connection to the Puerto Rican electric grid.

Although they couldn't afford a solar system, they did have enough capital—$300—to buy a 4,500-watt Black Max generator, and they used it for most of their needs.[2] But it was too noisy to run at night, so they also purchased a 2,100-watt, gasoline-fueled generator that was quieter. They used the smaller unit to power fans to keep them cool while they slept. The smaller unit cost another $650.[3] Thus, for about $1,000, they cobbled together an electric grid that met their needs. Their limited capital, in turn, determined the type of fuel they used to generate electricity. Wilfredo was buying gasoline every few days to keep the two machines running. The fuel was expensive and put a big crimp in the family's finances. But gasoline was readily available, which meant Wilfredo and Iris could be assured of having electricity when they really needed it. If they forgot to buy fuel, or they ran out, or they couldn't afford to buy the gasoline they needed, their grid would fail.

What was true for the tiny grid that provided electricity to Wilfredo, Iris, and their three girls in Puerto Rico holds true for electrification efforts everywhere. Every grid requires integrity, capital, and fuel, and of those three, integrity is the most important. By integrity, I mean that the system in which the grid is operating doesn't leak too much. Electric grids are like buckets: if they leak, they become useless.

Theft is the enemy of light. Leakage—whether it's people stealing money from the company running the grid or marijuana growers who have illegally tapped a distribution line to reduce their electricity costs—must be kept to manageable levels. Put yet another way, for an electric grid to work and work well, there must be some *esprit de grid*. That is, the people who operate the grid, as well as the people who rely on it, need to have some sense of responsibility for it. Or, if they don't feel responsible for the grid, they have to fear getting caught and punished for stealing electricity.

Since the time of Edison and Insull, the trend in electrification has been to achieve economies of scale by making grids bigger, with bigger power plants, more customers, and more lines. Larger generators consume less fuel per unit of electricity produced than smaller ones. Those big machines allow utilities to spread their capital and fuel costs over more customers and thereby reduce the cost of electricity. Smaller grids are generally less efficient—both in terms of fuel costs and capital deployment—than larger grids. But as is evident with Wilfredo and Iris's grid, it's also far easier to assure the integrity of a small grid than it is for a large grid. Indeed, as integrity in a given society declines, electric grids tend to shrink.

Understanding why integrity is so critical only requires a look at the overhead electric wires in your neighborhood or city. Keeping electricity flowing to consumers requires putting enormous lengths of electric wire in the air. That wire—along with the necessary number of poles, towers, and insulators—must be kept up in the air, no matter the weather. Keeping those transmission and distribution wires high enough that they're safe from people—and people safe from them—is no small feat. It requires bucket trucks, rigging, slings, cranes, ladders, protective clothing, insulated tools, and lots of brave men and women who are ready and willing to go out in all kinds of weather at every hour of the day or night to keep the electricity flowing. If a distribution line or transformer cannot be fixed because someone stole the bucket truck, or maybe just the battery from the bucket truck, then electricity reliability in the service area will necessarily decline. Electricity use in the affected region will cease until the wire or transformer can be repaired.

The importance of integrity can be illuminated by correlating corruption and electricity use. As shown by the graphic below, the less corrupt a country is, the more electricity it uses; the worse the corruption, the lower the electricity use.

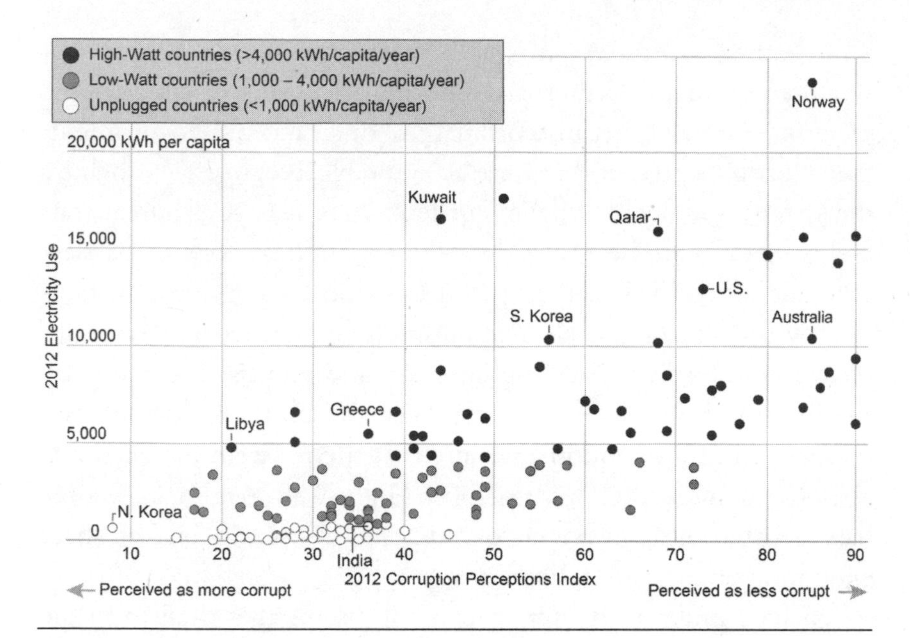

Corruption and Per-Capita Electricity Use, 2012

Sources: Transparency International, World Bank.

In most High-Watt countries, integrity is maintained because governments enforce the rule of law. That integrity in governmental and judicial matters is essential. The countries where humans are the freest and the richest are the ones where factions share political power in order to have a trustworthy government that enforces the rules. Credible governments, then, are key to successful electrification. In their 2012 book, *Why Nations Fail*, Daron Acemoglu and James A. Robinson provide a convincing argument for why political and economic power must be shared for a society to achieve sustained prosperity. Impoverished countries are often poor, they explain, because the elites have "organized society for their own benefit at the expense of the vast mass of people." By contrast, wealthy countries are rich because "their citizens overthrew the elites who controlled power and created a society where political rights were much more broadly distributed, where the government was accountable and responsive to citizens, and where the great mass of people could take advantage of economic opportunities."[4]

In other words, Acemoglu and Robinson are saying that wealthy countries—which are also the places where electricity is plentiful—are rich because ordinary people, and the governments they elect to represent them, are accountable. Regular folks believe the system is legit. That belief creates a virtuous cycle: bureaucrats and businesses are held accountable because beat cops in Bartlesville and on Wall Street keep the thieving to a minimum. People pay their taxes and electric bills. The courts work. Trash gets picked up. Hospitals have the lights on and enough medicine. By sharing political and economic power, the elites, politicians, and businesses in those countries assure that their system has accountability and integrity. Thus, in most High-Watt countries, people have reliable electric power because the people in charge share economic power and political power.

Societal integrity matters because it discourages theft from the grid. Electricity theft is a crime without an apparent victim. It doesn't require sticking a Glock in someone's face and making off with their Rolex or Rolls-Royce. It only requires willingness to go around the meter. This could be called the "broken windows" theory of the grid: the more people steal electricity, the worse the problem gets. If consumers know the local electric provider isn't watching out for thieves, the more they'll be inclined to get "free" electricity for themselves. If civil society isn't working—if the cops on the beat aren't enforcing the law—people on the lower rungs of the social ladder quit believing in the system. Those people don't have a stake in society, and they aren't going to get a stake, so they might as well grab what they can at the first opportunity.

Integrity assures that an electric grid doesn't leak too much: the money at the top doesn't get siphoned off by crooked executives and the electricity being distributed on the wires in local neighborhoods doesn't get siphoned off by thousands of illegal taps.

Keeping theft and corruption at a minimum is imperative because theft robs the grid of the capital it needs. Sure, electric grids need fuel. But the thing that fuels the production and distribution of electricity—always, everywhere—is money. That brings me to the second imperative: capital. The business of generating

Electric and telecommunications wires above a market street, Delhi, India, 2016.
Source: Photo by author.

and distributing electricity requires constant infusions of cash. The electricity business is the world's most capital-intensive industry. In 2018, global spending on electricity infrastructure totaled $780 billion, an amount roughly equal to the GDP of the Netherlands.[5] By comparison, investment in the oil and gas sector—which for years has accounted for the biggest share of global energy spending—totaled some $760 billion.[6] The US electricity sector's capital requirements are similarly huge. From the mid-2000s to the mid-2010s, the American electricity business spent an average of $100 billion per year on new capital projects.[7] In 2016, that amount surged to $135 billion, which was spent on new generators, wires, poles, transformers, solar panels, and a myriad of other equipment.[8]

Given the vast amount of money needed to maintain and grow the electric grid, the need for integrity is obvious. Whether the grid is a tiny one with a few individuals or one serving millions of customers, someone, somewhere, has to come up with the capital to make the system work. Furthermore, the bigger the electric grid, the greater the capital requirements, which means someone has to borrow the money. With money at risk, banks and borrowers need lawyers to negotiate the contracts. Those banks and borrowers need courts to enforce the contracts and to penalize scofflaws. Repaying the loans requires electricity producers and distributors to be paid for the energy they provide. That requires sound money and secure payments. It also requires that the people who are collecting all that cash aren't stealing from the grid's owner.

Integrity and capital are closely related to the final imperative: fuel. Without fuel, you can't make electrons move. Edison's plant on Pearl Street used coal. Today, nearly 140 years later, utilities all over the world are still using coal. We are also using numerous other fuels, including uranium, natural gas, oil, wind, sunlight, biomass, falling water, and the heat of the earth. The choice of fuel depends on many factors, including geography, wealth, and availability. Wilfredo Roque used gasoline to fuel his generator because it was the only fuel available to him in rural Puerto Rico. Meanwhile, in Wyoming, America's biggest coal-producing state, about 90 percent of the electricity generated comes from coal-fired power plants.[9] In Djibouti, a tiny country on the east coast of Africa, the World Bank, African Development Bank, and others are financing the development of a new geothermal project that could be producing 100 megawatts of renewable electricity by 2020.[10]

Geothermal plants, wind-energy projects, and solar-energy projects don't have to buy fuel. But the operators of those facilities still need capital to keep their plants operating. Wind-turbine blades break. Solar panels need cleaning. Geothermal wells, pipes, and turbines must be maintained. Regardless of the fuel used, transmission lines and distribution lines that carry the electricity from generation stations to customers have to be regularly checked. Downed wires and old transformers must be repaired or replaced.

In the next few chapters, I will look at electric grids in Low-Watt and Unplugged countries to show how integrity, capital, and fuel are the key determinants of the quality, cost, and cleanliness of the electricity they consume. And one of the surest ways to hamstring the integrity of a society is to destroy its electricity infrastructure.

9

THE AMERICAN WAY OF WAR

We never had any intention of destroying 100 percent of all the Iraqi electrical power.

—GENERAL NORMAN SCHWARZKOPF[1]

By early 1991, General Norman Schwarzkopf, the commander of US and allied military forces in the Persian Gulf, was a media star. Over the previous few months, as the US military was amassing the hundreds of thousands of soldiers and hundreds of aircraft and ships it would use for its invasion of Kuwait and Iraq, Schwarzkopf had given numerous briefings to media outlets about the impending war. The coverage of the brawny tough-talking general, a West Point graduate whose nickname was "Stormin' Norman," had been almost uniformly positive, if not glowing. (The *New York Times* would later declare him "the nation's most acclaimed military hero since the midcentury exploits of Generals Dwight D. Eisenhower and Douglas MacArthur.")[2]

On January 30, 1991, Schwarzkopf conducted a media briefing in Riyadh, Saudi Arabia, to detail the extent of the damage that allied air and naval forces had inflicted on Iraqi targets over the preceding two weeks.[3] Military aircraft from the United States and other countries had flown more than 30,000 sorties. They had attacked Saddam Hussein's air-defense systems, combat aircraft, and targets associated with Iraq's leadership. And, as Schwarzkopf explained, "we also attacked power plants and telecommunication facilities."

About one-fourth of Iraq's electrical-generating facilities were "completely inoperative" and another "50 percent suffered degraded operations," the general told the press before continuing, "I think I should point out right here that we never had any intention of destroying 100 percent of all the Iraqi electrical power. Because of our interest in making sure that civilians did not suffer unduly, we felt that we had to leave some of the electrical power in effect, and we've done that."[4]

Stormin' Norman was being coy. A few days earlier, the US military had used a new blackout bomb that had proved amazingly effective at paralyzing the Iraqi grid. In what appears to be the first published reference to the new weapon, a January 18, 1991, article in the *New York Times* described it as "a top-secret warhead filled with carbon filaments designed to short-circuit electrical power grids." The warhead was mounted on Tomahawk cruise missiles that were used to attack Iraq's power stations during the early hours of the First Iraq War.[5] Known in military jargon as the BLU-114/B, the blackout bomb, just before it hits the ground, unleashes clouds of long strands of filament that then settle across power lines. The ensuing short circuit shuts down the local electric grid without damaging the transformers or generators.[6]

Blackout bombs were only part of the deadly arsenal unleashed on the Iraqi grid. In all, the US-led bombing campaign included 215 sorties aimed specifically at electricity infrastructure. Before the war, Iraq had about 9,500 megawatts of electricity-generation capacity. By the time the bombing stopped, that capacity had been reduced to about 300 megawatts. One analyst concluded that the attacks "virtually eliminated any ability of the Iraqi national power system to generate or transfer power."[7]

The US military's destruction of Iraq's electric grid was no anomaly. Destroying electricity infrastructure has become part of the American way of war. The common strategy of the US military in all of the major conflicts since World War II—Korea, Vietnam, and Iraq—has been to cripple the opponent's ability to produce and distribute electricity. As I will show in a moment, the obliteration of the Iraqi electric grid led to a humanitarian and public health crisis that resulted in the deaths of tens of thousands of Iraqi civilians. Intended or not, the US military's annihilation of Iraq's electric grid assured that Iraq would be a failed state for decades to come, where outbreaks of water-borne maladies like cholera would become common and tens of thousands of people would die from disease.[8] Furthermore—and the geopolitics of this outcome are truly stunning—by ruining Iraq's electric grid, the United States helped assure that Iraq grew closer to its longtime adversary, Iran.

•

THE KOREAN WAR PROVIDES THE first modern example of the US military's sustained attacks on electricity infrastructure. The first North Korean electricity project to be bombed was the Fusen hydroelectric plant, which was attacked on September 25, 1950—three months to the day after the Korean War began.[9] Two years later, American warplanes hit the Sui-ho dam, which at the time was the world's fourth-largest hydro plant, with 600 megawatts of generating capacity.[10] Of course, the US bombing was not limited to power plants. Dean Rusk, who later became secretary of state, said the United States bombed "everything that moved in North Korea, every brick standing on top of another."[11] By the time the bombs stopped falling, about 90 percent of North Korea's electric generation capacity had been destroyed. A memorandum written by air force officials in Washington in 1950 provided the rationale for bombing North Korea's electricity sector. Bombing "was expected to lower morale by putting out lights" and that it would "bring some electrically powered industry to a halt."[12] That same rationale was used to justify the bombing of power plants during the Vietnam War.

In 1965, the US military began a bombing campaign in North Vietnam called Operation Rolling Thunder, which would become the longest aerial-bombardment campaign in the history of American air power.[13] In the fall of 1966, the Joint Chiefs of Staff urged President Lyndon Johnson to begin "a concentrated attack" on electricity targets. Johnson approved the attacks, and by May 1967 some 85 percent of the generating capacity in North Vietnam had been destroyed. The bombing of North Vietnam's electricity infrastructure continued during President Richard Nixon's tenure in the White House with the bombing of the Lang Chi hydropower plant in mid-1972. In December 1972, as peace negotiations with North Vietnam began to stall, Nixon ordered even more attacks on electricity targets in the north. Some 166 bombing sorties were made against electricity targets, with special emphasis placed on power plants near Hanoi.

In a 1994 study of military attacks on electricity infrastructure, Thomas E. Griffith, a former air force pilot, said that theorists believed that attacking electric grids would reduce morale among the enemy and "inflict costs on the political leaders to induce a change." Instead, as Griffith explains, after the Vietnamese power plants were destroyed, the North Vietnamese leadership made sure that "priority users still had electricity. They did this through the use of some 2,000 portable generators and five underground diesel generating stations." Those generators allowed the North Vietnamese to continue the war effort in spite of the bombing. In all, by late 1972, about 90 percent of the generating capacity in North Vietnam had been destroyed. Despite this widespread destruction, the United States, of course, lost the war. Griffith explains, "As in Korea, the attacks on the North Vietnamese electrical power system did not prove decisive in achieving American policy goals."[14]

Although the bombing of power plants in North Korea and North Vietnam didn't prove decisive in either conflict, the US military would go on to target electricity infrastructure in Iraq. As it was in Vietnam, the destruction was almost total.

In June 1991, a few months after the US halted its military campaign in Kuwait and Iraq, the *Washington Post* published a piece by reporter Barton Gellman about the bombing campaign in Iraq and, in particular, why so much firepower had been aimed at electricity infrastructure. By that time, the First Iraq War had been over for four months. Hussein's forces were out of Kuwait, but the Baathist dictator was allowed to retain control in Baghdad. Gellman quoted one air force strategist who said that, by destroying Iraq's electricity grid, the United States had "imposed a long-term problem on the leadership that it has to deal with sometime. . . . Saddam Hussein cannot restore his own electricity. He needs help." An unnamed officer told Gellman that the United States targeted Iraq's electric grid because "it's a leveraged target set." Yet another unnamed strategist told Gellman, "Big picture, we wanted to let people know, 'Get rid of this guy and we'll be more than happy to assist in rebuilding. We're not going to tolerate Saddam Hussein or his regime. Fix that, and we'll fix your electricity.'"[15] Of course, Saddam Hussein didn't get deposed. And the United States didn't fix Iraq's electricity.

In October 1991, an international study team released a survey of Iraq's infrastructure. The team reported that "at least ten of 16 power stations visited were attacked on the first day of the war, and at least 14 were attacked multiple times." One power plant in Basra was attacked thirteen different times. The team also reported that blackout bombs had been used on "Beiji and Musayeb, the largest generating stations in Iraq."[16]

Before going further, let me stipulate that Saddam Hussein was a ruthless bastard. His reign of terror in Mesopotamia caused untold thousands of needless deaths, including up to 500,000 Iraqi soldiers killed during a futile eight-year war with Iran.[17] But it's also clear that Hussein was good at keeping the lights on. Between 1979 and 1990, per-capita electricity use in Iraq jumped by nearly 80 percent.[18] Over that same time period, according to World Bank data, infant mortality rates in Iraq declined by about 25 percent and per-capita GDP increased threefold.[19]

But after the First Iraq War, the country's electricity use plummeted. Shortly after the war, a United Nations report said the results of the aerial bombardment of Iraq were "near apocalyptic" and that Iraq had been relegated to a "pre-industrial age." Until the country had sufficient fuel and new electrical generation capacity, food "cannot be distributed; water cannot be purified; sewage cannot be pumped away and cleansed; crops cannot be irrigated; medicines cannot be conveyed where they are required; needs cannot even be effectively assessed."[20] Human Rights Watch also condemned the bombing, saying that the destruction of the grid "resulted in severe deprivation of clean water and sewage removal for the civilian population and paralyzed the country's entire health care system."[21]

The lack of clean water led to a cholera outbreak.[22] But cholera was only one of the maladies to hit Iraq. In the years following the First Iraq War, according to a report by the Center for Social Responsibility, Iraqi civilians "experienced one of the most rapid declines in living conditions ever recorded." The report said that Iraq's place on the Human Development Index "dropped from 96 in 1991 to 127 by the year 2000, on a par with the small southern Africa country of Lesotho. No other country has ever dropped so far, so fast."[23]

Iraq's decline was due to several factors, including ongoing sanctions against the country. But it's also clear that Iraqi civilians suffered due to the breakdown of civil infrastructure and, in particular, the lack of electricity. Water-purification and sewage systems were shut down, resulting in widespread water contamination. Hospitals ran short of power. The water contamination and other health-related problems resulted in a surge in civilian deaths, with credible estimates putting the number of Iraqis killed by disease at 70,000.[24]

A 1995 assessment by two scientists from the UN Food and Agriculture Organization produced an even higher number of dead civilians. Due to the war and economic sanctions, malnutrition and disease in the country had soared. The researchers estimated that as many as 575,000 Iraqi children had died since the end of hostilities in 1991.[25]

In 2003, the US military launched the Second Iraq War, and again Iraq's power plants took a beating. This time, the grid wasn't damaged by missiles or blackout bombs; instead it was crippled by saboteurs and looters who thrived in Iraq after the American invasion. Saboteurs toppled power lines to harass the government, while looters stole copper wire and sold it for scrap.[26] By mid-2004, saboteurs were attacking the country's high-voltage transmission lines an average of twice a week. The *New York Times* quoted a US military official who said that "insurgents in Iraq had begun to realize that with summer coming on, damaging the electrical and water infrastructure could sow widespread distrust and discontent with the occupation and its allies."[27] By 2007, four years after the Second Iraq War began, and near the height of the US troop buildup known as "the surge," the Iraqi grid was still a mess. More than $4 billion in US taxpayer money had been allocated to electricity projects, but Iraq's per-capita electricity use had fallen to about 750 kilowatt-hours per year, or roughly half of what it was prior to the second war.[28]

Iraq's electricity situation was embarrassing for the George W. Bush administration. In 2003, then vice president Dick Cheney famously claimed that the US military in Iraq would "be greeted as liberators."[29] Liberators or not, the US military and its provincial government could never produce enough electricity to meet the country's needs. Between 2003 and 2013, Michael O'Hanlon of the Brookings Institution published the Iraq Index, which repeatedly showed that electricity production consistently lagged demand. In 2011, a note published with the index said that the "available supply of electricity averaged about 56 percent of demand."[30]

In the post-Saddam, post-American-occupation years, Iraq has been wracked by civil war, sectarian violence, and corruption. By 2012, nine years after the Second Iraq War began, the Iraqi government had only been able to connect about 8,000 megawatts of generation capacity to the country's electric grid—that's 1,500 megawatts less than what the country had back in 1991, when the United States and its allies routed Iraq's military in one of the most lopsided wars of the past century.[31]

By the summer of 2018, Iraqis were suffering from electricity shortages so extreme that some consumers were getting as little as four hours of electricity per day.[32] Making matters yet worse was repeated sabotage of the country's electric grid. In 2018, gunmen planted explosives under a high-voltage power line near the Kasibah power plant, which caused blackouts in Baghdad.[33]

The breakdown in societal integrity contributed to soaring levels of electricity theft, which further exacerbated the shortages. By 2018, as much as 65 percent of all the electricity being generated by the Iraqi government was being stolen. Nor were Iraq's grid operators overly concerned with exerting their authority by stopping the thievery and collecting unpaid bills. That same year, the *Wall Street Journal* reported that "in 2015, its best year to date," the Ministry of Electricity collected just 12 percent of the revenue it was due from its customers.[34]

The ongoing electricity shortages led to mass protests in Basra and other Iraqi cities. The shortages also contributed to the country's ongoing public health problems. In August 2018, some 17,000 residents of Basra were admitted to hospitals due to illnesses caused by the lack of clean drinking water. The public health crisis led Basra's governor to demand that all of the energy companies in the oil-rich region should build water "desalination and sterilization treatment plants" if they wanted to continue working in the province.[35] About a month later, as the protests continued, Jane Arraf, a reporter with National Public Radio, visited Basra and quoted one resident, Ahmed Hussein, who said, "We don't have water, we don't have electricity, we don't have anything. Aren't we Iraqi? Aren't we people of oil?"[36]

The ongoing electricity shortages have forced Iraq to become more closely tied to Iran. In 2017, Iraqi officials signed a long-term natural-gas supply contract with the Iranians. The Iranian fuel was used to generate about 15 percent of Iraq's electricity.[37] In February 2019, Iran and Iraq signed a deal that will tie them even closer together. Despite objections from the Donald Trump administration, which was trying to impose sanctions on Iran, the Iraqi government signed an agreement with Tavanir, Iran's state-run grid operator, under which Tavanir will supply 1,200 megawatts of electricity to Iraq.[38]

The deal between Iraq and Iran shows how energy politics can trump old rivalries. During the 1980s, the two countries fought an eight-year war that left about one million people dead.[39] Despite that conflict, which ended in a stalemate, the two countries have tied their electric grids together, and it's apparent that neither country cares much about objections coming from the United States. About the time the electricity-supply deal was signed, Iraq's prime minister, Adel Abdul Mahdi, declared that "Iraq will not be part of the sanctions regime against Iran." The country's former prime minister, Haider al-Abadi, told the *New York Times* that the United States has to "look at the geopolitics of Iraq. We happen to be neighbors of Iran; the U.S. is not. We happen to have the longest border with Iran; the U.S. does not."[40]

Importing electricity from Iran has helped Iraq meet some of its needs. But even with the imports, ordinary Iraqis are still not getting enough electricity. Unable to rely on the main power grid, citizens in Basra, Baghdad, and other Iraqi cities often get the power they need from small neighborhood generators that provide a viable, if expensive, alternative to the state-run grid.[41] Those private electricity generators aren't unique to Iraq. They are also a defining feature of life in another country in the Middle East that, like Iraq, has been torn apart by decades of war, corruption, and factionalism.

10

BEIRUT'S GENERATOR MAFIA

> We pay two power electric bills. We pay one power electric bill to EdL and we pay the double of that to the private generators.
>
> —**JOSEPH EL ASSAD**, professor, Holy Spirit University of Kaslik[1]

About five minutes after we loaded our gear into the van that was taking us from Rafic Hariri Airport to the Lancaster Plaza hotel, our driver, Hussein Mousl, was explaining why Lebanon's electricity grid is such a mess. "I'm an American guy," he told us. Mousl used to live in Dearborn, Michigan. He had moved back to Beirut a few years earlier to take care of his parents. "What do you want to know? I will tell you exactly what is happening here."

When I asked him if he had heard of Lebanon's generator mafia, he smiled and quickly replied, "One hundred percent. Everyone knows about the generator mafia." Every month, Mousl, who lives in the Haret Hreik neighborhood near the airport, pays two electricity bills: $35 to Electricité du Liban (EdL), the state-owned utility, and $100 to the generator mafia, the people who own and operate generators and then sell electricity to customers in their neighborhoods. In Haret Hreik, he said, EdL only provides electricity for "six hours. Seven hours, maximum. The rest of the hours, we pay for the generator." Why not set up his own

Hussein Mousl, driver, Beirut, Lebanon, 2017

Source: Photo by Tyson Culver.

generator, or buy electricity from someone else? Mousl shrugged and gave a brief, resigned smile. "If I try to set up a generator to compete with them, they might kill me, or cut my wire."

Mousl had reason to be concerned about violence. Two weeks before we arrived in Beirut, a turf war had flared up in Sidon—a coastal town located about twenty-five miles (forty kilometers) south of Beirut—between rival generator mafias. Here's how the *Asharq Al-Awsat* newspaper reported it: "A dispute between owners of power generators escalated on Monday into an armed clash that left two people dead before the army intervened to contain the unrest." The story continued, "The incident occurred in the Barrad neighborhood in central Sidon when a verbal dispute erupted between a member of the Shehadeh family and Walid al-Saddiq, an owner of a power generator that provides electricity to subscribers." To quell the violence, the Lebanese army was deployed "to end the clash and the perpetrators were arrested."[2]

Throughout our visit in Lebanon, we kept hearing stories just like the one that Mousl had told us during our trip from the airport to the hotel. It didn't matter with whom we were talking: government employees, academics, former politicians. Nor did it matter where we were: Beirut, Byblos, Jounieh, or Kaslik. Nearly everyone in Lebanon pays two electric bills—one to EdL and another to the generator mafia—and no one, it seems, expects that to change.

The generator mafia thrives in Lebanon because of the ongoing economic and political disarray in the country. Indeed, the generator mafia provides a stark illustration of the fact that people will do whatever they have to do to get the electricity they need. If the government can't provide reliable electricity, someone else will; and for the people who need electricity, concerns about cost, air pollution, and climate change take a back seat to their need for power.

Understanding how the generator mafia became integral to everyday life in Lebanon requires an understanding of its tear-stained history. Lebanon is a war-torn cocktail of religious, political, and geographic alliances. It has eighteen officially recognized religious groups, including four Muslim sects and twelve Christian ones. About 27 percent of the population is Sunni Muslim, 27 percent is Shia, and 21 percent are Maronite Christians. Greek Orthodox make up 8 percent of the population and Druze about 5 percent, with the remainder of the populace split among Mormons, Jews, Catholics, and other sects. Making things more complicated is the 1943 agreement known as the National Pact. Under that arrangement, Lebanon's president must be a Maronite, the prime minister must be Sunni, and the speaker of parliament must be Shia.[3]

Lebanon's political-religious-military Rubik's Cube has been further complicated by the flood of refugees fleeing the war in Syria. Those refugees are putting additional strain on Lebanon's overtaxed power grid, which was already threadbare due to repeated conflicts with its neighbors. And while Lebanon's deadly civil war, which raged from 1975 to 1990, is officially over, the country continues to be paralyzed by its hyper-factionalized and deeply corrupt politics. A friend of mine, I'll call him Khaled, has lived in Beirut for many years. He has a doctorate in economics and has worked for the Lebanese government, as well as for private consulting firms and big international lenders. He described Lebanon as "a failed state dominated by populism, sectarianism, nepotism, and fascism."

That lack of integrity prevents Lebanon from improving its electric grid, which in turn imposes an energy tax that is felt across the entire Lebanese economy. And it's the generator mafia that collects much of that money. A 2016 analysis by Elie Bouri and Joseph El Assad, two academics at Holy Spirit University of Kaslik, found that blackouts are costing the Lebanese economy about $3.9 billion per year, or roughly 8.2 percent of the country's GDP.[4]

Blackouts are so common that many Lebanese use Beirut Electricity, an iPhone app created by savvy software engineer Mustafa Baalbaki. The app lets users know when the electricity will be off and will even send an alert to warn you when the blackout will start. During an interview in Beirut, Baalbaki, a shy, slightly built man who has a passion for writing software code and flying drones, told me, "I'm not fixing the electricity issue here in Lebanon. I'm just trying to help people to cope with it."

In an effort to understand why the generator mafia has gained so much influence in Lebanon, I made an appointment with Khaled Nakhle, a senior official at the Ministry of Energy and Water. When I met him at the ministry's office in central Beirut, I asked why the Lebanese government can't put the private generators out of business. He replied that EdL is losing some $1.3 billion per year, while the private generators are taking in as much as $2 billion per annum.[5] "It's a huge business," he said, "and it's very dangerous to interfere with this business." When I used the term "generator mafia" to describe the private generators, Nakhle quickly replied, "They are not all mafia. But some of them are and they have connections at all levels and they pay for those connections."

It was a remarkable moment. Nakhle, an official in the Energy Ministry, was admitting that the generator mafia bribes Lebanese politicians to make sure that EdL stays weak and blackouts persist. I heard the same thing from Maya Ammar, a model and architect in Beirut, who told me, "The one reason in Lebanon that we do not have electricity is corruption, plain and simple." Corruption "is so ingrained in the system that no matter what you do, you will find it." The electric grid, she continued, is "a microcosmic example of how this country runs."

Lebanon's weak government serves the interests of other countries and political players who are building their own bases of power in the country. Among the most powerful factions in Lebanon is Hezbollah, the Shia military-political entity that is backed by Iran. Hezbollah's influence in Lebanon helps determine who gets electricity. By one estimate, about 80 percent of people living in Hezbollah-controlled areas of Beirut get electric service from EdL, but they are not required to pay for that electricity.[6]

Hezbollah's military and political influence is so strong that in 2017 the leader of Hezbollah, Hassan Nasrallah, took a page out of George W. Bush's playbook and declared "mission accomplished" shortly after his soldiers helped the Lebanese army push the last group of ISIS fighters out of Lebanon and back into Syria. On its Twitter account, Hezbollah posted a video showing a group of buses that were to carry several hundred ISIS fighters and a few dozen of their injured comrades back to Syria.[7]

But Hezbollah's military and political power in Lebanon has also resulted in nearly constant conflict with Israel. And those conflicts have thrashed Lebanon's electricity infrastructure.

•

WHEN LEBANON'S CIVIL WAR FINALLY ended in 1990, the country was in shambles. Large sections of Beirut had been reduced to rubble. The electricity sector was particularly hard hit. Over the next half decade, with Syrian troops occupying much of Lebanon and the Israeli military occupying a southern swath, some progress was made in rebuilding the country's electricity infrastructure. That progress came to a halt in 1996, after Hezbollah launched rockets into Israeli territory. In response, the Israel Defense Forces launched a military operation in Lebanon. The attacks damaged the Jamhour power plant outside Beirut, as well as the Bsalim power plant northeast of the capital.[8]

In 1999, Hezbollah military forces launched a number of Katyusha rockets from bases in Lebanon into northern Israel, killing two civilians. In response, the Israeli military launched a series of air strikes on Hezbollah strongholds, as well as a number of bridges. The Israeli military attacked the Jamhour power plant.[9] They also bombed electrical facilities at Bsalim, Baalbeck, Bint Jbeil, and a power relay station near Sidon.[10] In 2000, the conflict flared up again and the Israeli military attacked Lebanon's electrical infrastructure again. According to a report by Human Rights Watch, attack helicopters fired missiles at the Bsalim electrical station, which destroyed three of the six transformers at the facility.[11]

In July 2006, there was more conflict between Hezbollah and Israel. From an electricity standpoint, the most destructive attack of that conflict was delivered by Israeli jets, which were likely American-made F-16s. After crossing into Lebanese airspace, the planes probably only needed about four minutes to reach their target: the 346-megawatt power plant at Jiyeh, which lies about fourteen miles (twenty-three kilometers) south of Beirut. Bombs from the jets hit several fuel storage tanks at the power plant. Over the next few hours, about 4.2 million gallons (16 million liters) of heavy fuel oil poured into the Mediterranean Sea.[12] The slick went on to pollute about 75 miles (120 kilometers) of the Lebanese coast.[13] A month later, Israeli forces hit Lebanon's electric infrastructure again, bombing the electric power plant at Sidon. That attack left some 25,000 residents in Sidon and Tyre without power.[14] The damage to the Lebanese power stations was estimated at more than $200 million.[15]

The decades of conflict between Hezbollah and Israel have had a clear result on the fuel needed to run the Lebanese electric grid. During my visit to Beirut, I spoke to Riad Chedid, an engineering professor at the American University of Beirut, one of the most prestigious universities in the Middle East. Before the civil war began in 1975, Chedid told me, "the state of the electric utility in Lebanon was so good that they were selling electricity to Syria. We didn't have any shortage." Prior to the civil war, Lebanon was getting about 50 percent of its electricity from oil and about 40 percent from hydroelectric facilities.[16] But over the past few decades, the country's hydro capacity has withered and by 2014 the country was only getting about 1 percent of its electricity from hydro.[17]

The result is that about 99 percent of the electricity now being generated in Lebanon comes from oil-fired generators.[18] That, in turn, causes extensive air pollution. A 2013 study by researchers from the American University of Beirut found that emissions from the diesel-fired generators owned by the generator mafia in the capital "may significantly increase inhalation exposure to harmful substances." It found that "carcinogen exposure from diesel generators is similar to that of smoking a few cigarettes per day," and that in areas of Lebanon where power cuts exceed three hours per day, the carcinogen exposure is "significantly higher." The same study found that electricity being supplied by the generator mafia in central Beirut is nearly three times as expensive as the electricity being supplied by EdL.[19]

Lebanon's heavy reliance on oil for electricity production makes the country an outlier when compared to the rest of the world. Since 1973, when about 21 percent of global electricity was produced by oil-fired generators, countries around the world have been sprinting away from petroleum because oil is usually more expensive than other fuels. Of course, a few oil-rich countries—like Kuwait—still generate more than half of their electricity from oil. But all over the world, electricity producers try to avoid using oil to produce electricity because of its high cost. Thus, by 2014, according to the World Bank, less than 4 percent of global electricity production was coming from oil-fired generators.[20]

A very large example of Lebanon's reliance on oil for electricity production can be seen at Jiyeh, south of Beirut. In 2013, the Lebanese government contracted with the Turkish firm Karadeniz Holding to deliver the *Orhan Bey* powership to Jiyeh and the *Fatmagül Sultan* powership to Jounieh, a short drive north of Beirut. The government is using the powerships because EdL is effectively bankrupt and therefore cannot borrow the money it would need to build the generation capacity necessary to meet demand. So it is

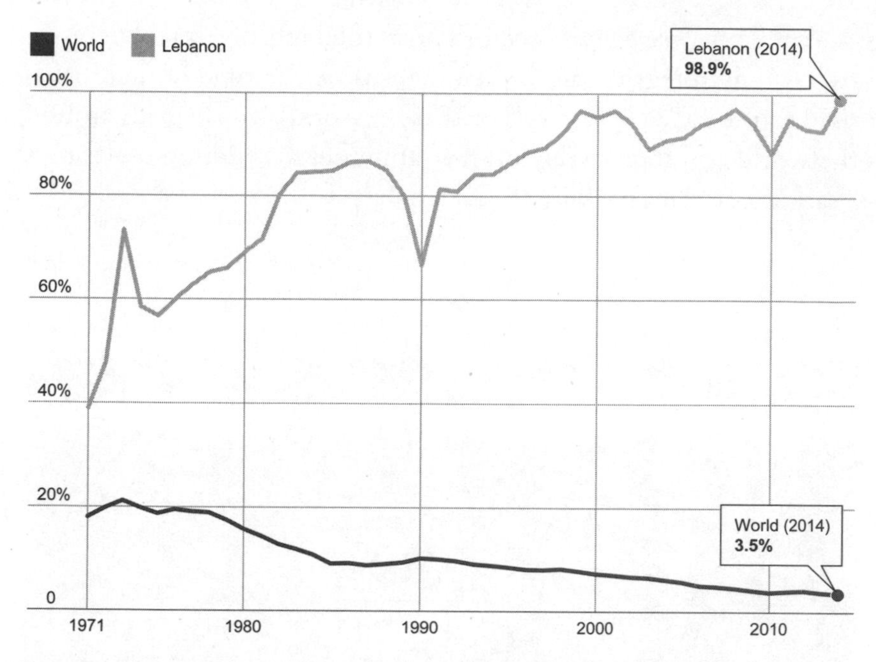

Percentage of Lebanon's Electricity Generated from Oil, Compared to World Average, 1971 to 2014

Source: OECD/IEA, "Electricity Production from Oil Sources (% of Total)," 2014, World Bank.

relying on the Turkish firm's ships, both of which are more than 130 meters (426 feet) long, to provide about 340 megawatts of electricity to the Lebanese grid.[21] The powerships produce electricity by burning heavy fuel oil, which sends clouds of dense, black smoke into the air. The pollution from the powerships' stacks is so noxious that the apartment buildings on the coastal hills above the power plant at Jiyeh have been abandoned. Similar pollution problems are apparent at Jounieh. There, I interviewed Khalid Al-Barouk, who owns a Range Rover repair shop that sits on a narrow street a few blocks north of the *Fatmagül Sultan* powership. "The pollution from the power plant is so bad, it destroys the paint on the cars around here," he told me in Arabic. Then, grabbing his nose, he said that on some days it's also difficult to breathe. Like other Lebanese, Al-Barouk has to pay the generator mafia to make sure his business has reliable electricity. Thus, he can see the plant. He can smell the plant. He just can't count on it to provide reliable electricity.

The Lebanese electric sector's near-total reliance on oil demonstrates how integrity and capital determine the type of fuel that is used to produce electricity. For these reasons, electric grids around the world continue to rely on the same fuel that Edison used at his Pearl Street plant back in 1882.

Since 2013, the powership on the right, *Fatmagül Sultan*, owned by the Turkish firm Karadeniz Holding, has been providing about 200 megawatts of generation capacity to the Lebanese electricity grid. Docked near the Jounieh power plant north of Beirut, the powership's generators burn heavy fuel oil, which causes local air pollution problems. Karadeniz Holding has deployed similar ships to Iraq, which like Lebanon suffers from ongoing electricity shortages.

Source: Photo by Tyson Culver.

11

IT'S NOT POSSIBLE TO KEEP THE LIGHTS ON WITHOUT COAL

> If we wait and do not act, then our factories will not be able to keep our people on the job with reduced supplies of fuel. Too few of our utility companies will have switched to coal, which is our most abundant energy source.
>
> —JIMMY CARTER[1]

Sanjay Kar Chowdhury didn't hesitate when I asked him about the importance of coal to India's electricity sector. Coal, he said, "is a lifeline. It is a lifeline of all the thermal power stations. Without coal you cannot survive."

When I met him on the campus of Jadavpur University in Kolkata, Chowdhury had been working at the Calcutta Electric Supply Corporation for about three decades. He was clearly proud of the company. He had worked in various parts of the organization and was now helping recruit and train employees. "Our company was formed in the year 1897 by royal charter by Queen Victoria. . . . We had our headquarters in London." Kolkata was "the second town or city to be electrified after London in the British Empire."

Chowdhury explained the electricity situation in Kolkata. The average resident of the sprawling, chaotic city of 4.5 million was consuming about 1,200 kilowatt-hours of electricity per year, or about 50 percent more than the Indian average. Not only were Kolkata residents using more electricity than many of their fellow Indians, Calcutta Electric Supply Corporation had not imposed any power cuts on its customers for the previous five years. Instead, Chowdhury said, the company had an electricity surplus. "This is winter time, we are having excess power and we are pushing the power into the market for power trading," he told me. He then added, "It's not possible to keep the lights on without coal."

About 75 percent of all the electricity in India is generated by coal-fired power plants, making it one of the world's most coal-dependent countries.[2] And its reliance on coal will likely continue for decades to come. In 2015, Jairam Ramesh, the former environment minister of India, told the *Washington Post*, "We cannot abandon coal." He went on, saying, "It would be suicidal on our part to give up on coal for the next 15 to 20 years, at least, given the need."[3] By early 2019, India had some 36,000 megawatts of new coal-fired capacity under construction.[4]

Whether all of that capacity gets completed remains to be seen. But India's continuing reliance on coal illustrates the growing bifurcation in the global electricity sector. Most of the wealthy countries are running away from coal at the same time that developing countries are building hundreds of gigawatts of new coal-fired generation capacity. This bifurcation—and coal's persistence in the global electricity mix—likely poses the single biggest challenge to the politicians and climate activists who want to see dramatic cuts in global carbon dioxide emissions.

The persistence of coal in India and other countries shows, again, that people will do whatever they have to do to get the electricity they need. It also shows that climate change concerns are not as important to decision makers as reliable electricity. Roger Pielke Jr. has dubbed this "the iron law of climate policy," which says, "When policies on emissions reductions collide with policies focused on economic growth, economic growth will win out every time."[5] In fact, Pielke's idea should be extended specifically to electricity and dubbed the iron law of electricity: when forced to choose between dirty electricity and no electricity, people will choose dirty electricity every time.[6]

Pielke, a professor at the University of Colorado as well as a highly regarded author on the politics of climate change and on sports governance, has since elaborated on the iron law. During an interview in Boulder, he explained it to me thusly: "The iron law says, we're not going to reduce emissions by willingly getting poor. Rich people aren't going to want to get poorer, poor people aren't going to want to get poorer." He continued, "If there is one thing that we can count on it is that policymakers will be rewarded by populations if they make people wealthier. We're doing everything we can to try to get richer as nations, as communities, as individuals. If we want to reduce emissions, we really have only one place to go and that's technology."

India's leaders have said they want to reduce the country's reliance on coal. But India cannot abandon the carbon-heavy fuel because it faces the world's biggest challenge when it comes to poverty and electrification. According to the Indian government, about 30 percent of all Indians are living in poverty and about three hundred million—a population nearly the size of the United States'—are living without electricity. Another ninety-two million don't have access to safe drinking water.[7]

Energy poverty in India is so extreme that it must be seen and smelled and coughed out of your own lungs to be fully comprehended. Trust me, if you spend much time in India's cities, you will be coughing. India's economy isn't so much coal fired as it is wood, twig, and dung fired. Everywhere in the country, it seemed, someone was burning something. Open fires in the markets and along the roadsides were being used for cooking. Even at office buildings in expensive sections of Delhi, cooking was being done outside over open fires. Delhi's air quality was so bad that visibility—even from the top of a ten-story hotel—was no more than a few hundred meters. During two weeks of travel, I saw men, women, girls, and boys cadging for scraps of wood, straw, and refuse—anything that would burn. Pity the trees of India. They are a beleaguered lot. In parks and along roadways, trees were under regular assault by children and adults looking for wood to burn. There were no fallen tree branches in the parks or along the roads. All of the fallen wood had been scooped up to use as firewood. In rural villages in West Bengal, I saw women using rice straw to cook their food. On a dusty road in Uttar Pradesh, west of the city of Agra, I saw a woman following a cow along the left side of the road. On top of her head, in a shallow basket, was the still-warm dung of the beast in front of her. On the houses in the nearby village, many roofs were covered with drying cow dung that would be used for home cooking and heating.

In addition to providing electricity, India's coal sector remains an integral part of the economy. It employs more than one million people and is India's second-largest employer, behind only the railroads. The black fuel matters to India's railroads too. Coal transport is the Indian railroads' main source of revenue.[8]

Thus, despite growing concerns about greenhouse gas emissions and climate change, electricity producers in India and other countries around the world are still relying on coal to fuel their generators. By 2017, more than 6,600 coal-fired power plants, with a combined capacity of about 2,000 gigawatts, were operating around the globe.[9] That coal-fired capacity accounted for nearly one-third of all global generation capacity.[10] Not only that, coal's share of global electricity production has remained nearly constant, at about 40 percent, since the mid 1980s.[11]

Coal's persistence continues despite efforts by environmental groups, super-wealthy donors, and politicians to reduce or eliminate the use of the carbon-heavy fuel. For instance, the Sierra Club has a Beyond Coal campaign that has been funded by Bloomberg Philanthropies, a charity started by former New York mayor Michael Bloomberg. In 2017, the charity announced that it was giving the Beyond Coal effort an additional $64 million. In all, since 2011, it has given $110 million to the campaign.[12] In its 2018 annual report, Bloomberg Philanthropies said it "increased our commitment to the Beyond Coal campaign in the US and

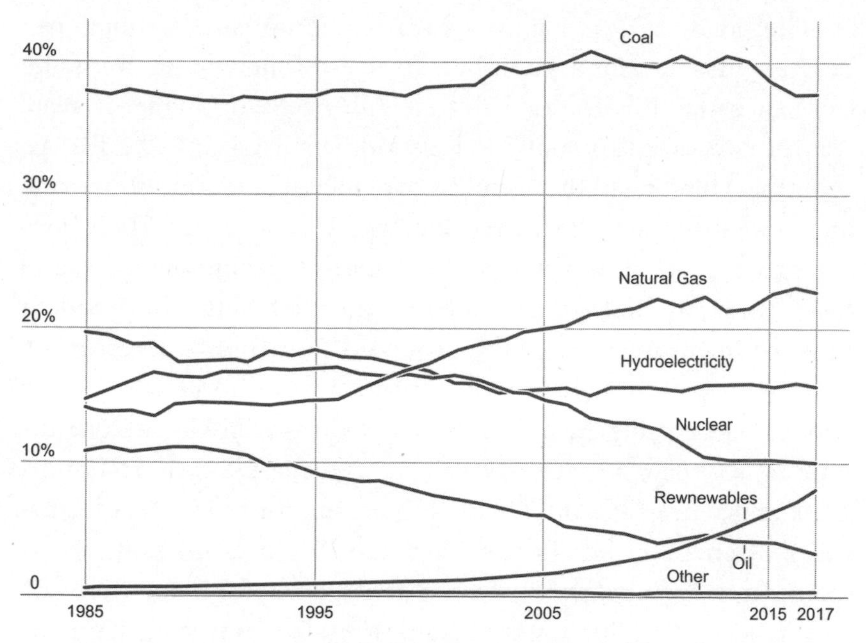

Share of Global Electricity Generation, by Fuel, 1985 to 2017

Source: BP Statistical Review of World Energy 2018.

expanded it to combat coal-fired power globally, starting first in Europe—where we've already helped to close some of the continent's dirtiest plants. In the US, through this effort, more than half of all coal plants have announced planned closures since 2010." To help in its anti-coal efforts, Bloomberg Philanthropies coproduced a feature-length documentary, *From the Ashes*, which premiered at the Tribeca Film Festival in 2017. According to the charity, the film "was shown at more than 300 screenings around the world, with many hosted by local community groups. *From the Ashes* was also released internationally by the National Geographic Channel."[13]

In June 2019, Bloomberg announced that his charity was upping the ante and would spend $500 million on advocacy, legal fights, and electoral strategies to close all of the coal plants in the United States. In a statement, Bloomberg Philanthropies declared the $500 million pledge was the "largest-ever philanthropic effort to fight the climate crisis."[14]

The anti-coal messaging from environmental groups can be seen in a headline published in a 2017 newsletter from the Oregon-based Environmental Law Alliance Worldwide. It read: "Burning coal kills people."[15] Environmental groups like ELAW and others have been successful in convincing lenders to stop loaning money for new coal-fired electricity projects. In 2013, several groups, including Friends of the Earth, Greenpeace USA, and the Center for Biological Diversity, sent a letter to US president Barack Obama, which said the proposed Thai Binh 2 power plant, a 1,200-megawatt coal-fired facility in northern Vietnam, would "emit unacceptable air pollution that will worsen climate disruption." A few days after the letter was sent, the US Export-Import Bank announced it would halt its financing for the project.[16] That announcement came shortly after the World Bank said that it would limit financing of coal-fired generation projects to "rare circumstances."[17] In 2018, World Bank president Jim Yong Kim announced that the bank had abandoned the last coal project on its books, the Kosova e Re plant in Kosovo, and would not support any more such projects.[18]

Given all this opposition and coal's heavy carbon footprint, why has the fuel been so durable? There are several reasons. First and foremost, it's cheap. For Asian countries, coal is about one-half to one-third of the price (on an energy-equivalent basis) of imported liquefied natural gas.[19] Second, coal prices are not affected by any OPEC-like entities. That means no single country, or group of countries, can reduce supply and therefore cause price spikes. Third, coal deposits are widely dispersed geographically. Fourth, the world has gargantuan coal deposits. At current rates of consumption, global coal reserves are projected to last another 134 years. The United States and Australia both have more than three hundred years' worth of coal reserves in the ground. Russia has nearly four hundred years' worth.[20] The large number of countries that produce and export coal allows buyers to compare prices from a number of suppliers and therefore get the best quality and price.

Finally, there is little technology risk. Coal-fired power plants have been in use for decades all over the world. Those many

Street scene, Kolkata, India, 2016.
Source: Photo by Tyson Culver.

decades of operational experience have helped coal-fired technology to become common around the world. The result: numerous engineering and construction companies from countries like Japan, China, and Malaysia can design, build, and finance coal-fired power plants. That, in turn, makes it easier to raise the capital needed to build those power plants.[21]

India isn't the only country that leans on coal to keep the lights on. In 2017, the United States produced more electricity from coal than India did. So did China, which produced more than 4,300 terawatt-hours of coal-fired electricity, or nearly four times as much as was produced in India that year.[22] By itself, China accounts for half of global coal consumption, and its consumption of the carbon-heavy fuel is unlikely to decline significantly anytime soon.[23]

Other Asian countries are also ramping up their coal-fired generation capacity. In 2016, Bangladesh announced plans to build a 1,320-megawatt coal-fired power plant. The $2.5 billion project will be built by two Malaysian state-owned firms. The fuel for the project will come from collieries in South Africa, Australia, and Indonesia.[24] In early 2018, Pakistan, which is chronically short of electricity, had some 3,200 megawatts of new coal-fired capacity under construction. Meanwhile, Vietnam was building nearly 11,000 megawatts of new capacity, Indonesia more than 12,000 megawatts, and the Philippines more than 4,500 megawatts.[25]

Booming coal consumption in Asia is occurring at the same time that utilities in the United States and Europe are slashing their reliance on the black fuel. In April 2017, England's electricity sector grabbed headlines in newspapers around the world when it went twenty-four hours without burning any coal. The *Guardian* newspaper declared that it was "Britain's first ever working day without coal power since the Industrial Revolution." Coal use had been declining in Britain, "accounting for just 9 percent of electricity generation in 2016, down from around 23 percent the year before, as coal plants closed or switched to burning biomass such as wood pellets. Britain's last coal power station will be forced to close in 2025, as part of a government plan to phase out the fossil fuel to meet its climate change commitments."[26]

In late 2017, Canada, France, the United Kingdom, and several states, provinces, and cities formed the Powering Past Coal Alliance, which pledges to eliminate the use of coal for electricity production by 2030.[27] The new alliance gained significant media coverage, but, as the *New York Times* noted, the members of the group "account for less than 3 percent of coal use worldwide."[28] While wealthy countries in the West are using less coal to generate electricity—or even trying to eliminate it—Low-Watt and Unplugged countries are increasing their coal-fired capacity.

The bifurcation of the electricity sector is particularly obvious in two Low-Watt countries: Turkey and Poland. In late 2017, a group of Turkish and Chinese companies announced plans to build a 1,320-megawatt coal-fired power plant in Adana, in southern Turkey.[29] Between 2006 and 2017, Turkey's electricity production from coal doubled.[30] The country now gets about 34 percent of its electricity from coal.

Poland also relies heavily on coal and it is planning to add more coal-fired capacity. That expansion demonstrates how history and geography affect the fuels used to generate electricity. Poland has a long history with Russia, nearly all of it bad. Given that history, it's not surprising that Poland's leaders have repeatedly said they will use their country's coal to generate electricity rather than rely on Russian natural gas. Poland is home to one of the world's biggest coal-fired power plants: Elektrownia Belchatow, a 5,258-megawatt plant that burns lignite, a low-rank coal that produces higher emissions during combustion than other forms of coal. Belchatow is also the largest single producer of carbon dioxide emissions in the European Union.[31]

In 2015, Prime Minister Ewa Kopacz said flatly, "Polish energy security is based on coal, and that is our priority." Poland gets about 85 percent of its electricity from coal, and it was expected to bring four new coal-fired power plants online in 2019.[32] Poland is also pushing to diversify its fuel mix to include more natural gas, as long as that gas doesn't come from Russia. In mid-2017, Poland began importing liquefied natural gas (LNG) from the United States, and in 2018 the Polish government announced a deal with Houston-based Cheniere Energy to supply it with LNG through 2042. The fuel will be unloaded at a Polish terminal on the Baltic Sea.[33]

In all, by early 2018, some 209 gigawatts of new coal-fired power plants were under construction around the world.[34] That estimate comes not from the coal lobby but from ardent coal opponents: the Sierra Club, Greenpeace, and CoalSwarm. To put those 209 gigawatts into perspective, that is roughly equivalent to all of the power plants—of all types—now operating in Germany.[35]

In fact, Germany illustrates coal's remarkable staying power. After the disaster at the Fukushima Daiichi power plant in Japan, Germany's environmental groups convinced the government to shutter all of the country's nuclear reactors. That, in turn, has forced Germany to rely more heavily on its coal-fired power plants. In 2017, the country's lignite-fired power plants had the same share in Germany's electricity mix as they did in 2000.[36] Germany's continuing reliance on coal has meant that the country—which has the largest economy in Europe—has not come close to achieving the emissions cuts targeted under the *Energiewende*, the name for the country's plan to overhaul its energy and power systems.

Between 2000 and 2017, Germany spent about $222 billion on renewable energy subsidies as part of its efforts to slash its greenhouse gas emissions. The country has pledged to slash those emissions by 40 percent compared to 1990 levels by 2020, and by 95 percent by 2050.[37] The total cost of the *Energiewende* could total more than $500 billion by 2025, and that figure only accounts for the investment needed in the electricity sector.[38] Despite the massive costs, in 2017, Germany's greenhouse gas emissions were at roughly the same level as they were in 2009. In 2018, the German government was forced to admit that it would not meet its 2020 emissions-reduction targets.[39] Also in 2018, under pressure from the Federation of Germany Industries, the German government stopped efforts by the European Commission to accelerate the pace of emissions cuts. Rather than make pledges to achieve bigger cuts, German chancellor Angela Merkel said, "we should first stick to the goals we have already set for ourselves. I don't think permanently setting ourselves new goals makes any sense."[40] In early 2019, the German government announced that it would close all of its coal-fired power plants by 2038 and that it instead will be relying on renewables for 65 to 80 percent of its electricity.[41] But it remains to be seen if that target will actually be achieved.

Japan also demonstrates coal's enduring popularity. In 1997, Kyoto was the site of the negotiations for the first international treaty to reduce emissions, the Kyoto Protocol. But rather than abandoning coal, Japan, like Germany, continues to rely heavily on the carbon-heavy fuel. In the wake of the disaster at the Fukushima Daiichi nuclear plant in 2011, Japan shuttered all fifty-four of its nuclear plants. By 2018, it had restarted just seven of those fifty-four. The loss of that nuclear capacity, combined with the need to keep electricity prices in check, has induced Japan to turn to more coal-fired electricity. Between 2016 and 2018, Japan opened eight new coal-fired generation plants, and the country has plans to build about thirty more plants with a total capacity of nearly 17 gigawatts.[42] By 2030, Japan expects to be getting 26 percent of its electricity from coal-fired power plants. (Under prior pledges, the government had said it would cut coal's share of electricity generation to 10 percent.)[43]

All of Japan's new coal plants will be built using ultra-supercritical combustion, which converts more than 40 percent of the heat energy in coal into electricity. (Coal-industry officials call the new plants "high-efficiency, low-emission," or HELE, projects.) Older coal-fired plants commonly use subcritical combustion, which converts about 33 percent of the heat energy in coal into electricity.[44] Although the higher-efficiency coal plants will help Japan reduce some of its emissions, the country's emissions have not fallen significantly. In 2017, they were at about the same level as they were in 1997, the year that Japan hosted the climate meeting in Kyoto.[45]

In short, coal has persisted since the days of Edison and Insull because it can be used to produce the vast quantities of electricity the world demands at prices consumers can afford. As concerns about climate change get more attention, the coal business will be pressured by environmental activists and governments. But even as those pressures mount, the need for abundant, reliable electricity will continue to be a priority for countries and corporations all over the world.

In the next section, I will look at some of the world's biggest and fastest-growing industries, and show how all of them depend on abundant, reliable electricity.

PART THREE

The View from on High-Watt

12

THE NEW (ELECTRIC) ECONOMY

Our computer and communications systems and operations could be damaged or interrupted by fire, flood, power loss.

—**AMAZON**, in its 2017 Form 10-K filing, discussing "risks related to system interruption"[1]

On August 2, 2018, Apple became the first American publicly traded company to attain a stock-market valuation of $1 trillion.[2] About a month later, Amazon became America's second $1 trillion company.[3]

Apple and Amazon—along with Alphabet (the parent company of Google), Facebook, and Microsoft—are among the richest and most powerful companies of our time. Those companies, which I call the Giant Five, not only dominate the stock market, they also infiltrate nearly every aspect of our digital lives, providing us with TV, movies, music, news, software, hardware, and a myriad of other gizmos and must-haves. The Giant Five are the biggest and most powerful players in the new economy, an economy dominated not by production of physical goods like steel or automobiles, but one dependent on digital information, and therefore on electricity.

While Wall Street stock pickers—as well as regulatory agencies from the United States and Europe—have focused on the Giant

Five's dominance, the essentiality of electricity in their astonishing growth has been largely overlooked. The Giant Five dominate the Information Age because all of them are able to produce, manipulate, transfer, and store gargantuan quantities of digital information. These five companies operate some of the world's biggest and most powerful data centers—the computer-packed warehouses that are the digital smelters of the Information Age—and every one of those data centers needs ultrareliable electricity.

Electricity is the fuel of the Information Age. The Giant Five can't do business without it. Therefore, over the past few years, the five colossi have spent tens of billions of dollars deploying their own private electric grids, so that their data centers won't be affected if—or rather, *when*—the local electric grid experiences a blackout.[4] By creating their own private grids, the Giant Five have become, in effect, electric utilities. They've done so because the rule of the new economy is simple: the bigger your network, the more valuable it is to those who own it and use it.[5] Big data rules. The bigger your data set, the more it is worth. That means the Giant Five must continue building new data centers to keep up with the inexorable growth of Internet traffic, and in particular with the amount of video being consumed online. Each new data center requires additional on-site electric generators to ensure uninterrupted supplies of electricity.

By 2017, about four hundred hours of video—everything from *Leave It to Beaver* reruns to Manu Ginobili highlight clips—were being uploaded to YouTube every minute.[6] Networking giant Cisco has estimated that the amount of data created by digital devices will quadruple between 2016 and 2021. That, in turn, will require global data-center capacity to quadruple over that time period.[7]

Crunching large amounts of digital data has always required large quantities of electricity. The earliest computers were energy hogs. In 1943, the US military agreed to provide funding for ENIAC, short for Electrical Numerical Integrator and Computer, which was the world's first general-purpose electronic computer. Built at the University of Pennsylvania by a team headed by J. Presper Eckert and John W. Mauchly, the massive machine was a wonder of the pre-transistor, pre–integrated circuit era. ENIAC contained 17,468 vacuum tubes, along with 10,000 capacitors, 1,500 relays, 70,000 resistors, and 6,000 manual switches.[8] In 1946, when ENIAC was finally completed, it required 174,000 watts (174 kilowatts) of power. The computer consumed so much electricity that when it was switched on, it allegedly caused lights in parts of Philadelphia to momentarily dim.

Since 1943, companies in Silicon Valley and around the world have drastically reduced the electricity intensity of computing. But the Information Age is still about the merger of bits and electrons. Sure, the ever-growing efficiency of computers—making them smaller, faster, lighter, and cheaper—has drastically reduced the amount of electricity needed for computation, but those efficiency gains keep colliding with ever-increasing volumes of digital data. Crunching lots of bits requires harnessing lots of electrons, and the key technologies of the twenty-first century—artificial intelligence, video-on-demand, and robotics—require handling staggering quantities of both bits and electrons.

The importance of electricity to the Giant Five's businesses can be seen by looking at how their electricity use has grown in near lockstep with their overall value. Between 2012 and 2017, the combined electricity use by those five companies jumped by 146 percent. Over that same time period, their combined market capitalization jumped by 228 percent. By the end of 2017, the Giant Five were worth some $3.4 trillion.[9]

The Giant Five have become so valuable and politically powerful that governments around the world are struggling to regulate and tax them. In a 2017 article in the *New York Times*, Farhad Manjoo wrote that we are at a "great turning point in the global economy," in which the Giant Five (Manjoo calls them the Frightful Five) will "control the technological platforms that will dominate life for the foreseeable future." Manjoo's piece, titled "Can Washington Stop Big Tech Companies? Don't Bet on It," claims that, despite their dominance in the marketplace, the tech giants "have been spared from much legislation, regulation and indeed much government scrutiny of any kind."[10] In fact, the Giant Five aren't just too big to fail. They may be too big to regulate.

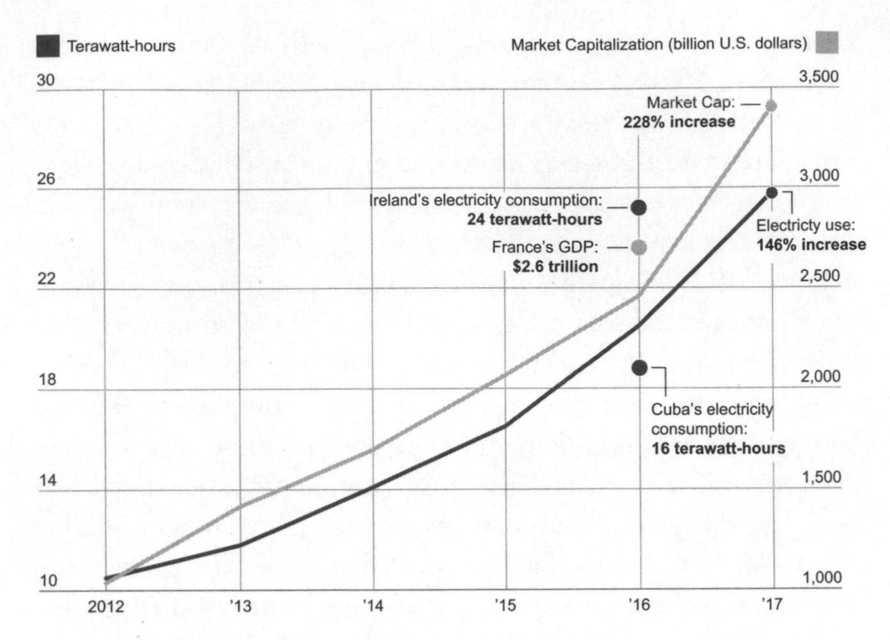

Combined Electricity Consumption Increases and Market Capitalization Increases of the Giant Five, 2012 to 2017

Sources: Company filings, Macrotrends.net, author estimates, CIA.

In mid-2017, the European Commission hit Alphabet with a $2.8 billion fine for abusing its near monopoly on Internet search to "give illegal advantage" to its own shopping service. The EU's competition commissioner claimed that the company "denied other companies the chance to compete" and that it had promoted "its own shopping service in search results" while "demoting those of competitors. What Google has done is illegal under EU antitrust rules."[11] Shortly after the European Commission announced the fine, Alphabet said that it would appeal the penalty, a move that will likely delay any final decision on the case for several years.[12] Even if Alphabet does eventually pay the fine, it will likely be only a minor setback. Why? Google controls about 80 percent of the market in desktop searches. Its nearest competitor, Bing, owned by Microsoft, has just 7.5 percent. Alphabet is even more dominant in mobile and tablet searches, with a staggering 92 percent market share.[13]

While Alphabet dominates the search business, Apple dominates in iPhones, music sales, and—it appears—in not paying taxes on the hundreds of billions of dollars it collects from its customers. By April 2019, Apple was sitting atop a cash hoard of $225 billion, a sum that exceeds the GDP of numerous countries, including places like Vietnam and Portugal.[14] Apple's enormous profits and cash stockpile have led it to stash huge amounts of money in tax havens such as the island of Jersey. In 2017, the BBC reported that Apple used loopholes in tax laws that allowed the company to funnel all of its sales outside of the Americas "through Irish subsidiaries that were effectively stateless for taxation purposes and so incurred hardly any tax." The BBC report estimated that in 2017 Apple "made $44.7 billion outside of the US and paid just $1.65 billion in taxes to foreign governments, a rate of around 3.7 percent. That is less than a sixth of the average rate of corporation tax in the world."[15]

Amazon's ultraconvenient shopping experience has been a boon for consumers. But its dominance is hurting small- and big-town retailers alike. As one analyst wrote, Amazon could "destroy more American jobs than China ever did." Retailers like Macy's, The Limited, Best Buy, and others are being hammered by Amazon's low prices and ease of use. Some 6.2 million people work in stores that sell furniture, appliances, books, and general merchandise; those are the same stores that are being undermined by Amazon.[16] By mid-2019, retailers had announced plans to close more than 7,000 retail stores in the United States.[17]

Facebook may let you keep track of your friends, but it won't let you block Mark Zuckerberg.[18] Zuckerberg's company can do pretty much what it wants because it dominates the social-media landscape. Some 76 percent of Facebook users in the United States log on to the company's site every day.[19] That loyalty has made the company a giant in both advertising and news. Along with Alphabet, Facebook controls about 70 percent of the digital advertising business in the United States, an industry worth some $73 billion per year.[20] The company's role in advertising may not bother consumers, but its growing importance as a news outlet provides plenty of cause for concern. Facebook tailors the news feeds delivered to individual users based on algorithms that predict what those users want. That means that Facebook users seldom see news stories on the site that offer a contrary perspective. As Roy Greenslade put it in an article published by the *Guardian*, "The Facebookisation of news has the potential to destabilize democracy by, first, controlling what we read and, second, by destroying the outlets that provide that material."[21]

Facebook's enormous growth has come at the expense of newspapers. In 2016, US newspaper advertising revenue totaled just $18 billion. That's a sharp drop from $50 billion a decade earlier. In response to the rise of Facebook, some of America's biggest newspapers—the *Washington Post* (which is owned by Amazon founder Jeff Bezos), *Wall Street Journal*, and *New York Times*—along with a group of smaller print and online publications have banded together to try to convince Congress to give them an antitrust exemption that would allow them to more effectively compete against Facebook.[22] While such an exemption will be difficult to obtain, the fact that rival newspapers are working together to take on Facebook is indicative of the threat that they face.

The dominance of the Giant Five in their respective markets is as great, or greater, than the monopolies of the past. The entrepreneurs who created the Giant Five—Jeff Bezos, Bill Gates, Mark Zuckerberg, Sergey Brin, and the late Steve Jobs—have become household names and will be remembered alongside such dynastic families as Carnegie, Vanderbilt, and Rockefeller. They are familiar to us not only because of their staggering wealth (in 2018, Bezos was worth some $166 billion), but also because of their influence over our everyday lives.[23] In 2016, demographer and author Joel Kotkin called the big technology companies "a fearsome threat whose ambitions to control our future politics, media, and commerce seem without limits." Yes, Kotkin admits, they "may be improving our lives in many ways, but they also are disrupting old industries—and the lives of the many thousands of people employed by them."[24]

In 2018, Greg Ip, a reporter for the *Wall Street Journal*, wrote that "Standard Oil Co. and American Telephone and Telegraph Co. were the technological titans of their day, commanding more than 80 percent of their markets. Today's tech giants are just as dominant," with Alphabet's Google handling 89 percent of all Internet searches. About 95 percent of young adults on the Internet use a Facebook product. By the end of 2018, Amazon had captured nearly half of the US e-commerce market.[25] "The monopolies of old and of today were built on proprietary technology and physical networks that drove down costs while locking in customers, erecting formidable barriers to entry," Ip explained. "Just as Standard Oil and

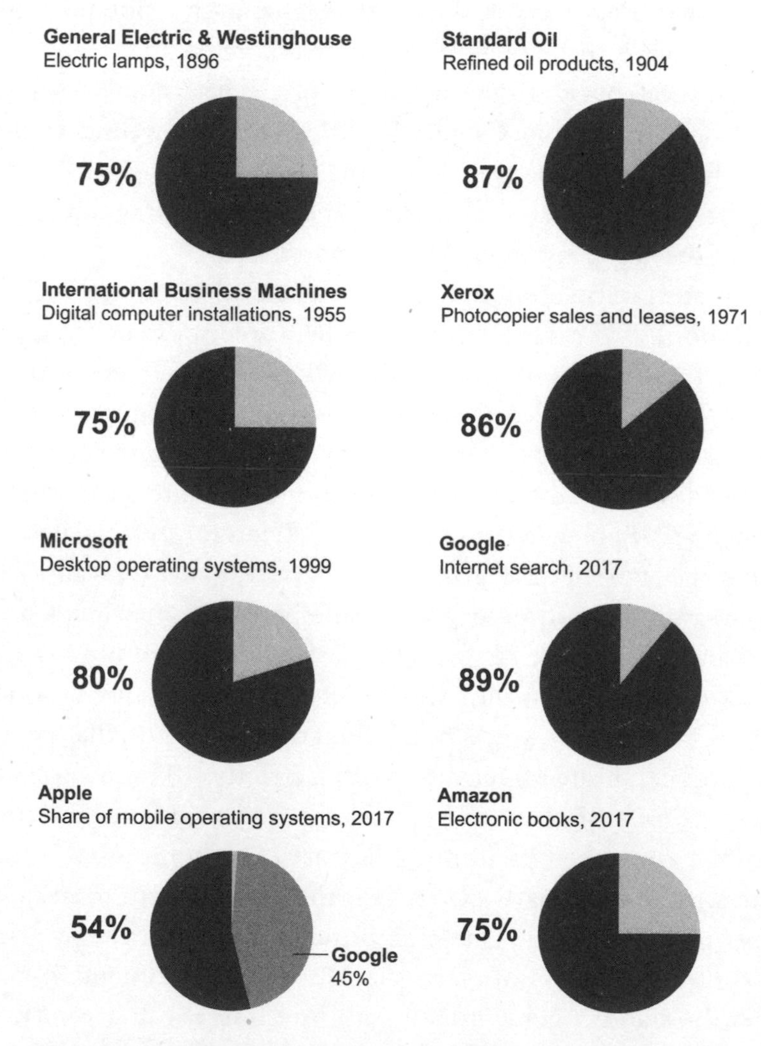

Comparing the Monopolies of Old with the New Tech Giants

Source: Wall Street Journal.

AT&T were once critical to the nation's economic infrastructure, today's tech giants are gatekeepers to the internet economy."[26]

For the Giant Five to remain the gatekeepers of the Internet economy, they have to do all they can to make sure their data centers have what computer wonks call "maximum uptime." Amazon's customers want to shop for everything from Crest to Levis, and they want to be able to order those products 24/7/365. The same is true for Google's search business and Apple's App Store. Facebook's advertisers want to know that whenever a potential customer is online, they are never far from a targeted ad.

Even short outages, due to the loss of electricity or computing power, can cost millions. On July 16, 2018, Amazon's website crashed in the middle of Prime Day, one of the company's biggest sales days of the year. Analysts later calculated that the sixty-three-minute outage cost the company about $100 million in sales.[27]

The push for maximum uptime can be understood by looking at Facebook's data center in Prineville, Oregon, which requires about 28 megawatts of power.[28] To make sure its servers don't go dark if the local grid experiences a blackout, the company needs 28 megawatts of on-site generation that can be up and running at full speed in a minute or two. Very large data centers can require as much as 100 megawatts of capacity.[29] Therefore, each data center has its own electric grid. These private grids typically have several large diesel-fired engines made by companies like Caterpillar and Cummins. For instance, a popular Cummins engine has sixteen cylinders, displaces ninety-five liters, and produces about 4,000 horsepower. When bolted to a generator, that engine can produce about 3 megawatts of electricity.[30] The data-center grids also have big banks of batteries. (Lead-acid batteries, similar to the ones used in automobiles, are commonly used in data centers.) The grids have switch gear that allows them to instantly supply power from the batteries to the computers, thereby assuring continuous flows of electricity during the moments it takes to get the standby generators up and running at full capacity. Finally, each data center usually has a large tank filled with diesel fuel that allows the on-site generators to provide the electricity

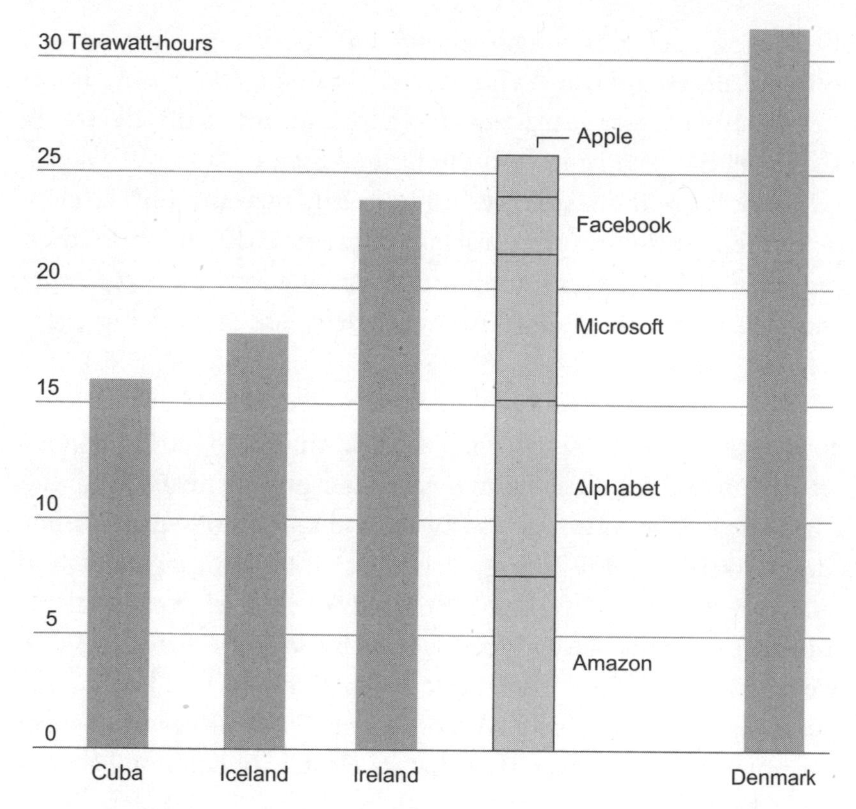

Electricity Consumption of the Giant Five—Alphabet, Amazon, Apple, Facebook, and Microsoft—in 2017

Sources: Company reports, CIA, author estimate.

so that the data center can operate independently from the local grid for hours, or even days.[31]

How much electricity do the Giant Five consume? In 2017, by my count, the Giant Five's electricity use totaled 26 terawatt-hours.[32] That's about as much as what's consumed by Ireland, a country of 4.8 million people.[33] But keep in mind that 26 terawatt-hours is a minimum number. I got that sum from energy-consumption data published by the companies themselves. But Amazon, which has more data centers than the other four companies, doesn't disclose its electricity use.

To put that electricity use in perspective, consider this: all of the data centers in the United States use about 70 terawatt-hours of electricity per year, which amounts to about 2 percent of total domestic electricity use.[34] Thus, the Giant Five's 26 terawatt-hours of electricity use per year is equal to about a third of the electricity used by all the data centers in the United States.

Not only are the Giant Five using Ireland-size amounts of electricity, their consumption is soaring. Between 2012 and 2017, their combined electricity consumption grew by about 25 percent per year. That means the Giant Five's electricity use is doubling every three years or so.[35]

Their rising electricity demand has led the Giant Five to move into the electricity-generation business. In 2015, Microsoft announced a $200 million data-center project near Cheyenne, Wyoming.[36] A year later, the software and technology giant struck a deal with Black Hills Energy that allows it to share its generation capacity with the utility. During periods of peak electric demand, Microsoft's generators will feed electricity into the grid. The deal saves the utility money because it doesn't have to build new generation capacity. In return, Microsoft gets more utilization out of its generators and an assurance that its electricity costs will be relatively stable.[37]

Microsoft is also building new electric generation capacity in Europe. In 2017, the company announced that it would build 18 megawatts of new gas-fired generation to provide the electricity it needed for a data center it was building near Dublin, Ireland. According to the *Irish Independent*, Microsoft had to build the new capacity "because the local transmission network hasn't been upgraded quickly enough to meet a surge in demand" from new data centers.[38]

While all of the Giant Five are massive users of electricity, Amazon almost certainly uses more than any of the others. I am qualifying my claim because Amazon doesn't disclose its energy use. Therefore, in my calculations, I assumed that Amazon's electricity use was the same as that of Alphabet, the next largest electricity user in the Giant Five. In 2017, Alphabet consumed 7.6 terawatt-hours of electricity.[39] That's about as much electricity as what's consumed by the country of Zimbabwe.[40]

Although Amazon does not disclose its electricity-usage data, even a cursory analysis of its business shows that it must be using far more electricity than the other members of the Giant Five. Amazon could be using two, three, or even four times as much electricity as what is being used by Alphabet.[41] Why? Amazon Web Services (AWS) has become an Internet-and-electricity colossus. Over the past few years, at the very same time as Amazon has taken over the online-shopping business, AWS has become the world's biggest provider of cloud computing and storage, a business known as infrastructure as a service (IaaS).

In 2016, Synergy Research Group estimated that AWS controlled 45 percent of the global IaaS market, meaning it had twice the market share of the next three companies—Alphabet, IBM, and Microsoft—combined.[42] By 2018, AWS itself was placing its share of the global cloud business at 52 percent.[43] That dominant market share means that if Amazon's data centers were to all go offline at the same time, huge chunks of the Internet would simply disappear. Amazon's servers handle the data needs of thousands of companies, including brands like Netflix, Pinterest, Lamborghini, the National Football League, and Major League Baseball.[44] In addition, AWS has become a major provider to the US government. In 2013, it won a $600 million contract to provide services to the Central Intelligence Agency.[45] In 2017, the company announced the formation of AWS Secret Region, a business designed to provide computing power to US government intelligence agencies, including the CIA. In its announcement, AWS claimed that it was the "first and only commercial cloud provider to offer regions to serve government workloads across the full range of data classifications, including Unclassified, Sensitive, Secret, and Top Secret."[46] By early 2019, AWS was considered a leading contender for a $10 billion computing contract with the US Department of Defense.[47]

In addition to all the standby generation capacity that they have installed at their data centers, the Giant Five are also trying to burnish their green credentials by buying oodles of renewable-energy capacity. But the Giant Five aren't buying renewables because the electricity from wind and solar projects is superior to conventional electricity; they are doing it as part of their marketing campaigns. The Giant Five can't afford to let public opinion go against them when it comes to their massive energy consumption. Buying renewable energy helps them in the court of public opinion.

Amazon has built or contracted for several hundred megawatts of wind-energy capacity and says that it is "committed to using 100 percent renewable energy."[48] Microsoft has purchased more than 500 megawatts of wind-energy capacity and claims it has been "carbon neutral" since 2012.[49] In 2017, Alphabet announced that it had signed twenty agreements to purchase nearly 2.6 gigawatts of renewable-energy capacity in the United States, Europe, and South America.[50]

When Apple opened its new "spaceship" campus in Cupertino, California, it claimed that the 175-acre facility would be run completely by renewable energy, including 17 megawatts of rooftop solar capacity.[51] Apple has also claimed that its Maiden, North Carolina, data center runs on solar energy. That claim didn't convince James Hamilton, a top engineer at AWS. In 2012, Hamilton scrutinized Apple's claims and estimated that, for the data center to run completely on solar panels, each square foot of data center would require 362 square feet of solar panels. In all, Hamilton estimated that powering Apple's 500,000-square-foot data center would require about 6.5 square miles (16.8 square kilometers) of solar panels, an amount that Hamilton concluded is "ridiculous on its own."[52]

While renewable-energy claims may help the Giant Five put some green spin on their operations, the reality is that the electrons that are delivered by the grid are indistinguishable from one another. The Giant Five's corporate offices and data centers are energized by whatever electricity is being delivered through the wires of the local grid. As one energy analyst explained in a 2018 article on energy demand and information infrastructure, "Regulatory legerdemain allows one to pretend that a purchase of wind or solar power somewhere else is, by fiat not fact, attached to a particular building. This is no different than the once-popular idea of paying to plant a tree somewhere as an indulgence for taking a trip on an aircraft." He continued, "No amount of vigorous PR changes the fact the aircraft necessarily burns aviation fuel. And no similar 'credit' changes the fact of a data center's 24x7 [*sic*] need for power."[53]

Although some of the Giant Five's green claims are suspect, there's simply no denying that they are electricity behemoths. By 2018, Amazon had about 3,656 megawatts of conventional generation capacity.[54] In addition, the company has contracted for about 1,016 megawatts of renewable generation capacity.[55] Thus, Amazon's total generation capacity was about 4,672 megawatts.

Amazon's secrecy makes calculating exactly how much generation capacity the Giant Five own particularly difficult. The companies don't include their diesel-fired generation in their corporate reporting. They only brag about their renewable portfolios. Nevertheless, based on their corporate and environmental reports, I estimate that by the end of 2018 the Giant Five's generation capacity—conventional and renewable—was about 12,100 megawatts.[56] That's more than five times as much as the Hoover Dam, which has 2,080 megawatts of capacity.[57]

To put that 12,100 megawatts of electric generation capacity into context, consider this: The Los Angeles Department of Water and Power has about 7,600 megawatts of generation capacity.[58] Austin Energy, the city-owned utility in my adopted hometown, has about 4,000 megawatts of generation capacity, which allows the utility to provide electricity to about one million people.[59] Thus, when adding up all of their conventional and renewable generation capacity, the Giant Five have more generation capacity than the cities of Austin and Los Angeles *combined*.

In summary, the scale and influence of the Giant Five—regardless of whether the issue is their electricity use, their market value, or their influence over the media—stagger our ability to comprehend. The old model of monopolies and capitalism—the one that depended on oil, coal, and manufacturing—hasn't gone away. Instead, it has been superseded by one in which companies gain market share and trillion-dollar valuations because they are able to manage massive quantities of digital information requiring terawatt-hours of electricity. The oil and steel trusts wanted to control the market for their commodities and therefore control profits. The five colossi want to know everything about you so they can sell you anything and everything you might want. They want your money too. All five have their own payment systems. Take your pick: Microsoft Pay, Apple Pay, Google Pay, Amazon Pay, or Facebook Messenger.[60]

The Giant Five's foray into payments is only one example of the electrification of our money. That trend—the merger of electrons, bits, and cash—has been underway for decades.

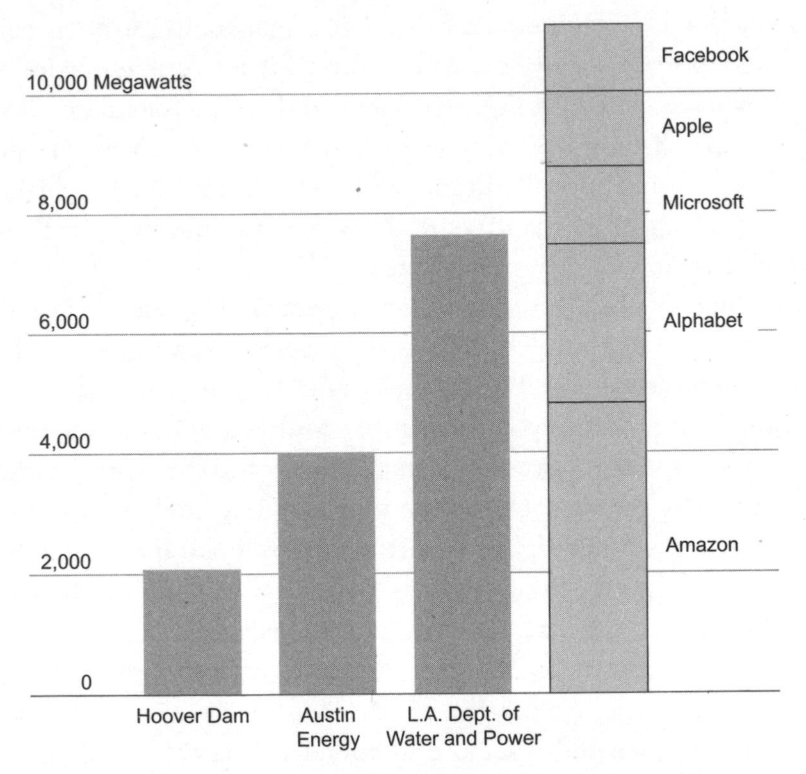

Total Renewable and Conventional Electricity Generation Capacity of the Giant Five, 2018. To assure that their data centers have highly reliable electricity, the Giant Five technology companies have installed thousands of megawatts of generation capacity. They have also bought or contracted for thousands of megawatts of renewable energy capacity.

Sources: DataCenterfrontier.com, Greenpeace, Bloomberg New Energy Finance, author estimates.

Chapter 13
ELECTRIFIED CASH

Electricity is really just organized lightning.

—GEORGE CARLIN

In its 2015 annual report, Visa Inc.—the world's largest credit card company—made a remarkable admission: Visa told its investors that the company's key asset isn't the credit card at all.[1]

"People often think of the card as the innovation," the company explained. But "the physical card was just the mechanism to access the real innovation—the network—which provided the ability to conduct and grow commerce safely and easily in partnership with our clients worldwide."[2] In 2011, Rick Knight, Visa's head of global systems operations and engineering, put it more simply: "Most people think of us as a financial institution, but the network is the brand."[3]

By 2018, there were some 3.3 billion Visa cards in use around the world.[4] To make sure all of those cards can connect to its network, the company has built the digital equivalent of Fort Knox. Somewhere on the East Coast—for security reasons, Visa won't reveal the exact location—the card giant operates a heavily fortified data center known as Operations Center East. The roads leading to the facility are equipped with hydraulic bollards that can stop a car traveling at fifty miles (eighty kilometers) per hour. The place even has a moat. The building that holds the company's computer servers is designed to withstand earthquakes and winds of up to 170 miles (274 kilometers) per hour.[5] In case of a grid failure, Visa keeps about 14,000 heavy-duty batteries and a battalion of diesel-fired generators with about 56 megawatts of capacity at the ready.[6] If the electric grid fails, Visa has enough diesel fuel stored on-site to keep Operations Center East humming for a week.[7]

Whether it's Visa's network or the emergence of cryptocurrencies like Bitcoin, we are seeing the rapid electrification of money. Whenever we charge anything, we are relying on electronic information that's sent through wires to computer systems that then confirm—or deny—our ability to pay for a Whataburger (no onions, please) in Corpus Christi or an espresso in Enid. In Visa's case, maintaining the network requires ultrareliable electricity so the company can maintain always-on digital links to more than 14,000 financial institutions and 2.3 million automated teller machines. Those entities are connected to Visa's computer system, which can process about 56,000 transactions per second. In 2015, those computers handled transactions worth some $7.4 trillion.[8] That's roughly equal to the GDP of Japan and Germany combined. Any interruption in the flow of data—any outage or security breach—would result in a loss of confidence in the card and therefore decrease customers' willingness to use it in place of cash or checks.

The price of anything—whether it's a physical object or a service—is what a buyer is willing to pay for it at any given time. The purpose of currencies and payment systems is to make transactions happen as easily and quickly as possible. Sure, paper money can be used for those transactions, but paper currency can also be a liability. It has to be secured in a safe or a bank vault so it cannot be stolen. It has to be protected from fire, water, and insects. By electrifying money—by turning Benjamins into bits—the number and speed of financial transactions can grow enormously.

The electrification of money can be seen by looking at the decline of paper stuff that looks likes money. Between 2000 and 2012, the number of checks used in the United States fell by more than half as consumers switched to credit cards and electronic banking.[9] For decades, financial institutions, corporations, and individuals have been moving more and more money via wire transfers. In 1987, the amount of money moved through the Fedwire Funds Service, the wire-payments network operated by the US Federal Reserve, totaled $152 trillion. By 2017, the Fedwire Funds Service was transferring some $740 trillion annually, or nearly $2 trillion per day.[10]

Another example of the electrification of money can be seen by looking at the surge in mobile and phone-based payments. Apps like India's Paytm, China's Alipay, and US-based outfits like Apple Pay, Venmo, and Google Pay are convincing increasing numbers of consumers to ditch their cash and credit cards and instead manage transactions with their phones. In 2016, the global mobile payments business was valued at about $594 billion. By 2022, that market is expected to be worth more than $3 trillion.[11] Few countries demonstrate the potential of mobile payments better than Kenya. Some 93 percent of all Kenyans utilize mobile payments, and nearly half of the country's GDP is processed over M-PESA, the mobile payments platform that dominates Kenya's banking system. The popularity of M-PESA has led to a decline in the number of consumers using cash. According to one mid-2018 report, the number of ATMs in Kenya has fallen by about a third over the past few years.[12] Why? People don't need as much paper money.

The Indian government has embraced mobile payments as a way to fight corruption. In late 2016, Prime Minister Narendra Modi suddenly banned the use of 500 and 1,000 rupee banknotes. Those banknotes—which at the time were worth about $7.50 and $15, respectively—made up about 86 percent of all the cash in circulation in India. The move, known as demonetization, sent the Indian economy into a tailspin and some 1.5 million jobs were reportedly lost. But Modi insisted the move was necessary if the Indian government was going to increase tax collection. Tax avoidance, which was facilitated by the use of large sums of currency for transactions, was costing the Indian treasury enormous amounts of tax revenue. At the time, less than 3 percent of Indians were paying income taxes.[13]

By outlawing the large banknotes (which were later replaced with new banknotes), Modi's government effectively forced Indians to use credit cards, mobile payments, or other banking methods that can be tracked, and therefore taxed.[14] The result of Modi's crack-down: in the year after demonetization, mobile payments in India more than doubled.[15]

Mobile payments, credit cards, and wire transfers are all examples of the electrification of money. Cryptocurrencies take that trend a step further.

•

WHEN I GOT OUT OF the car a short distance east of Keflavik International Airport, the noise, vibration, and rush of wind made it feel like I was standing on the tarmac of a busy airport. But there were no airplanes in sight, only a vast treeless plain and four low-slung, metal-clad warehouses that were protected by a stout metal fence. From each of the warehouses came a dull roar of rushing air, similar to the sound of a large propeller-driven airplane.

Helmut Rauth, the director of Genesis Mining's Icelandic operations, apologized for the noise as he shook my hand on the lumpy, gravel-surfaced parking lot. "Yes, it's loud," he said as he leaned forward so I could hear him. A thin, energetic man with graying hair, Rauth wasn't dressed for the cold. As I shivered in the freezing morning air, he seemed perfectly comfortable in a light sweater and scarf. Leading me through the security gate that separates the parking lot from the warehouses—the gate required both a pass card and a numeric code—Rauth explained that Genesis, a Hong Kong–based company, is one of the world's largest cryptocurrency miners. Genesis is known as a cloud miner, meaning that customers pay it to mine cryptocurrency for them. They can buy contracts to mine cryptocurrencies like Bitcoin, Ethereum, Litecoin, Cardano, Ripple, Monero, and others. Genesis built its first large-scale Bitcoin-mining facility in Iceland in 2014 and had been expanding its operations in the country ever since. The company was also mining cryptocurrency in Bosnia and Russia.

Once we were past the security gate, Rauth led me through a door—which also required a pass card and a numeric code—into one of the warehouses. We strained to open the door as the cold morning air rushed past us. Once inside, two things became apparent. First: it was just as cold inside as it had been outside, maybe colder. Second: Genesis was putting massive amounts of computing horsepower to work. On both sides of the warehouse—which was roughly 75 meters (246 feet) long—from the floor to a height of about 2.5 meters (8 feet), were row upon row of blinking, whirring computer servers. Myriad power cords and Ethernet cables connected the servers to control panels at the end of the warehouse. Overhead metal canals suspended from the roof bore the weight of all the cables. The long rows were divided into numbered racks. Servers with LED lights and humming fans were stacked on nearly every shelf. A handful of racks didn't have any servers on them and were cordoned off with plastic sheets or panels, so that more cool air would be directed onto the machines that were mining cryptocurrency.

"The outside air is being pulled past the servers by those fans," Helmut explained, talking directly into my ear. Even with him nearly yelling, it was difficult to hear over the din of the six huge, high-speed fans overhead. The meter-wide fans were pulling the air through screened vents on the sides of the building, past air baffles and hot computing gear, and pushing it out through large cupolas that were evenly spaced along the peak of the roof. "What's the power rating on those fans?" I asked. "Thirty-five kilowatts each," Helmut replied.

No wonder it was so damn windy inside. Those six overhead fans were using 210 kilowatts of electricity (equivalent to about 280 horsepower). But the power being used by the overhead fans was only a fraction of the electricity being consumed by the computers. In the two warehouses that Rauth showed me, Genesis was operating about 4,700 servers. All of them were mining Ethereum, and each one was drawing 1,100 watts. Thus, those two warehouses were consuming about 5.2 megawatts of electricity. That electricity demand was a key reason why Genesis was mining cryptocurrency in Iceland.

What is cryptocurrency? There are several answers to that question. The most basic one is this: cryptocurrency is a digital or virtual currency that uses cryptography for security. The cryptography makes cryptocurrencies difficult to counterfeit. It also makes them difficult to produce, because the cryptography requires enormous amounts of computing power and therefore electricity. Cryptocurrency production (or mining) is, at root, an arbitrage on the price of electricity. One definition puts it almost exactly that way: "Cryptocurrency is electricity converted into lines of code with monetary value."[16]

In their 2015 book *The Age of Cryptocurrency*, Paul Vigna and Michael J. Casey explain that the idea of money has always been about trust. Big banks, insurance companies, credit card companies, and others have created a system where they are the trusted keepers of money because they manage secure networks and ledgers that keep track of who owes what to whom. Cryptocurrencies take control of the ledgers from centralized financial institutions and give it "to a network of autonomous computers, creating a decentralized system of trust that operates outside the control of any one institution." Vigna and Casey go on: "Cryptocurrencies are built around the principle of a universal, inviolable ledger, one that is made fully public and is constantly being verified by these high-powered computers." The result, they posit, is that "in theory . . . we don't need banks and other financial intermediaries to form bonds of trust on our behalf."[17] And that's the crux of the whole idea: cryptocurrency can become a new method of exchanging value that leaves banks and other intermediaries on the sidelines. But that idea has yet to be proven.

Cryptocurrencies haven't shown that they can process transactions as quickly and efficiently as outfits like Visa and MasterCard. Furthermore, there are ongoing questions about the security of the companies that handle cryptocurrencies. In 2014, Japan-based Mt. Gox, the largest Bitcoin exchange at the time, revealed that it had been hacked. The cyber bandits made off with 850,000 bitcoins worth an estimated $450 million and some $27 million in cash. Although some of the heisted bitcoins were recovered, the thieves were never caught.[18] In 2017, hackers invaded the cryptocurrency platform Enigma and made off with some $470,000 worth of Ethereum.[19]

Despite the many unanswered questions, by late 2017 cryptocurrency production was booming and cryptocurrency miners were searching for locations that could provide them with cheap electricity. Chinese Bitcoin miners had set up shop in Tibet, where they were using cheap hydropower to fuel their operations.[20] Others had settled in the Pacific Northwest in the United States, which also has cheap hydropower. Still others were finding their way to Iceland.

Genesis and other cryptocurrency companies were flocking to Iceland because the weather is almost always cool. Even during July, the average temperature stays at about ten degrees Celsius (about fifty degrees Fahrenheit). That means that Genesis and other companies that require huge amounts of computing power don't need air conditioning to keep their machines from overheating. Instead, they use ambient air. Iceland also has ultra-cheap electricity, all of which comes from renewable sources. In 2016, industrial customers in Iceland were paying about $25 per megawatt-hour.[21] By contrast, in the United States, industrial customers were paying about $68.[22] In Germany they were paying about $170, and in France they were paying about $105.[23] The combination of cold air and low-cost electricity means that it costs about half as much to operate a data center in Iceland as it does in the United States, and far less than when compared to Germany or France.[24]

For decades, that cheap electricity has been attracting industry to the island country. Iceland's biggest product is aluminum, which accounts for 39 percent of its exports. (Seafood is second.)[25] Iceland's aluminum smelters are responsible for about 70 percent of the country's electricity consumption and explain why the country has the world's highest per-capita electricity use, more than 53,000 kilowatt-hours per year.[26]

During my visit to Iceland, cryptocurrency was all the rage. In late 2017, the price of Bitcoin hit an all-time high of nearly

$20,000. As the prices of Bitcoin and another popular cryptocurrency, Ethereum, soared, numerous analysts began looking at the amount of electricity being used to produce them. One analyst estimated that production of those two cryptocurrencies alone required the use of about 20 terawatt-hours of electricity per year.[27] (That's about as much electricity as is consumed by the country of Iceland, by the way.) Another estimate claimed that, in 2017, Bitcoin mining was using 35 terawatt-hours per year, or nearly as much as New Zealand.[28] After walking through Genesis Mining's warehouses, I asked Rauth to explain why cryptocurrency mining requires so much electricity and why that electricity use would continue rising. He explained that the number of bitcoins that can be mined is limited. The difficulty of mining additional bitcoins is "programmed in the algorithm of the cryptocurrency," he said. "So the more bitcoins are already mined the more difficult it gets to mine one bitcoin. That's the challenge." As the difficulty of mining Bitcoin increases, more computational power will be needed, and that added computing power will, of course, require yet more electricity.

But the volatility of cryptocurrency prices had made the future electricity-demand question more difficult to answer. By late 2018, the price of a single bitcoin had fallen to about $3,200. By mid-2019, it had recovered to about $7,600.[29] Ethereum, which had been trading for about $1,400 in 2017, had, by late 2018, fallen to under $100. By mid-2019, its price had recovered to about $230.[30]

In 2018, economist Nouriel Roubini, a professor at the New York University Stern School of Business, dismissed cryptocurrencies, calling them "nothing but a fad." Perhaps Roubini is correct and the "cryptopocalypse" is coming. But there is also a chance that Roubini (whose nickname is Dr. Doom) is wrong and Bitcoin or another cryptocurrency survives and becomes a reliable store of value. Even if that happens, credit card companies like Visa and MasterCard and other payment systems like PayPal and Venmo are not going to give up their places in the financial ecosystem without a fight. Indeed, it's unlikely that cryptocurrency or electronic money will ever fully unseat cold hard cash. Paper money—and in particular $100 bills—remains a key store of value in countries all over the world for both licit and illicit transactions. According to one report, the number of $100 bills in circulation around the world doubled between 2008 and 2019.[31] Paper money has a long history. It is a trusted means of exchange. It's familiar. It's portable. And it doesn't have to be plugged in.

Nevertheless, the electrification of money—the merger of electrons, bits, and cash—shows no sign of slowing. Electricity simply makes exchanging money easier. Whether we are using a credit card, making a payment with our mobile phone, or mining cryptocurrency, we are using electricity to access a network that makes commerce faster, easier, and more secure. The electrification of money is only part of our ever-increasing reliance on electricity. We can also see our growing need for electricity in the recreational drug business.

14

WATTS INTO WEED

> The electricity from cannabis businesses represented about 4 percent of the city's total electricity use.
>
> —**EMILY BACKUS,** Denver Department of Public Health and Environment

Daniel was happy to show us his black-market cannabis operation. But by the time we arrived at the house he was renting in an unfashionable neighborhood in north Denver (it was about 7:30 p.m. when we pulled up), he was concerned about the schedule. "It's time to put the girls to bed," he told us.

Daniel was not living in high style. His house was sparsely furnished and neat. He was wearing old sweatpants, a well-worn T-shirt, and a pair of Crocs. Straight ahead of the front door was a modest kitchen. The air in the place was humid and smelled faintly of mold. The entire house, which probably covered about nine hundred square feet (eighty-four square meters), was vibrating—a feature I soon learned was due to the fact that several high-speed fans were attached to the floor joists. After dispensing with a few pleasantries, Daniel, a friendly man who was about my age and height—I met him through a friend from Tulsa—gave us a soup-to-nuts tour of his operation.

Black-market marijuana grow operation, Denver, 2017
Source: Photo by Tyson Culver.

The first stop was one of the bedrooms, where Daniel was cloning marijuana seedlings under a small set of lights. He then grabbed a headlamp and led us down a steep set of narrow, unlit stairs to the basement. The "girls" were a set of about two dozen foot-high cannabis plants that were in the veg cycle, meaning they were adding new leaves. The plants were thriving beneath a set of 600-watt metal-halide lights. They'd already had eighteen straight hours of light and he wanted to keep them on their regular schedule of eighteen hours of light and six hours of darkness. "They'll be okay for a few more minutes," he said. "But I'm going to have to turn off those lights pretty soon."

The basement was cramped, loud, warm, and even more humid than the upstairs living area. The air was filled with the smell of soil and chlorophyll. The floor joists were just high enough overhead that I could stand upright. My friend Phil, who is a bit taller than me, couldn't.

The basement had been partitioned with sheets of plastic and plywood into three miniature greenhouses. The wall on the right was covered with a neat array of electric ballasts, switches, and outlets, all connected to networks of yellow Romex electric cable that snaked through the rooms. "I'm pretty well versed in all this stuff," Daniel said as he pointed to the wiring. "I put all the lights on 220-volt circuits so that they won't draw as many amps." Walking into the next room, he proudly showed us the basement's new electric subpanel, which was filled with circuit breakers. He had installed all of the wiring himself. Daniel had been growing illicit marijuana in Colorado for more than a decade. He was growing weed underground for two reasons: First, he didn't have enough money to set up a legal growing operation. Doing so, he estimated, would probably cost about $500,000. Second, even if he could afford to go legit, he wouldn't have been able to get a permit from the Colorado authorities because he had a drug-related conviction. Daniel volunteered that he only smokes marijuana occasionally. His drug of choice was heroin.

As he led us through the grow rooms, he explained the basics of his business. Every two months, he produced about eleven pounds of cannabis, which he was selling to buyers who were driving in from neighboring states. Gross revenues were about $150,000 per year. But weed prices had been soft lately and expenses—for soil, fertilizer, and rent—were high. Plus, his house was using about ten times as much electricity as a typical home in Colorado, and that was a major operational cost: his electricity bill, he told me, averaged $650 per month.[1]

His lighting setup was standard for cannabis growers: 600-watt metal-halide lights for the veg stage, during which the plants mature, and 1,000-watt high-pressure sodium lights for the flowering stage. High-pressure sodium lights are pretty much irreplaceable to marijuana farming because they produce light at the right temperature—about 2,700 degrees Kelvin, on the red segment of the light spectrum—that forces the plants to flower.[2] The sun, of course, is an ideal light source for growing marijuana. But outdoor grow operations aren't as secure or predictable as indoor ones. Growing weed indoors gives producers both greater control over their crops and the ability to grow more product. Outside growers are usually limited to one crop per year. Indoor growers can produce up to six crops per year.[3] In addition, some of the governments that are permitting legal weed cultivation have prohibited outdoor cultivation.[4] But growing indoors requires lots of energy, with electricity accounting for about a third of the cost of an indoor operation.[5]

Electricity-hungry high-pressure sodium lights have become so important to cannabis production that commercial grow operations are often compared using a single metric: the number of 1,000-watt lights they have. That number determines output. And what's that output number? "It's 1,000 watts per pound," Daniel said. What about growers who are claiming they can produce more than that? I told him that a few weeks earlier I'd read about a Denver-based grower who was claiming his technique could produce as much as three pounds per 1,000-watt light.[6] "He's full of shit," Daniel replied with a dismissive wave of his hand.

While some growers may be producing more than one pound of pot per 1,000 watts and others may be producing less, a few things are clear: regardless of whether the grower is a black-market operator like Daniel or a legit one, indoor cannabis production is all about turning watts into weed.

While it's easy to associate marijuana farmers with hippies and stoners twisting up a doobie on the street corner, pot production is among the world's most electricity-intensive agricultural businesses.[7] Turning watts into weed is also one of the fastest-growing sources of electricity demand in the United States, as well as a major source of carbon dioxide emissions. According to one estimate, the carbon dioxide emissions associated with American cannabis production are roughly equal to the emissions from three million automobiles.[8] Talk about a buzzkill. Furthermore, the power density of indoor marijuana cultivation is on par with that of power densities found inside data centers like those operated by the Giant Five.

In 2018, the investment research company Morningstar named cannabis production as one of the biggest sources of new electric demand in the United States. Although domestic electricity consumption has "flatlined during the last decade," Morningstar said, "we think it will spring to life." Morningstar said cannabis production—as well as new demand from electric cars and data centers—will account for about "6 percent of total U.S. electricity demand by 2030, offsetting energy efficiency gains and resulting in 1.25 percent total annual electricity demand growth through 2030."[9] Morningstar predicts electricity use for pot production will jump from 15 terawatt-hours in 2018 to 65 terawatt-hours per year by 2030. If that happens, weed production could account for as much as 1.5 percent of the electricity used in the United States.[10]

The soaring demand for electricity can be seen by looking at Colorado, the first state to legalize recreational weed. In 2012, 55 percent of Colorado voters voted in favor of Amendment 64.[11] At about the same time Coloradoans began getting a new—or rather, *newly legal*—Rocky Mountain high, the state's regulators and utilities began seeing a spike in electricity use.

In 2018, Emily Backus, a sustainability adviser at the Denver Department of Public Health and Environment, told me that electricity use by the cannabis industry in Denver had been growing by about 34 percent per year.[12] In 2013, indoor pot farms in Denver were using about 100 gigawatt-hours of electricity per year. By 2016, that number had nearly tripled to about 275 gigawatt-hours of electricity per year, which was nearly 4 percent of the city's total electricity use.[13] To put that in perspective, cannabis production in the city of Denver was using about the same amount of electricity as countries like Burundi and Somalia.[14]

Since 2012, numerous states have followed Colorado's lead. By the end of 2018, thirty-three states and the District of Columbia had passed laws legalizing medicinal marijuana and ten states and the District of Columbia had legalized recreational use.[15] The spread of legal weed (which is favored by more than 60 percent of voters) reflects soaring consumer demand. Between 2016 and 2017, legal retail cannabis sales in the United States grew by 33 percent, to some $9.7 billion.[16] That torrid growth looks likely to continue. ArcView Group forecasts that legal marijuana sales in North America will grow by 26 percent per year through 2021. By that year, ArcView expects that North American legal marijuana sales will exceed $21 billion.[17] Other projections put US retail marijuana sales at more than $30 billion by 2021.[18]

Growing all that weed requires enormous amounts of electricity. In 2012, in perhaps the most famous academic paper on marijuana and electricity demand, Evan Mills, a researcher at the Lawrence Berkeley National Laboratory, estimated that US weed growers were using about 20 terawatt-hours of electricity per year. But Mills's paper is a tad confusing in that he also put domestic cannabis electricity use at "approximately 1 percent of national electricity consumption." According to the Energy Information Administration (EIA), the United States consumes about 3,800 terawatt-hours of electricity per year.[19] Therefore, if the cannabis industry was using 1 percent of all US electricity, it wasn't consuming 20 terawatt-hours; it was consuming about 38 terawatt-hours per year.[20] At that level of consumption, domestic cannabis production requires about as much electricity as Peru, a country with thirty-two million people.[21] Whichever number is correct, we are still talking country-size quantities of electricity. As mentioned above, by 2030, Morningstar thinks the US cannabis sector will be using 65 terawatt-hours, or about as much electricity as is now consumed by Iraq, a country of some thirty-eight million people.[22]

The cannabis sector's electricity demand has led to a corresponding crackdown by regulators. Few places have taken the issue more seriously than Boulder, arguably the epicenter of America's legal cannabis business. Boulderites love their artisanal, bespoke, *sativa-indica* hybrid weed, and they buy it in a staggering variety of forms: from fragrant buds and pre-rolled joints to cannabis-infused gummy bears and coconut oil. In 2017, Boulder had about two dozen weed dispensaries. That means the city had more retail weed outlets than it had Starbucks and McDonalds *combined*.

But about 60 percent of Colorado's electricity comes from coal-fired power plants.[23] Given the state grid's dependence on coal, the city of Boulder passed an ordinance that requires dispensaries to prove they're growing polar-bear-friendly ganja. Boulder's cannabis regulations require the city's pot growers to use 100 percent renewable electricity or buy carbon offsets.[24] Government officials in Boulder County have taken similar steps. Pot growers operating in the county must offset their electricity use with renewable energy or pay the county a fee of 2.16 cents per kilowatt-hour.[25]

In 2018, the Massachusetts Cannabis Control Commission issued a seventy-three-page ordinance that limits the amount of electricity growers can use, thereby effectively forcing them to use LED lights instead of high-pressure sodium ones.[26] LED lights have advantages. They don't produce as much heat and are far more efficient than high-pressure sodium lights. But they are also five to ten times more expensive.[27] In addition, crop yields for marijuana plants grown under LEDs are substantially lower than what can be obtained from high-pressure sodium bulbs. A 2018 study published by the Sacramento Municipal Utility District reported that yields from LED-grown plants were 35 to 40 percent lower than what could be obtained with high-pressure sodium lights.[28] That study's findings agree with what I heard from Daniel, the black-market grower in Denver, who told me, "LED doesn't work. It just doesn't have the punch."

As part of their effort to reduce electricity use, Massachusetts regulators also limited the power density—that is, the amount of electricity used per square foot or square meter—of marijuana grow operations to no more than 535 watts per square meter. That's far lower than what I saw in Daniel's basement grow operation. After leaving his house, I did some calculations based on the size of his basement and his electricity use. I concluded that the power density in the flowering room of Daniel's grow operation was about 835 watts per square meter. That means the electricity intensity (when measured by the amount of electricity used per square meter) of a cannabis-flowering operation is twenty-two times as much as what's needed by a hospital and about 130 times that of an average home.[29]

But my calculations for Daniel's basement grow operation may be too low. Evan Mills put the power density of marijuana cultivation at 2,000 watts per square meter, or 2.5 times the density of what I saw in Daniel's basement.[30] If Mills is right, indoor weed production requires power densities similar to what is found inside data centers operated by companies like Apple or Facebook.[31]

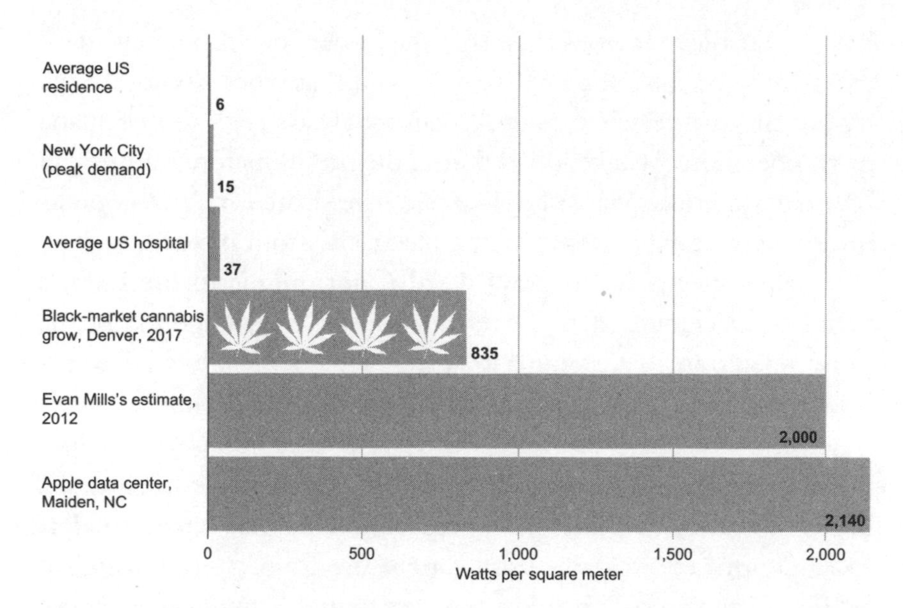

High-Watt Weed: Power Density of Cannabis Grow Houses Compared to Homes, Hospitals, and Data Centers

Sources: Energy Information Administration, Evan Mills, author calculations.

The cannabis industry's insatiable hunger for electricity has led to another result: a surge in electricity theft by black-market growers. By one estimate, about 88 percent of all US marijuana sales happen in the black market, a sector that had some $46 billion in revenue in 2016.[32] To feed the insatiable demand for weed, small growers and organized crime rings have set up clandestine indoor weed-production operations in residential neighborhoods from Florida to California.

To supercharge their profits, many of those clandestine growers are stealing electricity. In 2017, I began tracking media reports about marijuana cultivation and electricity use. Over a thirteen-month period, according to the news stories I catalogued, law enforcement officials discovered or made arrests on seventy-one illegal marijuana-production facilities. Out of those seventy-one busts, forty-seven occurred in California. An April 2017 bust in Merced was typical. Brianna Calix of the *Merced Sun-Star* reported that police confiscated about 350 marijuana plants from an illegal grow house after they received a tip that the house was stealing electricity. Calix quoted Merced police, who said that when the local utility investigates electricity theft, "it almost always leads to marijuana grow operations." Calix added that in the month before finding the house in question, Merced police had investigated six grow operations where growers were stealing electricity from the grid.[33]

Stealing electricity to grow weed is not unique to the United States. In December 2017, Spanish police busted a large marijuana grow operation in Catalonia that was set up in industrial warehouses that had been repossessed by banks. A local media outlet reported that "authorities suspect the ring stole $506,000 of electricity to grow the plants."[34] In February 2018, police in Australia busted a grower near the city of Adelaide who was forced to alert local authorities after his house caught fire from faulty wiring on his illegal operation. Police investigators found that not only was the man a lousy electrician, he was also stealing energy from the grid.[35]

Even if illegal cannabis operations aren't stealing electricity, their enormous energy demand can lead cops to their doorstep. Out of the seventy-one marijuana busts that I tracked, more than half a dozen operations were discovered due to their outsize electricity use. In October 2017, Larimer County, Colorado, investigators found nearly two hundred pounds of marijuana, as well as 173 cannabis plants, at a house in the town of Berthoud. According to a local news report, investigators got suspicious after finding that the house "was using large amounts of water and electricity, which is common for marijuana grow operations." Five men were arrested in connection with the case.[36]

In a case that deserves a medal for sheer audacity, black-market growers in California set up their cannabis-production facility inside the old Pacific Bell building in San Bernardino at the corner of Sixth and E, less than a block from the rear entrance of the San Bernardino Police Department's headquarters. At eight o'clock on the morning of December 13, 2017, a contingent of San Bernardino police got into their cruisers, drove a full city block, and raided what they described as a "very, very sophisticated operation." They seized nearly 25,000 marijuana plants from the Pacific

Bell building and two other locations in the city. Eight people were arrested in the bust. Police investigators made the raid after looking up the building's utility records and finding that the supposedly unoccupied building was racking up monthly electricity bills of $67,000.[37] After the bust, Lieutenant Mike Madden, a spokesman for the San Bernardino Police Department, told a local media outlet, "It's pretty bold that they would have their operation so close to the station."[38]

So many stoners are growing indoor weed that it has caused electrical problems even in places where marijuana is legal. In Oregon, in the first four months after recreational marijuana was made legal in 2015, Pacific Power, the Portland-based electric utility, experienced seven local blackouts that were caused by indoor growers who had overloaded the grid in their neighborhoods. The offending growers were each fined about $5,000 for having "overburdened and damaged local equipment."[39]

Regardless of whether blackouts are caused by weed growers or weather, they can be costly, even deadly—a fact that mischief makers and saboteurs know all too well.

15

THE BLACKOUT WILL NOT BE TELEVISED

> Short of a nuclear bomb, the most crippling affliction that can befall a modern metropolis is a total power failure.
>
> —***NEW YORK TIMES*** editorial, 1965

The April 16, 2013, attack on the Metcalf Transmission Substation in rural Santa Clara County, California, still has law enforcement officials stumped. Beginning at about 1 a.m., an unknown number of saboteurs fired 120 AK-47 rifle rounds into the transformers at the substation. They stayed for less than an hour and left no matchable boot prints. Their shell casings had no fingerprints and their vehicles left no tire tracks. Outfitted with night-vision scopes and heavy wire cutters, they cut fiber-optic cables and were able to briefly disable the local 911 emergency system and telephone lines.

Shortly after the shooting stopped, investigators from California law enforcement agencies and the FBI descended on the substation and began searching for evidence. The saboteurs knew the substation's layout. Small rock mounds were found near the positions from which the shooters fired at the transformers.[1] But the investigation quickly ran dry and no arrests were ever made.[2]

In all, seventeen transformers at the Metcalf substation were damaged by the AK-47 rounds. Fortunately, the attack did not cause a blackout. The California Independent System Operator, the entity that operates the state's grid, was able to reroute electricity from other power lines and increase the amount of electricity being generated. Nevertheless, it took workers from Pacific Gas & Electric nearly a month to bring the substation back into service.

About a year after the Metcalf attack, Rebecca Smith, a reporter for the *Wall Street Journal*, published an article that claimed the United States "could suffer a coast-to-coast blackout if saboteurs knocked out just nine of the country's 55,000 electric-transmission substations on a scorching summer day." Smith wrote that an analysis by the Federal Energy Regulatory Commission found that if the nine substations in question were taken off-line, the country could be plunged "into darkness for weeks, if not months."[3] Although some security analysts discounted Smith's story, the sabotage of the Metcalf substation highlighted the vulnerability of the US electric grid.

In the preceding pages, I have underscored how important electricity is to America's powerful tech businesses, the financial sector, and the weed business. Our High-Watt lives depend on electricity. But our near-total reliance on electricity also makes us vulnerable. Indeed, the one thing that scares the pajamas off of security experts is an extended blackout. A short blackout, one of, say, twelve hours or less, won't cause much damage. By thirty-six hours, things start to get dicey. After seventy-two hours, when much of the available ice has melted and food and water supplies start running short, things get messy in a hurry. The situation gets particularly bad in hospitals and nursing homes because modern health care demands reliable electricity. In fact, when the lights go off and stay off for more than a day or two, people die.

On August 31, 2005, about forty-eight hours after Hurricane Katrina began pummeling New Orleans, the on-site electric generators at Memorial Medical Center stopped working. Dr. Ewing Cook, one of the hospital's most senior physicians, later recalled that it was the "sickest sound" he had ever heard. For the previous two days, the medical professionals at the hospital had been scrambling to care for and evacuate hundreds of patients. That challenge was tough enough when the generators were working. Without electricity, the ventilators and incubators quit. The lights, computers, air conditioners, and elevators were all useless.

Before Katrina hit New Orleans, the administrators had been warned that the electrical system at Memorial Medical was vulnerable to flooding. Those warnings were ignored.[4] A few hours after the hospital's generators quit working, and as the wait for rescue boats and helicopters dragged on, the doctors at the hospital convened to decide which patients were most likely to live and which ones would probably die. Sheri Fink, the author of *Five Days at Memorial: Life and Death in a Storm-Ravaged Hospital*, explained in an interview that "a group of doctors . . . got together and they quickly made this choice. . . . They wrote 1, 2, and 3, on pieces of paper, or on the patients' gowns." She continued, "Ones were the relatively healthy. Twos were somewhat more sick. Threes were the very ill." Many of the patients given threes were also ones who had do-not-resuscitate orders in their records.[5] The patients with threes would be the last to be evacuated from the hospital.

As the weakest patients began deteriorating in the stifling heat inside the darkened hospital, Cook and the nurses began the triage process. One of the patients tagged with a three was a female in the intensive care unit who had advanced uterine cancer and kidney failure. The woman, who was heavily sedated on morphine, was bloated from fluid retention. Cook estimated she probably weighed more than three hundred pounds. That meant it would be nearly impossible for the staff to carry her down six flights of stairs so that she could be evacuated. Furthermore, four nurses were still caring for the woman. Cook decided on the spot that euthanasia was the best choice. He instructed the attending nurse to increase the amount of morphine being given to the woman until she died. Cook later told Fink, "To me, it was a no-brainer, and to this day I don't feel bad about what I did. . . . I gave her medicine so I could get rid of her faster, get the nurses off the floor."[6]

Cook wasn't the only one to order or administer deadly doses. In the day or two after the generators at Memorial Medical Center quit working, several other patients were given deadly doses of morphine by hospital staff. Dr. Anna Pou, one of the doctors who gave lethal injections to some of the patients, was later arrested by law enforcement officials. But a local grand jury refused to issue an indictment.[7]

Doctors euthanizing their patients is an extreme outcome. But the calamities at Memorial are just one example of how essential electricity has become to our medical system. In 2017, a dozen years after Katrina walloped Louisiana and the Gulf Coast, Florida was hit by Hurricane Irma. Shortly after the storm made landfall, power outages swept across the state. At about 3 p.m. on September 10, the Rehabilitation Center at Hollywood Hills, a nursing home in the town of Hollywood, lost power.[8] Employees at the nursing home called 911, the local utility, Florida Power and Light, state hotlines, and even the cell phone of the governor, Rick Scott, trying to get help. In the three days following the power outage, a dozen people who were being cared for at the nursing home died due to the high temperatures, which reached ninety-nine degrees Fahrenheit.[9]

While the power outages at Memorial Medical Center and the Rehabilitation Center at Hollywood Hills proved deadly, the death toll in those two instances is relatively small when compared to the calamities that hit Puerto Rico after it was slammed by Hurricane Maria. A report by researchers at George Washington University found that nearly 3,000 people on the island died due to the hurricane. The peak of the dying—known as excess mortality—came in January 2018, four months after the storm, when numerous people, many of them aged and infirm, began succumbing to the lack of medical treatment.[10]

In an extraordinary bit of reporting undertaken by the Associated Press, Puerto Rico's Center for Investigative Journalism, and the news site *Quartz*, a group of analysts and statisticians identified 487 of the people felled by Hurricane Maria. The investigators attributed 166 of those deaths to the lack of electricity. Some of those 166 couldn't get the medicines they needed due to the blackout. Others, like cardiorespiratory patient Eladia Dávila, depended on a mechanical ventilator to breathe. Dávila didn't die immediately after the storm. Instead, she died two months after the hurricane, on November 15, 2017.[11]

Of course, the hurricane that hit Puerto Rico was a relatively rare event, and the island's grid was fragile long before the storm delivered its deadly destruction. But saboteurs are continually probing electric grids, and the companies that make them work, for vulnerabilities. Saboteurs could, in theory, shut down a grid, or significant parts of it, using little more than a computer and an Internet connection. In 2015, and again in 2016, cyber saboteurs caused blackouts in Ukraine. In the 2015 attack, which was the first known instance of hackers taking down a power grid, the cyberattackers infiltrated the networks of three Ukrainian power companies by sending targeted emails to the companies' employees. Malicious code was embedded in Word documents attached to the emails. When the Word documents were opened, the code allowed the hackers to access the companies' networks. After the attack, which was attributed to hackers in Russia, one cybersecurity analyst said the hackers could have caused more extensive damage than they did, but the attack was "more like a demonstration of capabilities."[12]

In 2018, American Electric Power (AEP), one of the largest electricity utilities in the United States, experienced more than 1,400 cyberattacks on its computer systems. In early 2019, a company official said AEP was spending more than $100 million on cyber and physical security to protect its facilities from saboteurs.[13]

In addition to cyberattacks, the electric grid is vulnerable to an electromagnetic pulse, or EMP. An EMP is the result of a nuclear explosion. It sends a powerful burst of electromagnetic energy into the atmosphere that can fry electronics and disable the electric grid. William R. Forstchen's 2009 disaster novel, *One Second After*, is a chilling tale of the desolation that could occur if rogue states were able to detonate multiple EMPs at the right altitude.

Forstchen's novel describes thousands of modern vehicles, all of them made inoperable due to the fact that their onboard electronics were damaged by the EMP. In the novel, one character says that, after the EMP attack, he and his friends were suddenly as isolated "as someone in Europe 700 years ago when there was a rumor, just a rumor, that the Tartars were coming or there was a plague in the next village." In Forstchen's novel, food and medicine shortages hit almost immediately after the United States is attacked with an EMP. A *Mad Max* scenario ensues, complete with a climactic battle scene in which the hero and his fellow residents of Black Mountain, North Carolina, battle and destroy an invading horde of marauders who have grown fond of, yes, cannibalism.

Forstchen's novel projected that 80 to 90 percent of the American population would be dead within a year of a successful EMP attack.[14] While the apocalyptic scenario painted by *One Second After* is designed to scare readers, the real-life implications of a successful EMP attack are serious. An analysis by the Electric Infrastructure Security Council that looked at a 1962 EMP test done over Kazakhstan found that even a fairly small EMP could have devastating results. A comparable EMP, if it happened at high altitude over the United States today, "would likely damage about 365 large transformers in the U.S. power grid, leaving about 40 percent of the U.S. population without electrical power for 4 to 10 years."[15]

In 2004, a federal entity known as the EMP Commission—short for the Commission to Assess the Threat to the United States from Electromagnetic Pulse Attack—released a report that concluded that an EMP is "one of a small number of threats that can hold our society at risk of catastrophic consequences. . . . It has the capability to produce significant damage to critical infrastructures and thus to the very fabric of US society." The document continued, saying that one of the "primary avenues" to inflict "catastrophic damage" is through the electric grid. If the grid is compromised, our ability to get food, water, and medical services will also be compromised. "The recovery of any one of the key national infrastructures is dependent on the recovery of others. The longer the outage, the more problematic and uncertain the recovery will be."[16]

In 2017, a new EMP Commission assessed the threat again and came to a conclusion similar to that of the earlier effort, saying that the United States is facing "a present and continuing existential threat from naturally occurring and manmade electromagnetic pulse assault and related attacks on military and critical national infrastructures." It added that the EMP threat can be "exploited by major nuclear powers and small-scale nuclear weapon powers, including North Korea and non-state actors, such as nuclear-armed terrorists."[17]

Some terrorism experts believe the threat of an EMP attack has been overhyped. Michael A. Sheehan, a former member of the US Army and an expert on anti-terrorism who has worked in the Pentagon and for the New York Police Department, told me that the idea of an EMP attack "is the sexy thing right now." People forget that "nuclear bombs are bad enough even without an EMP." He continued, "Am I worried about an EMP attack from North Korea? No."[18]

Although the threat of an EMP may be overstated, grid-security analysts are also worried about the weaponization of drones. A drone fitted with ribbons of aluminum foil could be as effective—and far cheaper and easier to conceal—as the blackout bombs used by the US military in the First Iraq War in 1991 and in Serbia in 1999.[19] In 2019, the Kalashnikov Group, the Russian company that produced the AK-47 assault rifle, announced that it would begin selling a small suicide drone dubbed the KUB-UAV. The drone can fly for thirty minutes at up to 80 miles (129 kilometers) per hour and carry about 6 pounds (2.7 kilograms) of explosives.[20] Drones with that type of capability could be used to attack key transformer stations.

Blackouts aren't always the work of saboteurs or the military. Trees and squirrels are common causes of electricity failures. In 2003, tree branches in Ohio contributed to a blackout that left fifty million people in the dark for two days.[21] In 2013, the *New York Times* published a story that claimed that squirrels had contributed to fifty power outages in twenty-four states over a three-month period.[22]

Weather—both here on the earth and in space—can also cause blackouts. In 2012, Hurricane Sandy walloped the East Coast, flooding parts of New York, New Jersey, and other states. Some 8.1 million homes, spread over seventeen states, some of them as far west as Michigan, lost power. The New York Stock Exchange was shuttered for two days, the longest it had been closed since the blizzard of 1888. Total losses—due to the storm and ensuing blackout—were estimated at more than $25 billion.[23]

Solar storms can also wreak havoc. Also known as coronal mass ejections, these storms, which start on the surface of the sun, result in sudden releases of stored magnetic energy. If the sun and the earth are in the right positions, that solar energy collides with the earth's magnetic field, which can cause electrical surges on the grid and therefore blackouts.[24] In 1859, the earth was hit by a huge solar storm, known as the Carrington Event. The pulse of energy was so strong that telegraph operators were reportedly able to send and receive messages on their telegraphs even if they had disconnected the machines from their power supplies.[25] A 2017 report published by the American Geophysical Union estimated that an extreme solar storm like the Carrington Event could cause blackouts that would affect two-thirds of the US population, and that "daily domestic economic loss could total $41.5 billion plus an additional $7 billion loss through the international supply chain."[26]

In addition to the threats listed above, the grid is also vulnerable due to its near-total reliance on large, one-of-a-kind, high-voltage transformers. These machines aren't sexy, but they are difficult to build and service. More than 90 percent of the electricity consumed in the United States passes through high-voltage transformers. If saboteurs were able to damage or disable a significant number of those transformers, they could cause widespread power outages. A 2014 analysis of the transformer vulnerability issue by the Electric Power Research Institute explained that the "vulnerability is compounded by the fact that many U.S. high-voltage transformers are approaching or exceeding their design lives."[27]

Regardless of whether sabotage is increasing, the US electric grid is experiencing more frequent outages. According to data published by the US Department of Energy, in 2002, there were twenty-three "major disturbances and unusual occurrences" on the domestic electric grid. Those outages were caused by things like ice storms, fires, vandalism, and severe weather. By 2016, the number of disturbances and unusual occurrences had more than quadrupled, to 141. Just as interesting, the reports show an increase in the number of events listed as "sabotage" or "cyberattack."[28] Some of the increase in these events may be due to more scrupulous reporting by the utilities. Nevertheless, consumers are aware that the grid has become more vulnerable to outages. That awareness has spawned a booming business in standby home generators.

Generac Power Systems, the Waukesha, Wisconsin, manufacturer of electric generation units, said that, between 2010 and 2017, its sales of residential units more than doubled, to $785 million per year. Further, in its investor presentations, the company said that it expects to continue growing the residential backup generator business because of an "aging and underinvested grid, favorable demographics [and] heightened power outages." As for demographics, the company claimed that 70 percent of the buyers of home standby generators are age fifty or older.[29]

So, yes, blackouts have increased in frequency. And yes, many thousands of people have bought small generators. And yes, there are plenty of people in Moscow, Beijing, Tehran, and Pyongyang who'd like nothing better than to see you (and me, and millions more people) sweating in the dark on a hot August night because cyber saboteurs have managed to damage the electric grid with a bit of malicious computer code. But before you start stocking up on canned goods, *añejo* tequila, and shotgun shells—and making sure that your generator, laptop computer, and Nintendo machine are all safely ensconced inside a Faraday cage (putting electronics inside a metal housing, even a trash can, can shield them from the effects of an EMP)—remember that utilities and security experts understand the grid's vulnerabilities.

Since the gunfire attack on the Metcalf substation in California in 2013, and in particular since the Ukrainian cyberattack, electricity providers have gone through multiple assessments of their facilities and taken steps to make them less vulnerable to attack. In 2017, a spokesman for the Edison Electric Institute, a trade association that represents the investor-owned utilities (which serve about two-thirds of the US population) told me that many electric utilities have begun installing manual switches on some of their key equipment so it cannot be attacked via computer. Others are using dedicated fiber-optic lines between their operations centers and their key substations so they cannot be infiltrated by hackers on the Internet.

Of course, there is no guarantee that efforts made to prevent cyberattack and physical sabotage of the electric grid will be effective. A solar storm could wallop the earth at virtually any time and shut down the grid. Hurricanes, tornadoes, wind storms, wildfires, and suicidal squirrels could also bring it down. All of those threats are real and must be addressed, and they need to be dealt with at the very same time countries around the world must dramatically increase the amount of electricity they produce.

In the final section, I will look at the booming growth in global electricity demand and discuss how that soaring demand is likely to be met.

PART FOUR

Twenty-First-Century Terawatts

16

THE TERAWATT CHALLENGE

I love renewables. But I am also pro-arithmetic.

—**DAVID J. C. MACKAY,** physicist, University of Cambridge

The late Nobel laureate Richard Smalley called it the Terawatt Challenge. The world's most pressing problems, he explained, could only be addressed if the people of the world have plenty of energy. In the months before his too-early death from cancer in 2005, Smalley, who shared the 1996 Nobel Prize in Chemistry for his discovery of carbon molecules known as fullerenes, was one of the world's foremost energy evangelists.[1] During his last lectures, Smalley would show audiences this list:

TABLE 4

THE TOP TEN PROBLEMS FACING THE WORLD

1. Energy
2. Water
3. Food
4. Environment
5. Poverty
6. Terrorism and war
7. Disease
8. Education
9. Democracy
10. Population

While displaying that list, Smalley declared that if we can solve the first problem on the list—energy—then "the next four go away." But Smalley, who worked as a professor at Rice University in Houston, wasn't just promoting energy. He was laser focused on electricity. Electricity "is a much better answer" for meeting the world's energy needs because it "is a superb way to move energy from one place to another." He envisioned a sprawling high-voltage grid, where "the entire North American continent, all the way from the Arctic Circle down to Panama, would be wired together in a giant interconnected electrical energy grid." In addition, he imagined a system in which hundreds of millions of electricity consumers would have their own electricity storage units, "the equivalent of an uninterruptible power supply that not only gives a home computer a few minutes of power during an outage, but also can supply each of our houses or businesses with 12 to 24 hours of full operation."[2]

While Smalley's vision of a high-capacity, continent-wide grid backed up by hundreds of millions of batteries has not materialized, the chemist accurately foresaw the need for more electricity. In 2018, global electricity use jumped by 4 percent.[3] At that rate of growth, global electricity use will double in just eighteen years. That means that the amount of installed electricity generation capacity around the world will increase from roughly 6 terawatts today to about 12 terawatts by the late 2030s.[4]

Electricity demand is increasing for many reasons, including (as I discussed earlier) marijuana production, the expansion of digital commerce, and cryptocurrency production. Several other macro trends are also stoking electricity demand growth, including urbanization, population growth, air conditioning, water treatment, electric vehicles (EVs), and climate change. Let's take those in order.

For centuries, humans have been leaving the drudgery of rural farms for the bright lights and opportunity that can be found in cities. In 1500, less than 5 percent of the world's population lived in cities. By 1900, that figure had tripled to 15 percent. By 2000, it had tripled again to 47 percent. We are now living in an urban-majority world. That matters because, as Stewart Brand, the founder of the

Whole Earth Catalog, explained in a 2012 interview, "history is driven to a large degree by the size of cities." Brand said that "the dominant demographic event of our time is the screamingly rapid urbanization" now underway. The developing world, he explained, now has all of the world's biggest cities and those cities will continue growing. "The aggregate numbers are absolutely overwhelming: 1.3 million people a week are coming to town, decade after decade."[5] The result of this ongoing migration is this: by 2050, about 70 percent of the world's population will be living in cities.[6]

People are moving to cities and they are having babies. By 2050 or so, the world will likely add another two billion people, bringing global population to about 9.7 billion.[7]

Those billions of new residents will want to stay cool in their new apartments and houses. That will require billions of new refrigerators and air conditioners, all of which will need electricity. The booming demand for cooling can be seen by looking at India. By 2018, only about 5 percent of India's households had air conditioning. In the United States, that figure was 87 percent.[8] Given India's generally warm climate, it's no surprise that air conditioner sales in the country are soaring. A 2017 study by HDFC, India's largest bank, found that sales of air conditioners have been increasing by 13 percent per year.[9] At that rate, the amount of electricity needed just to fuel India's air conditioning units will double every six years or so. But India's demand represents only part of the global air conditioning trend.

In 2018, the International Energy Agency released a report called *The Future of Cooling*, which noted that of the "2.8 billion people living in the hottest parts of the world, only 8 percent currently possess air conditioners."[10] The report estimated that by 2050, the global stock of air conditioners will more than triple, meaning that there will be "10 new air conditioners sold every second for the next 30 years."[11] The amount of electricity that will be needed to keep all those air conditioners humming is staggeringly large. By 2050, the IEA estimates, some 6,200 terawatt-hours of electricity per year will be needed to drive the world's air conditioners.[12] That's a difficult number to imagine. So think of it this way: within three decades or so, global electricity demand solely for air conditioning will be nearly equal to the total amount of electricity now used by China.[13]

Increasing need for fresh water will also drive electricity demand growth. According to a recent estimate, global demand for clean water will increase by one third between now and 2050.[14] Meeting that demand will require enormous quantities of energy. Consider this fact: in the United States, drinking water and wastewater treatment accounts for about 2 percent of all energy consumption. For some water systems, energy can account for as much as 40 percent of total operating costs.[15]

As global freshwater sources are depleted, countries around the world are using electricity-hungry desalination plants to convert ocean water into drinking water. In 2017, Hexa Research estimated that global spending on water desalination will hit almost $27 billion per year by 2025.[16] In 2018, Adroit Market Research estimated that the global desalination business will grow by about 8 percent per year through 2025, meaning that the market is doubling every decade or so. That same report found that some of the biggest demand growth will likely occur in the Middle East and noted that Saudi Arabia gets nearly half of its fresh water from desalination plants.[17] All of those new desalination plants will require huge quantities of electricity. That can be seen by looking at the largest desalination plant in the United States, located in Carlsbad, California. The $1 billion plant, which began operating in 2015, provides about 50 million gallons (189 million liters) of fresh water per day to residents in and around San Diego.[18] Producing that much water requires 38 megawatts of electricity, or roughly the amount needed to supply 28,000 US homes.[19] As more desalination plants like the one in Carlsbad are built in the years ahead, demand for electricity will grow apace.

Increasing use of electric vehicles will also spur demand. By 2040, BP estimates, the global automobile fleet could include more than three hundred million EVs, all of which will need to be plugged in and recharged.[20]

Finally, whatever happens with the global climate—if it gets substantially hotter or it gets more volatile, with more extreme cold events and more extreme heat waves—we are going to need lots more electricity to keep us safe and comfortable. Proof of that came in June 2017, when the southwestern United States was hit by a merciless heat wave. Temperatures in Phoenix, Arizona, hit a record 119 degrees Fahrenheit (48 Celsius). The sizzling heat resulted in a spike in electricity demand, with several utilities in five western states recording all-time highs in electricity consumption.[21] Six months later, in early 2018, the United States was hit by a severe winter storm. During that cold snap, which was referred to as a "bomb cyclone," US utilities burned record amounts of natural gas to produce the electricity needed to keep their customers warm. Demand for natural gas in Boston was so great that spot prices for the fuel hit $35 per million Btu, or more than ten times the average price at the Henry Hub gas terminal in Louisiana. The price spike, reported *Forbes* energy writer Chris Helman, gave Boston the distinction of having the "priciest gas market in the world."[22]

In summary, there's no doubt that we will need vastly more electricity in the decades to come than we have now. Adding 6 terawatts of new generation capacity will be a huge challenge. To put that in perspective, recall that the United States currently has about 1 terawatt of generation capacity.[23] Therefore, over the next three decades or so, the countries of the world will have to add six grids the size of the existing US grid.

Without doubt, the biggest obstacle to the ongoing pace of electrification in the world, and to adding those six new grids of additional capacity, is the lack of integrity. Poor governance and corruption are rife in Low-Watt and Unplugged countries. Corruption is particularly problematic in Africa, where the majority of countries have failing grades on Transparency International's Corruption Perceptions Index.[24] Therefore, many countries are unlikely to add significant amounts of new electric capacity because their systems simply leak too much. Bankers and lenders will not commit the billions of dollars needed for full electrification if the countries seeking to borrow that money don't have integrity.

Meeting the Terawatt Challenge will also require gargantuan sums of capital. The World Energy Council has projected that meeting growing electricity demand will require as much as $1 trillion per year in new investment between now and 2060.[25] While we cannot know exactly how much capital will be needed, it is obvious that in the decades ahead we are going to need far more electricity than what we are producing today. In the next chapter, I will explain why, despite their widespread popularity, renewable energy sources alone will not be enough to meet the world's soaring demand.

17

THE ALL-RENEWABLE DELUSION

Men believe what they want to.

—**PUBLIUS TERENTIUS AFER,** Roman playwright

Since the 1970s, mainstream environmental groups and the Democratic Party have been united on one energy issue more than any other: that we should not be using nuclear energy and that we should be using far more renewables than we are now.

For instance, Greenpeace claims that renewable energy, "smartly used, can and will meet our demands. No oil spills, no climate change, no radiation danger, no nuclear waste."[1] A similar all-renewable-no-nuclear vision is being pushed by the Sierra Club, one of America's biggest environmental groups. The club has a Beyond Oil campaign, a Beyond Coal campaign, and a Beyond Natural Gas campaign. The group says, "We have the means to reverse global warming and create a clean, renewable energy future." Another big environmental group, the Natural Resources Defense Council, also opposes nuclear energy, claiming the technology poses too many risks and, until those issues are addressed, "expanding nuclear power should not be a leading strategy for diversifying America's energy portfolio and reducing carbon pollution."[2] The NRDC was among the environmental groups who negotiated the early closure of California's last remaining nuclear power plant, Diablo Canyon, which will be shuttered by 2025.[3]

In 2005, some three hundred environmental groups—including Greenpeace, Sierra Club, and Public Citizen—signed a manifesto that said, "We flatly reject the argument that increased investment in nuclear capacity is an acceptable or necessary solution. . . . Nuclear power should not be a part of any solution to address global warming."[4] In 2016, Michael Brune, the executive director of the Sierra Club, reaffirmed the club's position, saying it "remains in firm opposition to dangerous nuclear power." That's not a new position. In 1974, the Sierra Club said it "opposes the licensing, construction and operation of new nuclear reactors utilizing the fission process." It went on, stating that it will continue its opposition, pending "development of adequate national and global policies to curb energy over-use and unnecessary economic growth." The Sierra Club didn't delete that last part about "unnecessary economic growth" until 2016.

If nuclear is the outcast of American energy politics, solar energy is like apple pie: everyone loves it. In 2016, the Pew Research Center found that 89 percent of adults in the United States favor expanded use of solar energy. The same poll found 83 percent wanted more wind. Meanwhile, nuclear energy, hydraulic fracturing, and coal mining were favored by just 43, 42, and 41 percent, respectively.[5]

For four decades, the Democratic Party has either ignored or professed outright opposition to nuclear energy. The party's 2016 platform said that climate change "poses a real and urgent threat to our economy, our national security, and our children's health and futures." The platform contains thirty-one uses of the word "nuclear," including "nuclear proliferation," "nuclear weapon," and "nuclear annihilation." But it doesn't contain a single mention of the phrase "nuclear energy."[6] The last time the Democratic Party's platform contained a positive statement about nuclear energy was way back in 1972.[7]

The partisan schism over nuclear energy is evident in the polling data. A 2015 Gallup poll found that voters who identify as

Republican are twice as likely (47 percent to 24 percent) to support nuclear energy as are those who identify as Democrats.[8]

The most liberal members of the US Senate are pushing all-renewable policies. In 2017, four of them—Jeff Merkley (D-OR), Edward J. Markey (D-MA), Cory Booker (D-NJ), and Bernie Sanders (I-VT)—introduced the 100 by '50 Act, which calls on the United States to be completely free of fossil fuels by 2050.[9] The bill includes a "carbon duty" on any foreign goods that are made by energy-intensive industries.[10] The 100 by '50 legislation was immediately endorsed by a who's who of all-renewable obsessives, including actor and activist Mark Ruffalo, the Sierra Club's Michael Brune, and May Boeve, the executive director of 350.org.[11]

Also in 2017, New York governor Andrew Cuomo touted his renewable-energy goals and declared that his state was not going to stop "until we reach 100 percent renewable because that's what a sustainable New York is really all about."[12] That same year, more than fifty Massachusetts lawmakers—representing more than a quarter of the members of the state legislature—signed onto a bill that would require the Bay State to be getting 100 percent of its energy from renewable sources by 2050. The bill says the goal is to "ultimately eliminate our use of fossil fuels and other polluting and dangerous forms of energy."[13]

The 100-percent-renewable stance is also gaining traction with state and local governments. By mid-2019, more than 125 cities in the United States, along with eleven counties and five states plus the District of Columbia and Puerto Rico, had adopted goals to get all of their electricity from renewable energy.[14] (Note that many of these governments say they will be using "clean" or "carbon-free" energy, not just renewables.) In addition, more than 175 companies from around the world, including brands like Ikea, Visa, and Sony, had committed to getting all of their electricity from renewables.[15]

In early 2019, Representative Alexandria Ocasio-Cortez, a Democrat whose district includes parts of the Bronx and Queens, and Senator Ed Markey announced the Green New Deal, which calls for a complete overhaul of America's energy and power systems. Although the text of the resolution is silent about what types of technologies should be used to achieve that goal, Ocasio-Cortez has made it clear that she supports the all-renewable approach. Shortly after she was elected to Congress, Ocasio-Cortez told a group of supporters, "We don't have a choice. We do not have a choice. We have to get to 100 percent renewable energy in 10 years. There is no other option."[16] A few months later, she stated, "We need to declare our North Star, and our North Star is 100 percent renewable energy."[17]

In early 2019, some six hundred environmental groups submitted a letter to the House of Representatives that said that the United States must shift to "100 percent renewable power generation by 2035 or earlier." The same letter said that any "definition of renewable energy must . . . exclude all combustion-based power generation, nuclear, biomass energy, large-scale hydro and waste-to-energy technologies." It continued, saying that the new electric grid must have the "ability to incorporate battery storage and distributed energy systems that are democratically governed." Signers of the letter included groups like Food & Water Watch, Friends of the Earth, and the Environmental Working Group.[18]

Politicians, environmental groups, activists, and big business are getting traction with the all-renewable claim because, as one writer at *Vox* put it, the goal is "a clear, intuitive, and inspiring target, an effective way to rally public support and speed the transition."[19] In fact, one of America's most prominent climate activists, Bill McKibben, has admitted that he and his fellow activists are pushing the all-renewable agenda because it doesn't require them to muddy their message by defending nuclear energy.[20]

Despite the attractiveness of the all-renewable concept to voters, activists, politicians, and corporations wanting positive media coverage, here's the truth: Renewables aren't going to be enough to meet the Terawatt Challenge. Not by a long shot. Four factors will prevent renewables from taking over our energy and power systems: cost, storage, scale, and land use. Let's look at the cost issue first.

Germany provides a clear example of how renewable mandates push up electricity prices. According to Agora Energiewende, a think tank that focuses on Germany's transition toward renewables, residential electricity prices in Germany jumped by 50 percent between 2007 and 2018. The result: German residential customers now have some of the highest-priced electricity in Europe, about $0.37 per kilowatt-hour.[21] German industry has also been hit hard. Between 2016 and 2018, electricity prices doubled for the *Mittelstand*, the term used for the midsize companies that employ some twenty million German workers and account for a big part of the country's industrial output. A study of the *Mittelstand* found that a third of company leaders in the sector believed electricity prices were a threat to their businesses.[22]

Ontario, Canada, has also pushed hard for renewables. In 2009, the provincial government launched the Green Energy Act, which guaranteed long-term contracts to renewable-energy generators at prices that were well above market rates. To pay for the measure, Ontario, which is home to nearly 40 percent of Canada's thirty-six million residents, added surcharges to ratepayers' electric bills. The province also forced the closure of coal plants, claiming that doing so would improve public health. The result: between 2008 and 2016, residential electricity rates in the province jumped by 71 percent, which was more than double the average increase in the rest of Canada over that time period.[23] Soaring electricity prices led to a backlash from consumers, municipalities, and other electricity users. A 2017 article in the *Windsor Star* focused on the rural town of Kingsville, population 21,000, which saw its electricity costs nearly double between 2011 and 2014. The surging costs were forcing the town to raise taxes and spend more money on efficiency projects. Nelson Santos, the mayor of the town, said that even with "conservation efforts, the costs are going up."[24]

In 2018, the Fraser Institute, a free-market think tank, released a report that concluded that "soaring electricity costs in Ontario have placed a significant financial burden on the manufacturing sector and hampered its competitiveness." The same report found that phasing out coal had had "very little effect on [Ontario's] air pollution levels" and that the province's high electricity prices were "responsible for approximately 75,000 job losses in the manufacturing sector from 2008 to 2015."[25]

Voter disgust with sky-high electricity prices played a decisive role in the province's 2018 elections. Ontario's Progressive Conservative party, led by Doug Ford, drubbed the incumbent Liberal party, which had ruled the province for fifteen years. Ford campaigned on a pledge to reduce electricity rates by 12 percent.[26] That promise resonated. Ford's Progressive Conservatives won 76 of the 124 seats in the province's legislature, while the Liberals, who had been led by premier Kathleen Wynne, retained just seven seats and were reduced to rump-party status.[27] In an interview a few months after the election, Kenneth Green, an energy analyst at the Fraser Institute, told me, "The hottest issue was electricity rates. The election turned on that." Ford's new government quickly canceled 758 renewable-energy contracts on projects that were being developed in Ontario, claiming that the move would save ratepayers about $790 million.[28] In December 2018, Ford's government repealed the Green Energy Act.[29]

Or consider Australia, where electricity prices skyrocketed after the government imposed renewable-energy mandates and emissions caps on the electric sector. High electricity prices played a major role in the 2018 ouster of the country's prime minister, Malcolm Turnbull, a member of the center-right Liberal Party. The country's new prime minister, Scott Morrison, a conservative and staunch coal supporter, has pledged to bring down electricity costs by having the government intervene in the markets.[30]

Shortly after Turnbull was ousted, Australia's new energy minister, Angus Taylor, declared that the new government would be phasing out its renewable-energy targets and would "not be driven by ideology or grand gestures, but pragmatism."[31] While presenting the electricity overhaul plan to the Australian House of Representatives, Taylor said, "We have seen the experiment of 50 percent renewable energy targets in South Australia and the results were shocking. . . . In South Australia we now have prices at around 50 cents per kilowatt-hour. They are among the highest in the country." Taylor concluded his remarks, saying, "We are going to back investment in fair dinkum reliable generation because that's what this country needs."[32] (Fair dinkum is Australian slang for a fair deal.) In May 2019, despite numerous polls that predicted defeat, Morrison's center-right coalition was reelected. Morrison's challenger, Bill Shorten, the leader of the Labor Party, had based much of his campaign on pledges to do more about climate change.[33]

In the United States, California continues to be a leader in both renewable mandates and high electricity prices. In 2008, then governor Arnold Schwarzenegger signed an executive order that required the state's utilities to be obtaining one-third of the electricity they sell from renewables by 2020.[34] In 2015, Governor Jerry Brown signed a law that boosted the mandate to 50 percent by 2030.[35] Those moves had an effect on electricity prices. In 2018, Mark Nelson and Michael Shellenberger of the Berkeley-based think tank Environmental Progress released a report that showed that California's electricity rates rose at more than five times the rate of electricity prices in the rest of the United States between 2011 and 2017.[36]

Those high prices have led to a backlash. In April 2018, the Two Hundred, a coalition of civil-rights leaders, filed a lawsuit in state court against the California Air Resources Board (CARB), claiming that the state's climate policies discriminate against low-income and minority consumers. The 102-page lawsuit says California's

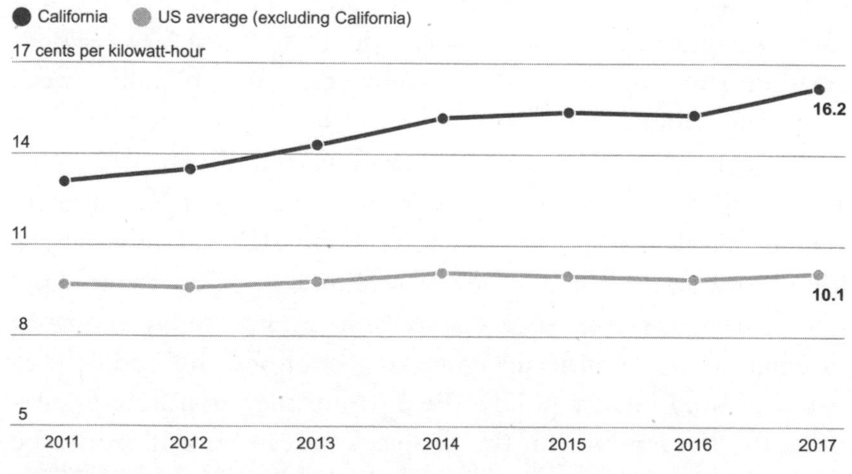

California Electricity Prices Compared to US Average, 2011 to 2017

Source: Environmental Progress.

"reputation as a global climate leader is built on the state's dual claims of substantially reducing greenhouse gas emissions while simultaneously enjoying a thriving economy. Neither claim is true."[37] The suit argues that the state's climate laws violate the Fair Employment and Housing Act because CARB's greenhouse gas emissions rules on housing units "have a disparate negative impact on minority communities and are discriminatory against minority communities and their members." The suit also claims the state's climate laws are illegal under the Fair Housing Act, again because their effect is felt predominantly by minority communities.

In addition, the lawsuit argues that California's climate regulations are a regressive tax that is hitting poor and working-class consumers harder than their wealthier counterparts. Since 2007, the suit says, "California has had the highest poverty rate in the country," with "over 8 million people living below the US Census Bureau poverty line when housing costs are taken into account." It continues, claiming CARB has "ignored" the state's "modest scale of greenhouse gas reductions, as well as the highly regressive costs imposed on current state residents by CARB's climate programs." The lawsuit focuses largely on the state's housing and transportation policies, but it also says that California's climate change policies increase "the cost of transportation fuels" and "further increase electricity costs." Those high costs, it claims, "have caused and will cause unconstitutional and unlawful disparate impacts to California's minority populations, which now comprise a plurality of the state's population."[38]

In September 2018, shortly after Governor Brown signed SB 100—a bill that requires the state to be obtaining at least 60 percent of its electricity from renewables by 2030—I interviewed John Gamboa, a member of the Two Hundred.[39] Gamboa told me that California's renewable energy policies are merely the latest example of how politicians ignore the poor and the middle class when making energy policy. "Every time they pass new regulations, the burden falls on the people who can least afford it," he told me. "That's the history of the environmental movement."

James Bushnell, an economist at the University of California, Davis, found that renewable-energy policies are forcing up the state's energy prices. In 2017, Bushnell published a short paper on the history of California's electric grid, which he calls "a long and gory one." Since the early 2000s, "the dominant policy driver" in California's electricity sector, he writes, "has unquestionably been a focus on developing renewable sources of electricity generation." Bushnell then explains that the renewable energy coming into the electricity market distorts prices because it favors wind and solar producers at the expense of traditional generators that rely on coal, natural gas, or nuclear. But because the electric grid still needs traditional generators to supply electricity when the sun isn't shining and the wind isn't blowing, the state's utilities must continue operating (and paying for) traditional generation units. The result, Bushnell explains, is that the utilities pass the cost of maintaining all that traditional generation capacity onto end users. In addition, consumers have to pay for billions of dollars' worth of new transmission lines needed to carry wind and solar electricity from remote regions into cities.[40] Bushnell specifically mentions the $2 billion Tehachapi Renewable Transmission Project, which will carry electricity from renewable generators in Kern County south to San Bernardino County.[41]

Transmission spending by California has skyrocketed. In 2016 alone, the state's utilities spent more than $20 billion on transmission projects. And, as Doug Karpa pointed out in a 2018 article for *Utility Dive*, between 2008 and 2018, spending on transmission projects in the state grew at an annual rate of more than 12 percent.[42] The cost of all of those projects will have to be absorbed by ratepayers.

Under an all-renewable scenario, ratepayers would also be stuck with big bills for electricity storage. A key reason why attempting to rely solely on renewables is so costly is that doing so would require enormous batteries to overcome seasonal fluctuations in wind and solar output. For instance, in California, wind- and solar-energy production is roughly three times as great during the summer months as it is in the winter. Storing summer-generated electricity and saving it until it's needed in winter months would require batteries, batteries, and more batteries. According to a 2018 analysis done by Stephen Brick, an energy analyst at the Clean Air Task Force, a Boston-based energy-policy think tank, for California to get 80 percent of its electricity from renewables, the state would need about 9.6 terawatt-hours of storage.[43]

It's difficult to get a handle on what that number means. Therefore, Brick put it into more easily digestible terms: the number of Tesla Powerwalls that would be needed to store that quantity of electricity. (In 2015, Tesla, the same company that makes electric cars, began producing lithium-ion battery packs for use in energy-storage systems. The newest model, the Tesla Powerwall 2, can hold about 13 kilowatt-hours of energy.)[44] By my calculations, storing the 9.6 terawatt-hours of electricity needed for California to get 80 percent of its electricity from renewables would require the state to install more than seven hundred million Powerwalls. That would mean every resident of California would need roughly eighteen Tesla Powerwalls. A full 100-percent-renewable electricity mandate would require even more batteries: some 36.3 terawatt-hours of storage, or about seventy-one Tesla Powerwalls for every resident.[45] At roughly $6,700 per Powerwall, that much storage would cost each Californian about $479,000.[46]

Brick's estimates are similar to the findings of four American energy analysts who published a 2018 report, which concluded that attempting to obtain all US electricity from renewables would require overcoming "seasonal cycles and unpredictable weather events," which in turn would necessitate having "several weeks' worth of energy storage and/or the installation of much more capacity of solar and wind power than is routinely necessary to meet peak demand."[47] The quantity of storage needed "would be prohibitively expensive at current prices."[48] How expensive? Using the cheapest batteries available, it would require spending roughly $1 trillion.[49] That would mean a bill of roughly $3,000 for every citizen of the United States. And remember, that sum doesn't include the cost of all the wind turbines and solar panels needed to charge those batteries. Finally, and this is no small matter, batteries have a relatively short life span, and that life span can be reduced if the batteries are charged and discharged frequently. Lead-acid batteries, like the ones commonly used in conventional automobiles, last three to five years. Tesla offers a ten-year warranty on its Powerwalls.[50] Thus, providing enough battery storage to offset the seasonal variation of renewable sources like wind and solar would also require an ongoing battery inspection and replacement system involving millions upon millions of individual batteries.

Storing enough electricity to fuel the entire US economy is yet more daunting. In 2019, Mark P. Mills of the Manhattan Institute noted that Tesla's $5 billion Gigafactory near Reno, Nevada, is one of the world's largest battery-manufacturing facilities. "Its total annual production could store *three minutes'* worth of annual U.S. electricity demand," Mills explained. "Thus, in order to fabricate a quantity of batteries to store two days' worth of U.S. electricity demand would require *1,000 years of Gigafactory production*" (emphasis added).[51]

Other credible studies have also concluded that the all-renewable scenarios being promoted by politicians and environmental groups are not based on sound science. In 2017, a study led by Australian academic and pronuclear activist Ben Heard found that, "while many modelled scenarios have been published claiming to show that a 100-percent-renewable electricity system is achievable, there is no empirical or historical evidence that demonstrates that such systems are in fact feasible."[52]

In an interview in Berkeley at the Environmental Progress office, Heard and I discussed the intermittent nature of renewables and how that requires grid operators to have sufficient amounts of backup generation capacity or large amounts of storage. Heard neatly summed up the problem. The "chaotic nature of renewable energy supply, particularly from wind and solar, is an enormous challenge because it's the opposite of what we actually want in the system we're trying to provide. We don't try to create something chaotic," he said. "We're trying to create something stable, predictable, that can give people what they want, when they want it . . . at low cost, every hour of the year. Preferably you wouldn't do that with something that was driven by the weather."

The generally higher cost of renewables has had a discernible effect: the bulk of global renewable-energy spending is concentrated in High-Watt countries, even though electricity demand in those countries is generally flat or declining. For instance, in both the United States and Germany, electricity production in 2017 was roughly the same as it was 2004.[53] Meanwhile, in the Low-Watt and Unplugged countries—where electricity tends to be scarce and demand is soaring—spending on renewables lags far behind that of the High-Watt countries.

In 2016, global investment in renewable energy projects totaled some $242 billion. Of that, some $106 billion, or 43 percent, was spent in Europe and the United States. China spent another $78 billion. Thus, the United States, Europe, and China together accounted for more than 75 percent of all global spending on renewables.

Meanwhile, spending on solar and wind in Africa, the Middle East, and India totaled just $17 billion. As shown in the graphic, in 2016, the United States spent about $144 per person on renewable energy projects. In India that figure was twenty times less: just $7 per capita. In Africa and the Middle East it was lower still, just $6 per capita, or twenty-four times less than renewable-energy spending in the United States that year.[54]

The vast disparity in spending on renewables is particularly apparent when it comes to solar energy. In 2017, all of the countries of Africa produced 6 terawatt-hours of solar energy. By contrast, the United States generated 78 terawatt-hours, or thirteen times as much. That's a remarkable contrast when you remember that Africa's population of 1.2 billion is nearly four times as large as that of the United States.

In addition to the cost and storage problem, renewables are not scaling fast enough to meet global electricity demand growth. Understanding the scale challenge requires only that we do the math. Between 1997 and 2017, global electricity production increased by an average of 571 terawatt-hours per year.[55] That's the equivalent of adding about one Brazil (which used 591 terawatt-hours of electricity in 2017) to the global electricity sector every year.[56]

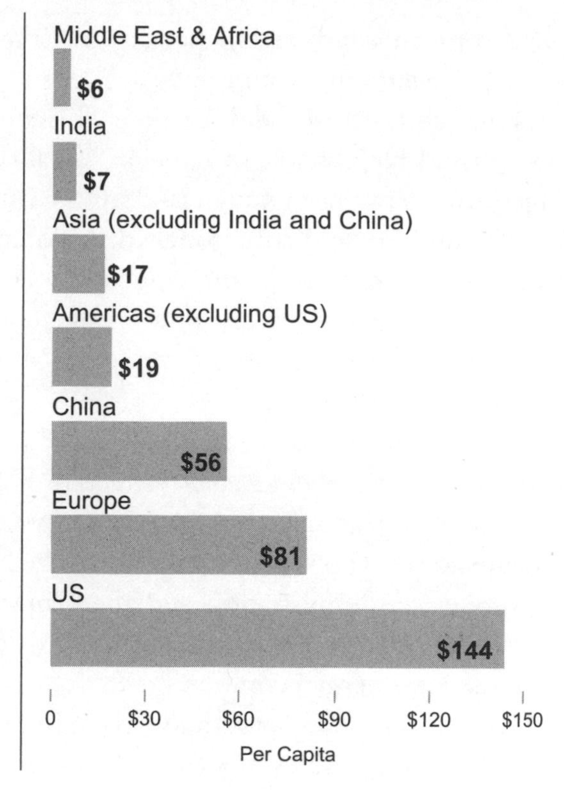

Per-Capita Renewable-Energy Spending Around the World, 2016

Sources: UN Environment Program and Bloomberg New Energy Finance

What would it take solely to keep up with that growth in global demand by using solar? We can answer that question by looking at Germany, which has more installed solar-energy capacity than any other European country, about 42,000 megawatts.[57] In 2017, Germany's solar facilities produced about 40 terawatt-hours of electricity.[58] Thus, just to keep pace with the growth in global electricity demand, the world would have to install fourteen times as much solar capacity as now exists in Germany, and it would have to do so every year.

Prefer to use wind? Fine. Let's look at China, which has more installed wind capacity than any other country—about 164,000 megawatts. (That's roughly twice the amount installed in the United States.)[59] In 2017, China produced about 286 terawatt-hours of energy from all that wind capacity. Recall that global electricity use is growing by 571 terawatt-hours per year. Thus, just to keep pace with electricity demand growth, the world would have to install twice as much wind-energy capacity as China has right now, and it would have to do so annually.

While cost, storage, and scale are all significant challenges, the most formidable obstacle to achieving an all-renewable scenario is simple: there's just not enough land for the Bunyanesque quantities of wind turbines and solar panels that would be needed to meet such a goal. In fact, deploying wind and solar at the scale required to replace all of the energy now being supplied by nuclear and hydrocarbons would require paving state-size chunks of territory with turbines and panels.

Proof of that can be had by looking at a 2017 paper published in the *Proceedings of the National Academy of Sciences* by an all-star group of American scientists. The paper, by Christopher Clack—a mathematician who has held positions at the National Oceanic and Atmospheric Administration and the University of Colorado—and twenty other top scientists, thoroughly debunks the idea of an all-renewable energy economy.[60]

Clack's paper repudiates the work of Mark Z. Jacobson, a Stanford engineering professor whose claims about the economic and technical viability of a 100-percent-renewable energy system have made him a celebrity. In 2013, Jacobson appeared on David Letterman's TV show to plug his claims.[61] He also became the darling of big environmental groups and climate activists. Jacobson became a hero despite the fact that his claims were based on Enron accounting, alternative facts, and technology hopium. His work contained numerous flaws. But his claims were politically popular and that, apparently, was enough. His academic papers sailed through peer review. In 2015, Jacobson published a paper, co-written with Mark Delucchi, a research engineer at the University of California, Berkeley, and two Stanford graduate students, Mary Cameron and Bethany Frew, in the *Proceedings of the National Academy of Sciences*. The paper, which claimed to offer "a low-cost solution to the grid reliability problem" with 100 percent renewables, went on to win the Cozzarelli Prize, an annual award handed out by the National Academy of Sciences. A Stanford website bragged that Jacobson's paper was one of six chosen by "the editorial board of the *Proceedings of the National Academy of Sciences* from the more than 3,000 research articles published in the journal in 2015."[62]

In the months following the publication of Jacobson's award-winning paper, Clack and his colleagues studied the paper's assertions and concluded that the numbers simply didn't add up. In their 2017 paper exposing Jacobson's claims, Clack and his coauthors—including Dan Kammen of the University of California, Berkeley, former EPA Science Advisory Board chair Granger Morgan, and Jane Long of Lawrence Livermore National Laboratory—concluded that Jacobson's work contained "numerous shortcomings and errors." His paper also used "invalid modeling tools, contained modeling errors, and made implausible and inadequately supported assumptions." Those errors "render it unreliable as a guide about the likely cost, technical reliability, or feasibility of a 100-percent wind, solar, and hydroelectric power system."

Among the biggest errors—and one that should force the National Academy of Sciences to withdraw the 2015 paper—is that Jacobson overstated the ability of the United States to increase its hydropower output by a factor of ten. While that was an egregious error, the inexcusable flaw in Jacobson's scheme involves the massive amount of land his plan would require. Jacobson's plan would necessitate installing nearly 2.5 terawatts of wind-energy capacity, with the majority of that amount onshore.[63] That's a staggering quantity. Recall that in 2016 the entire installed electric generation capacity in the United States, of all types—coal, gas, nuclear, hydro, wind, and solar—totaled about 1 terawatt.

Clack and his colleagues found that accommodating all of the wind turbines needed to achieve Jacobson's all-renewable vision would require "nearly 500,000 square kilometers, which is roughly 6 percent of the continental United States and more than 1,500 square meters of land for wind turbines for each American."[64]

But Clack's figure of 500,000 square kilometers—an area larger than the state of California—understates the actual amount of territory needed to accommodate the absurd number of wind turbines that would be required for an all-renewable energy system.[65] In 2018, Lee Miller, a postdoctoral fellow at Harvard University, and

David Keith, a physics professor at Harvard, published a paper in the journal *Environmental Research Letters*. Miller and Keith looked at 2016 energy-production data from 1,150 solar projects and 411 onshore wind projects. The wind projects in the study had a combined capacity of 43,000 megawatts, or roughly half of all US wind capacity in 2016.[66] They found that solar panels produce about ten times more energy per unit of land than wind turbines.[67] That finding alone makes their paper significant. But it was their analysis of wind energy's paltry power density that makes it newsworthy. Here's how Miller explained it to the *Harvard Gazette*: "We found that the average power density—meaning the rate of energy generation divided by the encompassing area of the wind plant—was up to 100 times lower than estimates by some leading energy experts." The problem, Miller said, is that most estimates of wind energy's potential ignore "wind shadow"—that is, how air flow through a given turbine disrupts the air flowing to turbines downwind of it.[68]

Miller and Keith determined that "meeting present-day US electricity consumption, for example, would require 12 percent of the continental US land area for wind." A bit of math reveals what that 12 percent figure means. The land area of the continental United States is about 2.9 million square miles (7.6 million square kilometers).[69] Twelve percent of that area would be about 350,000 square miles (912,000 square kilometers). Therefore, merely meeting America's current electricity needs would require a territory more than two times the size of California! The notion that the United States would be willing to cover two Californias with wind turbines—and remember, that much territory would be needed just to provide for our electricity needs, forget the liquid and gaseous fuels needed for transportation, home heating, and industry—is nonsense on stilts.

Before going further, I should note that Miller and Keith's 2018 calculations are almost identical to those done by Vaclav Smil in his 2010 book *Energy Myths and Realities: Bringing Science to the Energy Policy Debate*. Smil wrote that "relying on large wind turbines to supply all US electricity demand . . . [would] require installing about 1.8 terawatts of new generating capacity," which, he explained, "would require 900,000 square kilometers of land—nearly a tenth of the country's land, or roughly the area of Texas and Kansas combined."[70]

Miller and Keith's paper is important for three reasons: First, it uses real-world data, not models, to reach its conclusions. Second, it shows that wind energy's power density is far lower than what has been claimed by the Department of Energy, the Intergovernmental Panel on Climate Change, and academics like Jacobson. Third, it shows that wind energy cannot shrink its massive footprint. Miller and Keith write, "In summary, we find that while improved wind turbine design and siting have increased capacity factors (and greatly reduced costs) they have not altered power densities." That conclusion is remarkable. Over the past two decades or so, manufacturers like Vestas, General Electric, and others have invested untold millions of dollars improving the efficiency of the blades, gears, and generators they put on their turbines. They have also dramatically increased the size of their machines: the latest models stand some 853 feet (260 meters) tall. That's nearly three times the height of the Statue of Liberty.[71] But even with those improvements, the industry has not been able to wring more electric energy out of the kinetic energy of the wind. In physics terms, the wind-energy business is colliding with the Betz limit, named for German physicist Albert Betz, which sets the theoretical maximum efficiency for a wind turbine.[72] The newest and tallest wind turbines are coming close to this limit.

This means the wind sector is reaching the physical limits of the amount of energy it can produce from a given piece of land. As Miller and Keith—and Smil—have pointed out, the power density of a given wind turbine is roughly 0.5 to 1 watt per square meter. That means that if the wind sector is to make major increases in its energy production, it must cover more and more land with more and more turbines. But as those turbines get taller and cover more land, the more people will see them and the more people will object because they don't want them in their neighborhoods. Remarkably, that very point was made in late 2018 by Anne Reynolds, the executive director of the Alliance for Clean

Energy New York, a trade association that represents the wind industry. During a conference on renewable energy, Reynolds told attendees, "I personally think the arguments against wind energy are because people don't want to see the turbines."[73]

Perhaps the best work on the futility of all-renewable schemes was done by the late David J. C. MacKay, a physics professor at the University of Cambridge. In 2008, MacKay published *Sustainable Energy—Without the Hot Air*, one of the first academic books to take a serious look at the land-use impacts of renewables.[74] MacKay quickly became renowned for his hard-eyed calculations about land use and the vast quantities of energy needed by modern societies. For instance, he calculated that wind energy needs about seven hundred times more land to produce the same amount of energy as an oil and gas drilling site that uses hydraulic fracturing.[75] During a 2012 TED talk, he said that, when it came to renewables, "there's a lot of fluff, a lot of greenwash, a lot of misleading advertising, and I feel a duty as a physicist to try to guide people around the claptrap and help people understand the actions that really make a difference." He continued, "I love renewables, but I am also pro-arithmetic."[76]

In 2016, just a few days before his death at age forty-six from cancer, MacKay talked with British author and activist Mark Lynas about his life, his work, and his views on energy. During that interview, MacKay called the idea of relying solely on renewables an "appalling delusion."[77]

While academic assessments of all-renewable scenarios are valuable, observers need not comb through obscure journals to understand why they aren't feasible. They only need to look at the hundreds of rural villages, towns, and counties that are fighting the encroachment of large renewable-energy projects. That backlash can be seen from New York to California and from Ontario to Loch Ness.

18

THIS LAND IS MY LAND

> We feel this renewable energy push is an attack on rural America.
>
> —**K. DARLENE PARK,** resident of Frostburg, Maryland, and president of Allegany Neighbors & Citizens for Home Owners Rights[1]

Energy policy and land-use policy are inextricable. Regardless of whether the plan involves nuclear reactors, oil and gas drilling, pipelines, coal mines, wind-energy projects, or acres of solar panels, the end result is that something gets slated for construction somewhere and someone (or lots of someones) will object.

For instance, in 2016 and early 2017, thousands of protesters gathered near Cannon Ball, North Dakota, to oppose the Dakota Access Pipeline. Those protests got enormous amounts of media coverage, including front-page stories in the *New York Times*. More than seven hundred climate-change activists and others were arrested during the protest. They claimed that the pipeline, by crossing the traditional lands of the Standing Rock Sioux tribe, was violating the tribe's cultural and spiritual rights.[2] The pipeline was nevertheless fast-tracked for approval shortly after Donald Trump became president and began shipping oil in mid-2017.[3]

Or look at Colorado, where, in late 2018, voters rejected Proposition 112, a citizens' initiative endorsed by numerous environmental groups, including 350.org, Sierra Club, and Greenpeace, that would have prohibited oil and gas drilling activities within 2,500 feet (762 meters) of homes, hospitals, schools, and "vulnerable areas."[4] Had the initiative passed, it would have effectively banned new oil and gas production in Colorado, the sixth-largest natural-gas producer in the United States.[5]

Energy and land-use conflicts aren't just happening in the United States. In Germany, anti-coal protesters have been fighting for years against the expansion of an opencast lignite mine into the ancient Hambach Forest. In October 2018, a German court sided with environmental groups and ordered the mine's owner, RWE, to halt the expansion of the mine. A final ruling in the case may not be issued until 2020.[6]

Those land-use battles have received plenty of media coverage because they generally pit environmental groups against the hydrocarbon sector. But when the fights have to do with wind- or solar-energy projects, those same environmental groups—and most national media outlets—ignore them. Nevertheless, all across the United States, as well as in Europe and Australia, rural residents are mobilizing against large-scale renewable-energy projects, and that opposition is already limiting their expansion. This growing backlash against renewables illustrates what I call the vacant-land myth.

The vacant-land myth perpetuates the notion that there's an endless amount of unused, uncared-for land out there in flyover country that's ready and waiting to be covered with forests of renewable-energy stuff. The truth is quite different. All across the country, rural communities—even entire states—are resisting or rejecting wind projects, solar projects, and high-voltage transmission lines.

Before delving further into the rural backlash against renewables, I will gladly stipulate the obvious: the oil and gas industry has a big negative impact on the environment. Growing up in Oklahoma, I hunted quail in old oil fields that were littered with abandoned pipes, tanks, and pumps. I have personally seen the damage the oil industry can have on wildlife. In the early 1990s, I published several articles about how open oil pits in Texas, Oklahoma, and New Mexico were killing hundreds of thousands of birds per year and how the federal government was prosecuting oil and gas companies under the Migratory Bird Treaty Act to stop the carnage.[7] Accidents like the Deepwater Horizon blowout in the Gulf of Mexico, which resulted in an oil spill of more than 200 million gallons (757 million liters), can cause enormous damage to sea life, birds, and, of course, people. That 2010 accident resulted in the death of some 800,000 birds and 65,000 turtles. It also killed eleven men who were working on the drilling platform. Another seventeen workers were injured.[8]

Onshore, the oil and gas sector has had numerous accidents. In 2015, in California, a blowout at the Aliso Canyon natural-gas storage well forced thousands of residents from nearby Porter Ranch to be evacuated.[9] That accident and others have left the oil and gas sector facing legal and land-use battles on multiple fronts. Climate-change activists are opposing pipeline projects across the United States. Meanwhile, as the shale revolution has expanded, drilling has begun encroaching on urban and suburban residents. That has led to increased conflicts between drillers and residents of suburban neighborhoods that have been disturbed by heavy truck traffic and fifteen-story-high drilling rigs. By 2015, according to a tally done by the National Center for Policy Analysis, a conservative think tank, more than four hundred communities had restricted or banned hydraulic fracturing—the process used by energy producers to extract oil and gas from tight rock formations.[10] In addition, three states—Vermont, New York, and Maryland—have banned the process.[11]

Those bans have been promoted by national environmental groups, including Food & Water Watch, which in 2015 raised $17 million.[12] The group, which has seventeen offices in states across the country, has created a group called Americans Against Fracking, a coalition that "has over 275 organizations at the national, state and local levels united in calling for a ban on fracking and related activities."[13] The Natural Resources Defense Council has led fights against nuclear energy, fracking, and pipelines in communities from

New York to California. In 2017, the group raised a whopping $177 million.[14] It has launched a Community Fracking Defense campaign, which uses a policy and legal team to "craft effective local laws on fracking, defending those laws in court when challenged, and working at all levels to preserve and protect community rights and local control."[15] Another environmental group, Earthjustice, which took in $94 million in 2016, also promotes bans on fracking.[16] Earthjustice has offices in eighteen locations around the country.[17]

There's no doubt that hydrocarbon-extraction projects take a toll on the environment. They can pollute surface water, ground water, and drinking water. They can scar huge tracts of land. But renewable-energy projects are environmentally destructive too. They have significant negative effects on people, communities, and wildlife. Rural residents object to wind projects because they are protecting their property values and viewsheds.[18] They don't want to see the red blinking lights atop those fifty- or sixty-story-high wind turbines, all night, every night, for the rest of their lives. Nor do they want to be subjected to the noise—both audible and inaudible—that the giant machines produce.[19] (The range of sounds that occur between 20 hertz and 20,000 hertz is the audible frequency range. Sounds below 20 hertz, including deep bass tones, can be sensed by humans, but are not considered audible.)

It can be easy for Prius-driving city dwellers to dismiss the opponents of large wind projects as just NIMBYs—that is, "not in my backyard." But rural landowners have plenty of reason for concern. Numerous studies have shown the deleterious effect wind projects can have on property values and human health. And as rural residents learn more facts about the wind industry, more of them are fighting back.

In 2010, five people, including some members from the environmental group Earth First!, were arrested near Lincoln, Maine, after they blocked a road leading to a construction site for a 60-megawatt wind project on Rollins Mountain. According to the *Portland Press Herald*, one of the protesters carried a sign that read, "Stop the rape of rural Maine."[20]

In 2012, dozens of rural residents living near Utica, New York, filed a lawsuit against the owners of a $200 million wind project, claiming that the noise from the turbines was giving them headaches, disturbing their sleep, and hurting their property values. All of the plaintiffs lived within a mile of the Hardscrabble Wind Power Project, a 74-megawatt wind farm that began producing electricity in January 2011. Neighbors began complaining about noise from the turbines shortly afterward.[21]

In 2015, the Los Angeles County Board of Supervisors voted unanimously in favor of an ordinance banning large wind turbines in the county's unincorporated areas.[22] (One of the board members was Hilda Solis, who served as US secretary of labor in the Obama administration.[23]) During a hearing on the measure, Supervisor Michael D. Antonovich said, "Wind turbines create visual blight." In addition, he said the skyscraper-size turbines would "contradict the county's rural dark skies ordinance which aims to protect dark skies in areas like Antelope Valley and the Santa Monica Mountains."[24]

That wind turbines are a blight on the landscape—both day and night—is indisputable. And it's no small bit of irony that Los Angeles County politicos put a ban on them in 2015 at about the same time state legislators in Sacramento were passing a law requiring the state's electric utilities to get 50 percent of their power from renewables by 2030.[25]

Also in 2015, residents of Irasburg, Vermont (population: 1,100), held a town meeting on a proposed 5-megawatt wind project that was to be built just west of the village.[26] The meeting concluded with a vote. The tally: 274 against and just 9 in favor.[27]

In 2016, at the same time that thousands of protesters were gathered near Cannon Ball to oppose the Dakota Access Pipeline, officials in Billings County, North Dakota, rejected a proposed 383-megawatt wind-energy project that was to cover some 25,000 acres of land in the county. Chief among the county's concerns was the project's visual impact, including the fact that some of the turbines would have been visible from inside Theodore Roosevelt National Park, a local tourist attraction. During the meeting, Jim

Arthaud, a county commissioner, announced that he would vote against the wind project, saying there were "too many impacts to our county and to our citizens in different uses of our economy from ridgeline, to tourism, to being able to see it at the Painted Canyon, to the neighbors that are directly affected by it. . . . I just think the magnitude of this project in our county, the visual impacts it will have on western North Dakota is just more than the county can bear."[28]

In Iowa, a state that gets about a third of its electricity from wind, a three-turbine wind project being pushed by a company called Optimum Renewables was rejected by three different counties: Fayette, Buchanan, and Black Hawk.[29] In 2015, the Black Hawk County Board of Adjustment rejected the project after more than one hundred local residents expressed concerns.[30]

In New York, Governor Andrew Cuomo has mandated that the state be obtaining 50 percent of its electricity from renewables by 2030.[31] But three upstate counties—Erie, Orleans, and Niagara—as well as the towns of Yates and Somerset, have all been fighting a proposed 200-megawatt project called Lighthouse Wind, which aims to put dozens of turbines on the shores of Lake Ontario.[32] The same developer pushing Lighthouse Wind, Virginia-based Apex Clean Energy, also faced fierce resistance on a project in New York that aimed to put 109 megawatts of wind capacity on Galloo Island, a small island that sits off the eastern shore of Lake Ontario.[33] The project was opposed by the nearby town of Henderson for years, and, in the documents it filed with the state, Apex neglected to report that bald eagles have been nesting on Galloo Island.[34] That omission caused an uproar, and in early 2019 Apex withdrew its application for the Galloo project.[35] In April 2019, Apex announced it was also suspending work on the Lighthouse Wind project.[36]

The media's paltry coverage of the backlash against the wind industry is particularly obvious when looking at the Lighthouse Wind project. Even though the fight over the project raged for more than three years, and it was the highest-profile wind-energy project in New York, by mid-2019 the *New York Times* had not published a single story about the controversy.[37]

Although you won't read about it in the *New York Times*, the wind-energy business has become so unpopular in parts of rural America that the industry's biggest player has sued small towns that have resisted its projects. In 2017, Florida-based NextEra Energy filed lawsuits in both state and federal court against the town of Hinton, Oklahoma (population: 3,200).[38] NextEra, the world's biggest wind-energy producer, sued Hinton shortly after town officials approved an ordinance that deemed wind turbines "a public nuisance" and prohibited their installation within 2 miles (3.2 kilometers) of the town's borders. In a phone interview, Hinton's mayor, Shelly Newton, told me that the town passed the measure because "we were trying to give ourselves some elbow room." Newton said she had seen how other Oklahoma towns had allowed wind turbines to be built near their borders. The result was that all the new development in those towns was happening in areas that were furthest from the new turbines. "We wanted room to grow," she said.

NextEra also filed lawsuits against local governments in Michigan, Indiana, and Missouri that had passed measures restricting the encroachment of wind turbines. Just a few weeks after the company sued Hinton, it also filed suit against two small governments in Michigan—Ellington Township and Almer Township—both of which were opposed to Tuscola III, a 118-megawatt project that aimed to put more than four dozen wind turbines across thousands of acres of rural Tuscola County.[39] In Indiana, NextEra filed a state lawsuit after officials in Rush County denied a permit for a twenty-two-turbine project the company wanted to build.[40] In 2016, NextEra filed a state lawsuit against officials in Clinton County, Missouri, after that county passed a ban on wind turbines. In April 2019, the wind-energy giant sued Juniata Township, Michigan, after it revoked a land-use permit that could have allowed NextEra to install thirty-one wind turbines standing 490 feet high in the township.[41]

NextEra has also sued an anti-wind activist solely because she opposed one of its projects. In 2013, the company hired one of Canada's biggest law firms, Toronto-based McCarthy Tétrault, to file a defamation lawsuit in Canadian court against Esther Wrightman, a resident of tiny Kerwood, Ontario, who was fighting the company over its plans to put a 60-megawatt wind project next to her home. What did Wrightman, a mother of two young children, do to raise the company's ire? On her website, Ontario-wind-resistance.org, she called it "NexTerror" and "NextError."

When I wrote an update about the litigation for *National Review* back in 2015, NextEra refused to comment on the lawsuit. And no wonder. Shortly before McCarthy Tétrault filed the case against Wrightman, the Ontario Provincial Parliament was considering a law outlawing the very legal tactic, known as a SLAPP suit (short for strategic lawsuit against public participation), that NextEra was using in an attempt to silence her. Nearly thirty US states have passed laws prohibiting the use of SLAPP suits. Shortly after it filed suit against Wrightman, NextEra threatened to begin discovery in the case, all while claiming that any monetary damages it might collect from her would be donated to United Way. Then the company decided to let the case lie fallow. After NextEra built the 60-megawatt wind project, Wrightman moved her family out of Kerwood and settled in Saint Andrews, New Brunswick. But NextEra never dismissed the defamation case against her.[42] In June 2019, Wrightman told me the company's case against her is still pending. "You'd think it would have been decent for one of the World's Most Ethical Companies to tell me they were never going to pursue the suit, that may have even helped their reputation," she explained in an email.

NextEra's hardball legal tactics against Lilliputians like Esther Wrightman provide a telling counterpoint to the company's carefully burnished public profile. On February 26, 2019, the company issued a press release that noted that NextEra has been "named one of the World's Most Ethical Companies for the 12th time." It was given that distinction by the Ethisphere Institute, which was touted in the release as a "global leader in defining and advancing the standards of ethical business practices." The release quoted Ethisphere's CEO, Timothy Erblich, who said, "It's evident that high ethical standards and integrity are at the core of NextEra Energy's business strategy, and the company is helping to improve the communities they serve." The same press release noted that the company was also ranked "no. 1 overall among electric and gas utilities on Fortune's 2019 list of the 'Most Admired Companies' for the 12th time in 13 years."[43]

NextEra is dragging small towns and individuals into court because they stand between the wind giant and millions of dollars in subsidies. NextEra may be the biggest wind-energy producer in the world, but its real business appears to be subsidy mining.

In 2017, the Institute on Taxation and Economic Policy, a Washington, DC–based nonpartisan think tank, issued a report that named NextEra as one of the most subsidized corporations in America. Between 2008 and 2015, NextEra accumulated profits of $21.5 billion but ended up with a negative tax bill of $313 million. Put another way, over that seven-year period, NextEra had a *negative* tax rate of 1.5 percent. Only ten other companies—many of them industrial giants like Boeing and General Electric—outranked NextEra when it came to extracting money from taxpayers' pockets.[44] The Institute on Taxation and Economic Policy estimated that NextEra had collected $7.8 billion in subsidies. That sum is substantially higher than what has been reported by Good Jobs First, a Washington, DC–based nonprofit that tracks corporate welfare.[45] In 2019, Good Jobs First estimated that NextEra had collected some $5.8 billion in state and federal subsidies and loan guarantees.[46]

NextEra's proficiency at subsidy mining means it won't be paying federal corporate income taxes any time soon. In its 2018 10-K filing with the Securities and Exchange Commission, the company reported having $3.3 billion in tax-credit carryforwards on its balance sheet.[47] (A tax-credit carryforward allows a taxpayer, including individuals and businesses, to offset profits and thus reduce future tax liabilities.)

NextEra's multistate litigation effort—which is clearly aimed at intimidating cash-strapped rural governments—reflects the growing opposition that the wind industry is facing in rural America. My collection of published articles—nearly all of them from rural newspapers and TV stations—found that, in 2015 alone, sixty-five towns or counties in the United States passed measures to ban, restrict, or delay the construction of wind projects within their jurisdictions. By my count, between 2015 and mid-2019, some 230 government entities had moved to restrict or reject wind projects.

Among the examples of wind restrictions that occurred in 2018: Henry County, Indiana, where seven communities passed resolutions establishing a four-mile buffer zone around their towns. In a November 1, 2018, article titled "County Towns Putting Up Walls Against Wind," Darrel Radford, a reporter for the New Castle *Courier-Times*, wrote that "there's still lots of anti-turbine activity" in Henry County and that "as many as half" of the incorporated communities had passed anti-wind measures.[48] In late 2018, the zoning board in Penn Forest Township, Pennsylvania, denied an application by a company called Atlantic Wind that wanted to build more than two dozen turbines on property owned by the Bethlehem water authority. A member of the zoning board, Paul Fogal, told a local news outlet, "We just don't feel it is right for the township."[49]

Kevon Martis, a Michigander who has been tracking the growth of the wind industry since 2009, has consulted with dozens of rural governments in the Midwest to help them craft and implement land-use regulations that will protect them from the encroachment of Big Wind. In a 2018 phone interview, Martis, the volunteer director of the Interstate Informed Citizens Coalition, a nonprofit group, told me that rural residents are increasingly wary of the wind industry. "A few years ago, it might take a year to get one hundred or two hundred people to attend meetings about possible wind projects. Now, local opposition groups are getting four hundred or five hundred people in only sixty to ninety days."

Martis believes the fundamental problem is that, by installing huge turbines, the wind industry is getting "uncompensated nuisance and safety easements" on the unleased property that sits nearby. The result is what he calls "trespass zoning." By establishing setback and noise limits from homes rather than property lines,

Martis says that the wind industry is getting a de facto subsidy from neighbors without compensating them. That, he argues, "is fundamentally unjust and flies in the face of sound zoning principles."[50] Martis asserts that the only way for the wind industry to continue expanding is for it to acknowledge the costs it is imposing on nearby landowners and to buy easements from them.

The growing anti-wind backlash can also be seen by looking at the wind-energy capacity that is *not* being built in California, which has some of America's most ambitious renewable-energy mandates. In 2017, California had about 5,600 megawatts of installed wind capacity—about 107 megawatts less than what the state had back in 2013.[51] In 2017, Rob Nikolewski of the *San Diego Union-Tribune* reported that, in addition to the ban on wind projects in Los Angeles County, three other counties—San Diego, Solano, and Inyo—had passed restrictions on wind turbines. Nikolewski then quoted the head of the California Wind Energy Association, who lamented the industry's inability to site new projects in the state. "We're facing restrictions like that all around the state. . . . It's pretty bleak in terms of the potential for new development."[52]

Looking north of the US border, Ontario has been a hotbed of anti-wind activism. In that Canadian province, ninety towns have declared themselves "unwilling hosts" to wind projects.[53] The anti-wind backlash is also obvious across the Atlantic. In 2010, the European Platform Against Windfarms had about four hundred members in twenty countries. By 2019, it had nearly quadrupled in size and counted some 1,471 member organizations in thirty-two countries.[54] The backlash is particularly apparent in the German state of Bavaria and in Poland. Both places have effectively banned wind turbines by implementing the so-called 10H rule, which requires turbines be located no closer than ten times their height from the nearest homes or other sensitive areas.[55] In 2016, after the government adopted the 10H rule, the head of the Polish Wind Energy Association lamented that "wind farms will disappear from the Polish landscape."[56]

There are dozens of other examples of the growing resistance to Big Wind in Europe. In 2011, 1,500 protesters descended on the Welsh National Assembly, demanding that a massive wind project planned for central Wales be halted.[57] One of the protesters was quoted by the BBC as saying that Welsh politicians "want to destroy magnificent mid Wales. . . . Are we going to let them turn rural Wales into one gigantic power plant?"[58] In 2015, the British government refused a permit for the 970-megawatt Navitus Bay offshore wind project, which was planned to be built in the English Channel near the Isle of Wight.[59] Among the reasons given for rejecting the project, which would have utilized 194 turbines, were its "seascape, landscape and visual impact."[60]

In Scotland, numerous wind projects have been rejected by planning authorities due to local opposition.[61] In 2015, after several wind projects were rejected, Fergus Ewing, Scotland's energy minister, said the plans had been rejected due to "unacceptable landscape and visual impacts in the local areas and these are not outweighed by any wider policy benefit."[62] In 2016, a proposed wind project near Scotland's famous Loch Ness was rejected by local authorities because of its potential impact on tourism.[63] Among the biggest critics of the push for wind turbines in Scotland is the group Scotland Against Spin. On its website, the group says that wind energy is "bad for the customer, bad for jobs and bad for the environment. Civil servants and politicians alike have been comprehensively hoodwinked by renewable energy spin-doctors."[64]

In 2018, *Politico* reported that proposed wind projects in France were facing increasing friction, with "appeals lodged against 70 percent of new projects." The same article quoted Antonella Battaglini, the CEO of the Renewables Grid Initiative, an association that promotes grid development. "We are behind with realizing the grid infrastructure that is needed to support the growing share of renewables," Battaglini said. Why the delays? "Almost every project has local opposition—you might not see it at the national or European level, but everywhere you build there is opposition." An analysis by the European Network of Transmission System Operators for Electricity, the agency that coordinates the European grid, found that out of 350 proposed transmission projects,

100 had either been delayed or rescheduled due to problems with "local acceptability."[65]

Rural residents are opposing wind projects because they are concerned about their property values. In 2015, energy analyst Jude Clemente published an article in *Forbes* citing five studies that found that wind projects reduce the value of nearby properties. Clemente concluded that wind energy's "impact on local property values can no longer be ignored because wind power is set to play a larger role in our electricity portfolio."[66] For instance, in 2010, Michael McCann, a Chicago-based real-estate appraiser, submitted testimony to members of the county board in Adams County, Illinois, which concluded that "residential property values are adversely and measurably impacted by close proximity of industrial-scale wind energy turbine projects." He continued, "Real estate sale data typically reveals a range of 25 percent to approximately 40 percent of value loss, with some instances of total loss as measured by abandonment and demolition of homes, some bought out by wind energy developers and others exhibiting nearly complete loss of marketability."[67]

A 2014 study by the London School of Economics looked at more than one million sales of properties located close to wind projects over a twelve-year period and found that houses located within 1.2 miles (2 kilometers) of large wind projects saw their values reduced by about 11 percent. The study, by Steve Gibbons, the director of the London School of Economics' Spatial Economics Research Centre, included 150 wind projects in England and Wales. Gibbons summed up the study by saying that "property prices are going up in places" where wind projects are not visible, "and down in the places where they are."[68]

In 2016, two researchers from Aachen University in Germany published a study in the journal *Energy Economics*, which found that "the asking price for properties whose view was strongly affected by the construction of wind turbines decreased by about 9 to 14 percent. In contrast, properties with a minor or marginal view on the wind turbines experienced no devaluation."[69]

In 2019, a study by the German think tank RWI found that the value of a single-family home "falls by an average of 7 percent when a wind turbine begins operation within 1 kilometer of the property." RWI's analysis was based on the asking prices on more than 2.7 million houses that were posted on the site of Germany's leading online real-estate broker between 2007 and 2015. The drop in property value disappears on homes that are eight kilometers or more away from the wind turbines. RWI attributed the value reductions to potential noise pollution from the turbines, as well as their deleterious aesthetic effect on the countryside. RWI researcher Manuel Frondel summed up the study by saying, "Wind power may be important for the success of the energy transition but the implications for property owners can be severe in some cases."[70]

Rural residents also have reason to be concerned about the noise pollution created by wind turbines. The wind industry—through its Washington-based lobby group, the American Wind Energy Association—has consistently claimed that wind turbines don't produce much noise and that the noise they produce is not harmful to humans. The facts show otherwise. There is plenty of evidence—both scientific and anecdotal—that shows that the audible and inaudible noise produced by the massive turbines can irritate humans, cause sleeplessness, and in some cases make people sick.[71]

There is also ample evidence that wind-turbine noise adversely affects animals. A 2013 study by Polish researchers found that geese raised in close proximity to wind turbines gained less weight and exhibited "some disturbing changes in behavior" when compared to those raised farther away from the turbines.[72] A 2015 study of pigs done by another group of Polish researchers found that pigs raised in close proximity to wind turbines had lower-quality meat than those raised in control groups. The paper concluded with this blunt assessment: "It is crucial to reduce the exposure of animals to noise generated by wind turbines in order to avoid negative effects on meat quality."[73]

A 2016 study by British researchers from the Royal Veterinary College, Zoological Society of London, and Scottish Oceans Institute on badgers living near wind projects found that those living within one thousand meters of wind turbines were under greater physiological stress than those in a control group. The study found that the hair of badgers living close to turbines had cortisol levels nearly triple the levels of those in the control group. (Cortisol is a steroid hormone produced by the adrenal glands that helps the body regulate metabolism and respond to danger or stress. It is sometimes referred to as the "stress hormone.") The researchers concluded that "the very high levels of cortisol detected in hair from badgers living near wind farms" indicate that the badgers were suffering from a "chronic increase in their hypothalamo-pituitary-adrenal (HPA) axis activity, and thus can be described as stressed."[74]

My collection of news media articles about humans complaining about wind-turbine noise goes back to 2009. That year alone, rural residents in Texas, Oregon, New York, Minnesota, Wisconsin, Canada, New Zealand, Australia, France, and England complained about the noise from wind turbines and, in particular, the issue of sleep deprivation. In 2010, Dr. Michael Nissenbaum, a radiologist in Fort Kent, Maine, surveyed about two dozen residents who lived near the Mars Hill wind project in northeastern Maine. That same year, he published his findings: 82 percent of the residents he surveyed who were living within about 3,500 feet (1,100 meters) of the wind turbines complained of sleep disturbance. Nissenbaum also surveyed about two dozen people in a control group, all of whom lived at least three miles (five kilometers) away from the turbines. Nissenbaum found that only 4 percent of the people in the control group complained of disturbed sleep. He concluded that "there is absolutely no doubt that people living within 3,500 feet of a ridgeline arrangement of turbines 1.5 megawatt or larger turbines in a rural environment will suffer negative effects."[75] In an interview, Nissenbaum told me that the wind industry is "intentionally neglecting the issue of sleep disturbance."[76]

Depriving humans of sleep can make them sick. Nissenbaum made that point during a press conference in Montpelier, Vermont, in 2010, shortly after he completed his initial survey of the residents at Mars Hill. "Annoyance leads to sleep deprivation and illness as day follows night," Nissenbaum said. The people who suffer from the noise pollution, Nissenbaum added, don't need psychological help; "they need the turbines placed further away from their home."[77] In 2012, Nissenbaum, along with two coauthors, published his findings in the journal *Noise & Health.* The article concluded that "the adverse event reports of sleep disturbance and ill health by those living close to industrial wind turbines are supported."[78]

In 2011, in a peer-reviewed article in the *Bulletin of Science, Technology & Society*, Carl V. Phillips, a Harvard-trained PhD, concluded that there is "overwhelming evidence that wind turbines cause serious health problems in nearby residents, usually stress-disorder type diseases, at a nontrivial rate."[79]

Alec Salt, a research scientist at the department of otolaryngology at the Washington University School of Medicine in Saint Louis, has written extensively about the health effects of wind-energy projects and has concluded that noise from wind turbines "can be hazardous to human health." Salt said that the wind industry has "taken the position that if you cannot hear the infrasound, then it cannot affect you. . . . We disagree strongly."[80] In a 2012 paper, Salt and a colleague at Washington University, Jeffery Lichtenhan, concluded that "the physiological effects of low-frequency sounds are more complex than is widely appreciated. Based on this knowledge, we have to be concerned that sounds that are not perceived are clearly transduced by the ear and may still affect people in ways that have yet to be fully understood." They further conclude that infrasound and low-frequency noise can result in "localized endolymphatic hydrops," which is swelling of the inner ear. That condition can result in dizziness and loss of equilibrium.[81] Those two symptoms are common among people who complain about the noise generated by wind turbines. It appears that low-frequency noise and infrasound affect the body's vestibular system, which aids in balance.

In 2012, Peter Narins, a professor and expert on auditory physiology at the University of California, Los Angeles, published a paper in the journal *Acoustics Today*. In the paper, Narins and his coauthor, Hsuan-Hsiu Annie Chen, found that wind turbines generate "substantial levels of infrasound and low frequency sound," and therefore "modifications and regulations to wind farm engineering plans and geographical placements are necessary to minimize community exposure and potential human health risks."[82]

Other studies, from Denmark, Iran, Germany, and Portugal all came to similar conclusions. In 2014, Danish researchers found "that noise from wind turbines increases the risk of annoyance and disturbed sleep in exposed subjects in a dose-response relationship."[83] In 2015, researchers from Iran found that noise from wind turbines "can directly impact on annoyance, sleep and health."[84]

In 2017, German researchers concluded that "the construction of wind turbines close to households exerts significant negative external effects on residential well-being," and that those effects are felt by people living within about 2.5 miles (4 kilometers) of the wind projects.[85] A 2017 study by five Portuguese researchers concluded that "exposure to wind turbine sound significantly impairs individuals' well-being, because it strongly affects their decision to spend, or consider spending, resources in retrofitting their houses."[86]

For years, the wind industry has dismissed studies like the ones just listed. The industry has been aided in no small part by credulous reporters. For instance, a 2018 article in *Popular Science* said that "at a distance, wind turbines are about as noisy as a refrigerator." The article's source for that claim? General Electric, one of the world's biggest wind-turbine manufacturers.[87]

The wind industry and its allies also claim that there is a "nocebo" effect—implying that the people who are complaining are merely imagining their discomfort and the reasons for their sleeplessness. But if the nocebo effect is so strong, and noise pollution from wind turbines isn't a problem, why have so many people, in so many locations all over the globe, been complaining for so many years about the noise-pollution problem?

Furthermore, if wind-turbine noise isn't an issue, why have so many scientists from so many countries published papers that indicate it is a problem? And why has the wind industry bought out residents who have complained about noise?

In 2018, the Minneapolis *Star Tribune* reported that Wisconsin Power and Light bought the homes of "two persistent noise complainants, Bernie and Cheryl Hagen and Dave and Birgitt Langrud." The couples had complained for years about the noise coming from the Bent Tree wind project. A 2016 study done by the Minnesota Department of Commerce found that the noise complaints against the wind project were "substantial" and that the turbines on the project had exceeded state noise standards.[88] In 2017, before the utility agreed to buy his house, Dave Langrud told Mike Hughlett of the *Star Tribune* that ten wind turbines had been built within three-quarters of a mile of his home, and the closest one was just 1,150 feet (350 meters) away. "We can hear them inside our house—whoosh, whoosh, whoosh. It's hard to fall asleep and you don't get a restful sleep," Langrud said. "When I go out of town, I start catching up on my sleep." Langrud, like many other people who have had turbines built near their homes, said he often got dull headaches due to the noise. Furthermore, his property—including the interior of his house—was affected by shadow flicker, the stroboscopic effect caused by shadows of the rotating turbine blades. "It drives you nuts," Langrud said.[89]

If wind-turbine noise isn't a problem, why are towns in Massachusetts and Iowa forcing energy companies to dismantle and remove their wind turbines? In Falmouth, Massachusetts, a battle over noise forced the town to shut down a pair of four-hundred-foot wind turbines it had installed at the town's wastewater treatment plant.[90] In 2012, Neil P. Andersen and his wife, Elizabeth L. Andersen, sued the town of Falmouth, claiming the noise from the turbines had disrupted their ability to enjoy their property. The court agreed in 2017, and the two 1.65-megawatt turbines were ordered to shut down. The town of Falmouth estimated the shutdown will cost it more than $10 million, including about $1 million to dismantle and remove the turbines.[91] In 2018, Andersen described the turbine noise to a reporter at the *Falmouth Enterprise* newspaper thusly: "A pounding pulse that you can't hear, but it gets into your chest cavity. It's extremely distressing. It mimics your heartbeat and you cannot sleep; it just drives you nuts."[92]

In 2018, after myriad complaints about turbine noise, town residents in Fairbank, Iowa, won a three-year legal battle that forced wind-energy developers to remove three wind turbines. Cheyney Hershey, who lived near the 450-foot-high (137 meter) turbines, told the *Des Moines Register*, the noise from the machines was so bad, "you can't sit outside on the deck and have a conversation without the constant thumping of the blades going round."[93]

To be clear, not every wind project causes noise problems. Adding complexity is that the most problematic noise generated by the turbines—low-frequency sound (10 to 200 hertz) and infrasound (0 to 10 hertz)—has varying effects on people. Some individuals feel the effects of the noise quickly and compare it to motion sickness. Others may not feel it at all. Nevertheless, the harmful effects of low-frequency noise and infrasound are well known and have been known for decades. A 2001 report published by the National Institutes of Health said that exposure to infrasound can cause vertigo, as well as "fatigue, apathy, and depression, pressure in the ears, loss of concentration, drowsiness."[94]

In addition to the academic studies on wind-turbine noise and human health, over the past decade or so I have personally interviewed more than a dozen people who have experienced health problems or have abandoned their homes after wind turbines were built nearby. In 2010, I talked to Charlie Porter, a horse trainer whose farm near King City, Missouri, had been surrounded by a phalanx of giant turbines. He told me that "the overwhelming noise, sleep deprivation, constant headaches," and anxiety had forced him and his wife to abandon the twenty-acre (eight-hectare) farm where they'd lived for eighteen years. The closest wind turbine was built about 1,800 feet (549 meters) from their home. The noise from the turbines "just ruined life out in the country like we knew it," Porter told me. "We never intended to sell that farm. Now we couldn't sell it if we wanted to."

In 2010, I interviewed Janet Warren, who had been raising sheep on her five-hundred-acre family farm near Makara, New Zealand, when wind turbines were built nearby. The turbines, she told me via email, emit "continuous noise and vibration," which she said was resulting in "genuine sleep deprivation causing loss of concentration, irritability, and short-term memory effects." The turbines, installed about 2,900 feet (884 meters) from Warren's house, began generating electricity in July 2009. She said, "We started recording formal noise compliance complaints in August." In early 2010, she and her husband were forced to move out of their home due to the noise.[95]

Also in 2010, I interviewed Tony Moyer, a resident of Empire, Wisconsin, who was living with the noise created by three turbines that were erected within 1,400 feet (426 meters) of his home in late 2008. "If you get up at night, you can't go back to sleep because you hear those things howling." Moyer and his wife have tried white-noise machines, to no avail. They aren't able to sell. "The option to sell your home isn't there. I have wind turbines east, west, and north of me. If you talk to realtors, they can't sell homes near a wind farm."[96]

In 2016, I interviewed another victim of wind-energy noise: Michael Keane. In 2004, Keane and his wife, Dorothy, purchased a derelict cottage on 2.1 acres (0.85 hectares) in County Roscommon, Ireland, for 72,000 euros. They spent the next several years, and about 100,000 euros, renovating the place. In 2013, shortly after wind turbines were built nearby, they abandoned their home. In 2016, according to Keane, they sold the property for 55,000 euros.[97]

In an email, Keane told me that they left "after two years of noise torture from two 100-meter tall Enercon 2.3-megawatt turbines [that] were built 750 meters from our home. We took this drastic step on the advice of three medical doctors after we were diagnosed with sleep deprivation, chronic stress, and anxiety."[98] When I asked him about the wind industry's claims that the noise from turbines is not a problem, he replied: "In time the truth will come out" about how the wind industry "aided and abetted by world governments allowed people to be tortured, some to the point of suicide so that the pockets of a few could be filled by the suffering of many."[99]

Also in 2016, I met Dave and Rose Enz. In 2011, the Enzes abandoned their home near Denmark, Wisconsin, a small farming community located in Brown County, about one and a half hours' drive north of Milwaukee, because of the noise produced by

Anti-wind-energy sign near Denmark, Wisconsin, 2016.
Source: Photo by author.

the half dozen wind turbines that had been built near their home. The closest turbine was installed about 3,200 feet (975 meters) from their house, which sits on 41 acres (16.6 hectares) of rolling farmland. The couple bought their property in 1978. Shortly after the turbines at the 20-megawatt project owned by Duke Energy, known as the Shirley Wind Farm, began operating, the couple, both of whom were in their sixties, began experiencing numerous symptoms, including "headaches, ear pain, nausea, blurred vision, anxiety, memory loss, and an overall unsettledness," Mr. Enz told me.[100] Unable to stay in their home, the Enzes began living in their RV or staying with friends.

The Enzes weren't the only residents of Brown County to be afflicted by wind-turbine noise. As reported by the *Green Bay Press Gazette*, about twenty families in the county were adversely affected by the turbine noise and "three families have moved out of their homes rather than endure physical illness they blame on the low-frequency noise the wind turbines generate, according to Audrey Murphy, president of the board that oversees the Brown

County Health Department." In 2014, after investigating the issue, the board voted 4-0 on a motion that declared the turbines to be a public health risk. "We struggled with this but just felt we needed to take some action to help these citizens," Murphy told the *Gazette*.[101]

In 2015, the county's health director, Chua Xiong, effectively reversed the board's decision and concluded that there wasn't sufficient evidence to directly link the turbine noise to human health issues.[102] That same year, Xiong told a colleague that she got migraine headaches whenever she visited the areas near the wind turbines. Shortly after the reversal, Xiong resigned her position in Brown County to take another job.[103]

When I met the Enzes, they refused to meet me at their home in Denmark. Instead, we met at a Subway sandwich shop in nearby Wrightsville. I had spoken to Dave by phone several times in the previous years about their predicament, but had not met him, or Rose, in person. "We didn't expect any of this stuff," said Dave, who had spent more than thirty years working as a millwright at a paper mill in Green Bay. When I asked Rose and Dave why they hadn't sold their property, Rose replied, "We have a conscience. . . . If a family moved in there, I believe some of them will get sick. So how do I, in good conscience, sell it to them? They'd be stuck." After chatting for forty-five minutes or so, we decided to adjourn. As we stood in the parking lot saying our goodbyes, Dave turned to me and asked, "How can it be in America, a country where we are supposed to have property rights, that a company can build a wind project and force us off our land?"

The land-use battles also include fights over solar projects. In California, the 377-megawatt Ivanpah solar complex now operating in the Mojave Desert met fierce opposition from conservationists due to its impact on the desert tortoise, which is listed as a threatened species under the federal Endangered Species Act.[104] Construction on the project, which covers 5.4 square miles (14 square kilometers), was allowed to continue, but Ivanpah will likely be one of very few solar thermal projects built in the United States due to the size of its footprint. Meanwhile, in New York in 2016,

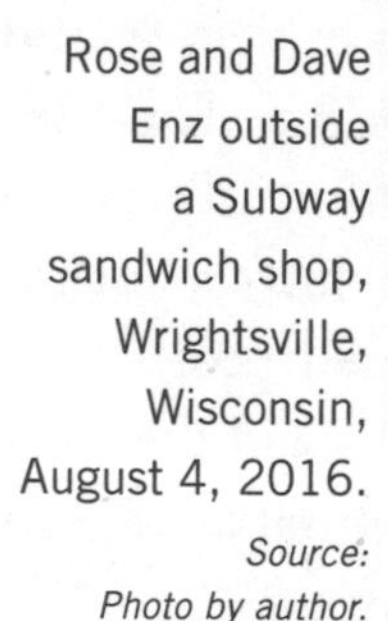
Rose and Dave Enz outside a Subway sandwich shop, Wrightsville, Wisconsin, August 4, 2016.
Source: *Photo by author.*

local environmental groups fought the clear-cutting of 350 acres of woodlands on Long Island to make way for a solar project. One of the opponents of the project, Dick Amper, told the Associated Press that "choosing solar over forests anywhere in the world is just plain stupid."[105]

In January 2019, residents of Spotsylvania County, Virginia, were fighting a proposed 500-megawatt solar project that, if built, would cover nearly ten square miles (twenty-six square kilometers). According to the Fredericksburg *Free-Lance Star*, local residents were opposed because they believed the solar project "is too big to be near homes and that it poses potential health and environmental risks. They also are concerned about impacts to property values."[106] That same month, the *Baltimore Sun* reported on a controversy that was raging in Charles County, Maryland, over a solar-energy project backed by Georgetown University. As reported by the *Sun*'s Scott Dance, the solar project, which requires clear-cutting 240 acres (97 hectares) of forest, is "a choice between a vanishing ecosystem and a push toward cleaner energy." Dance explained that the forest that will be cleared is among fewer "than three dozen areas in Maryland that the Audubon Society has deemed an 'important bird area,'" meaning it is a rare remnant of large contiguous forestland. A member of the southern Maryland chapter of the Sierra Club called the solar project "thoughtless of the future."[107]

In May 2019, the town board of Cambria, New York (population: 6,000), unanimously rejected the proposed 100-megawatt Bear Ridge solar project. If built, that $210 million project would cover about nine hundred acres with solar panels. A few days after the vote, I talked to Cambria town supervisor Wright Ellis, who has held that position for twenty-seven years. "We don't want it," Ellis told me. "We are opposed to it." The proposed project, he said, violates Cambria's zoning laws. In addition, Ellis said the solar project would result in a "permanent loss of agricultural land" and potentially reduce the value of some 350 nearby homes.[108]

The proliferation of wind energy is having a deadly impact on bats and birds. A 2013 study estimated that US wind turbines were killing about 888,000 bats and 573,000 birds per year. The bird kills include some 83,000 raptors.[109] That same year, raptor biologists at the US Fish and Wildlife Service published a paper reporting the deadly effect that Big Wind is having on bald and golden eagles. They found that, between 2007 and 2011, wind turbines killed eighty-five eagles, including six bald eagles.[110] Joel Pagel, the report's lead author, told me that that figure was "an absolute minimum."[111] Furthermore, the study showed that as wind capacity increased, so did eagle kills. In 2007, the United States had about 17 gigawatts of installed wind-energy capacity and the report confirmed two eagles were killed by turbines. By 2011, installed wind capacity had increased to about 47 gigawatts and the researchers found twenty-four eagles had been killed by turbines.[112] Thus, over a time frame when wind capacity tripled, the number of documented eagle kills increased by a factor of twelve.

Pagel's 2013 study on eagle mortality caused by wind turbines was published a few months after the Fish and Wildlife Service issued a report that concludes, "There are no conservation measures that have been scientifically shown to reduce eagle disturbance and blade-strike mortality at wind projects."[113]

In 2018, ecologists from the Indian Institute of Science in Bangalore studied the effect of wind turbines on wildlife in India's

Western Ghats mountain range, where wind projects have been operating for two decades. They found that wind farms can act like apex predators. "By reducing the activity of predatory birds in the area, wind turbines effectively create a predation-free environment that causes a cascade of effects on a lower tropic level," the report explained.[114] The researchers found almost four times as many birds of prey in areas without wind turbines. They also found that areas near the wind projects had far more lizards than those without wind turbines. The study concluded that wind farms have "complex ecological consequences" and that they "have emerging impacts that are greatly underestimated."[115]

Wind-energy promoters have repeatedly attempted to downplay the death toll by saying that buildings and cats also kill birds. That may be true. But house cats are not killing golden eagles and bald eagles—wind turbines are, and they are killing them by the dozens. Many of those same promoters claim that climate change is a bigger long-term threat to wildlife than are wind turbines. That may or may not be so. But allowing the immediate destruction of wildlife so that the animals might be saved from climate impacts at some point in the future makes no sense at all.

Wind turbines are particularly deadly to bats. In 2016, two scientists from the US Geological Survey, Thomas J. O'Shea and Paul M. Cryan, along with three colleagues, published a paper that said that wind turbines were the largest cause of mass bat mortality, and that they exceed the toll taken by white-nose syndrome, a fungal disease that afflicts bats.[116] In a discussion of the paper, Cryan said that the wind industry's harm to bat populations could have long-term negative effects. "Bats are long-lived and very slow reproducers," he said. "Their populations rely on very high adult survival rates. That means their populations recover from big losses very slowly."[117]

The deleterious effect of wind turbines on bat populations was further confirmed in 2016, when Bird Studies Canada, a conservation group, released a report on wind energy. According to the study, "across Canada, bat fatalities were reported more often than birds, accounting for 75 percent of all carcasses found." The report found that wind turbines in Ontario alone killed about 42,656 bats between May 1 and October 31, 2015, and each wind turbine had killed about eighteen bats over that time frame.[118] The bat fatalities in Ontario included several species of rare or endangered bats, such as the little brown bat and northern long-eared bat. The report also found that wind turbines in the province killed 462 raptors over that same six-month period.[119]

Bats are not as popular as eagles. But they are essential pollinators and insectivores. In Texas alone, economists have estimated that bats save the state more than $1 billion annually on pesticides.[120] In 2014, I talked to Merlin Tuttle, one of the world's foremost experts on bats, about the effect that wind-energy deployment is having on the only flying mammals. He told me, "Anyone familiar with bat population biology is deeply concerned about the impact of wind turbines on the long-term viability of a number of bat species." Tuttle, who founded Bat Conservation International, reiterated the point about bats' slow reproductive rates. "We are at great risk of needlessly creating new endangered species," he told me. "We risk losing the benefits of bats to natural systems and agriculture."[121]

The human and ecological impacts of electricity-generation facilities like wind turbines and solar projects are only one aspect of the land-use challenge facing large-scale renewable deployment. Land-use battles are also being fought over the high-voltage electricity transmission lines needed to carry solar- and wind-generated electricity from rural areas to customers in big cities.

In 2017, Iowa enacted a law that prohibits the use of eminent domain for high-voltage transmission lines. The move doomed the Rock Island Clean Line, a five-hundred-mile, $2 billion, high-voltage, direct-current transmission line that was going to carry electricity from Iowa to Illinois.[122] The opposition forced the project's developer, Houston-based Clean Line Energy Partners, to withdraw its application for the project in Iowa.

In early 2018, Clean Line Energy Partners also announced it was suspending its years-long effort to build a 720-mile, $2.5 billion transmission line across the state of Arkansas. The Plains and Eastern Clean Line aimed to carry wind energy from Oklahoma to customers in the southern and southeastern United States. But the project faced fierce opposition in Arkansas, where the state's congressional delegation opposed the deal.[123] After the project was canceled, the delegation issued a statement saying termination of the transmission project was "a victory for states' rights and a victory for Arkansas." It went on, saying that energy projects "should not cost Arkansas landowners a voice in the approval process."[124]

Also in 2018, New Hampshire regulators rejected a high-voltage electricity transmission project called Northern Pass Transmission that was to carry power from Quebec hydroelectric facilities to consumers in Massachusetts. But the 192-mile, $1.6 billion project—which was to go through the White Mountains—was vetoed unanimously by the New Hampshire Site Evaluation Committee.[125]

A similar high-voltage project, the $2.5 billion, 780-mile Grain Belt Express, has been delayed for years by opposition from rural residents in Missouri. First proposed in 2010, the 4,000-megawatt project is designed to move electricity from Kansas to Indiana and other states.[126] But in 2015, the Missouri Public Service Commission blocked the project after concluding the cost to the state's landowners exceeded its benefits.[127] The fight over the project was partially resolved in mid-2018, when the Missouri Supreme Court ruled in favor of the transmission line. But several counties in Missouri must still approve the project, and, by late 2018, the transmission line had only acquired about forty of the more than seven hundred easements it needs to acquire from private landowners.[128] It must also overcome legal challenges in Illinois.[129]

With the obvious exception of rooftop solar systems, all electricity-generation plants need transmission and distribution lines to carry the energy they produce to customers. But renewable-energy projects are particularly dependent on long transmission lines because the best wind, solar, and hydropower resources are in rural areas where electricity use is low. Moving electricity from those remote sites to urban areas, where demand is high, requires long transmission lines. The more renewable-energy capacity gets added to the grid, the more transmission capacity must be built. In 2012, the National Renewable Energy

Laboratory estimated that if the United States were to attempt to derive 90 percent of its electricity from renewable sources, it would have to double its high-voltage transmission capacity.[130] The United States currently has about 240,000 miles (386,000 kilometers) of high-voltage transmission lines.[131] Given the friction that high-voltage projects are already facing, it is highly doubtful that the United States will be able to double that capacity.

Of course, not all renewable-energy and high-voltage-transmission projects result in contentious land-use battles. But many of them do. And as those projects proliferate, more land-use conflicts are inevitable. Thus, if issues like land use and cost are already limiting the growth of renewables, what should electricity producers be using if they want to meet soaring global demand and reduce the growth of greenhouse gas emissions? The answer is nuclear energy.

19

THE NUCLEAR NECESSITY

> So far, the atom is a superb villain. Its power of destruction is foremost in our minds. But the same power can be put to use for creation, for the welfare of all mankind.
>
> —**HEINZ HABER**, *Our Friend the Atom* [1]

Nearly all of the machinery that generates electricity at the Indian Point Energy Center in Buchanan, New York, is hidden from view. Sure, the domed reactor buildings can be seen from miles away by boats traveling on the Hudson River, as well as from other vantage points. But the reactor building—where small amounts of uranium are fissioned to produce the heat that drives the generators—is strictly off-limits to visitors. The turbines—the bladed machines that are spun by the steam created by the reactors to convert that steam into electricity—are shrouded by big metal covers. The main generator, where the electricity gets produced, is similarly shrouded. Nevertheless, for visitors lucky enough to get inside the turbine hall of the Unit 2 reactor, a glimpse of the guts of the massive machine can be had.

Near the middle of the aircraft-hanger-size turbine hall, the shrouds, pipes, and safety barriers have been stripped away to expose a short segment of the solid-steel main turbine shaft. A light coating of oil gives it a faint glimmer as it spins at about 1,800 revolutions per minute. The shaft, which has a diameter of about eight inches (twenty centimeters), doesn't make any noise. Instead, its spinning motion unites with a faint hum and vibration that can be felt throughout the turbine hall.

The two-hundred-foot-long drive shaft is the steel spinal cord that allows the 3,200 megawatts of heat produced by the reactor to be converted into about 1,000 megawatts of electric energy.[2] That electricity drives much of the Big Apple. By itself, that single shaft energizes fully one-eighth of New York City and its 8.6 million residents. It's rather astounding to consider. Over one million New Yorkers—as well as one out of every eight lights, elevators, subway cars, and electric teapots in the five boroughs—depend on the electricity that is being generated thanks to that one finely balanced piece of spinning steel.

Since the time of Edison, the business of producing electricity on a large scale has depended on machines similar to the one in the turbine hall at Unit 2. The objective has remained the same: turn a drive shaft that spins a copper coil inside a clutch of magnets and, in doing so, produce electric current. And when it comes to spinning a drive shaft, Unit 2 and Unit 3 have few peers. The two reactors at Indian Point are among the largest engines in the United States. They produce so much heat that during the hottest days of the summer, each of the reactors requires as much as 840,000 gallons (3.2 million liters) of Hudson River water to be pumped through its condensers every minute to keep it at the proper operating temperature.

During my career in journalism, I've visited factories, mines, refineries, and numerous power plants. Indian Point leaves all of them in the shade. It is a marvel of engineering, architecture, and ingenuity that should be appreciated alongside other iconic American landmarks like the Hoover Dam, Gateway Arch, and Washington Monument. Alas, it is not. Instead, Indian Point is a relic from another age.

Indian Point was launched when government and companies were thinking big. In the wake of the New Deal and World War II, the federal government and state governments took on big public works projects, like interstate highways, bridges, inland waterways, and hydropower projects. During the 1930s, '40s, '50s, and '60s the United States built lots of dams—structures that architecture critic Lewis Mumford would later call "democratic pyramids."[3] Work on Indian Point began in 1956, the same year President Dwight Eisenhower launched the interstate highway system.[4] (Two years earlier, in 1954, work began on the first commercial nuclear reactor in the United States, at Shippingport, Pennsylvania.)

Indian Point equals or surpasses any of the great dams built in America. But unlike the sprawling reservoirs that are impounded by those dams, Indian Point represents the apogee of densification.[5] The twin reactors perfectly illustrate what may be nuclear energy's single greatest virtue: its unsurpassed power density, which, in turn, allows us to spare land for nature.

To illustrate that point, let's compare the footprint of Indian Point with the footprint needed to accommodate renewables. Indian Point covers less than 0.4 square miles (1 square kilometer).[6] From that small footprint, Indian Point reliably pumps out about 16.4 terawatt-hours of zero-carbon electricity per year.[7] To put Indian Point's footprint into context, think of it this way: you could fit three Indian Points inside New York City's Central Park.

Now, let's compare Indian Point's footprint and output with what would be required to replace it with electricity produced by wind turbines. Based on projected output from offshore wind projects, producing that same amount of electricity—16.4 terawatt-hours per year—would require installing about 4,005 megawatts of wind turbines.[8] That much capacity will require hundreds of turbines spread over some 515 square miles (1,335 square kilometers) of territory.[9] Thus, from a land-use, or ocean-use, perspective, wind energy requires about 1,300 times as much territory to produce the same amount of electricity as is now being produced by Indian Point.[10]

Those numbers are almost too big to imagine. Therefore, let's look again at Central Park. Recall that three Indian Points could fit inside the confines of the famed park in Manhattan. That means that replacing the energy production from Indian Point would require paving a land area equal to four hundred Central Parks with nothing but forests of wind turbines.

Despite its tiny footprint, despite its importance to New York City's electricity supply, despite its zero carbon emissions, Indian Point is headed for premature shutdown. By 2021, the drive shafts at Unit 2 and Unit 3 will stop spinning. The reactors are not being shuttered due to decrepitude. They could continue operating for

Land area required to produce 16.4 terawatt-hours of electricity per year.

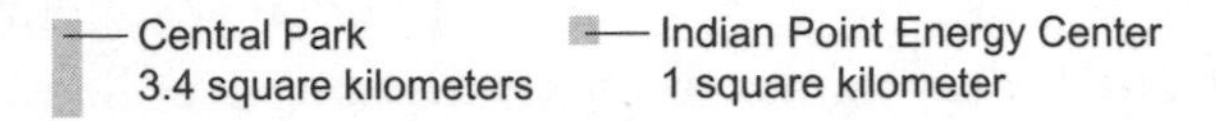

Nuclear
1 square kilometer or **0.3** the size of Central Park.

Wind
1,335 square kilometers or about **400** times the size of Central Park.

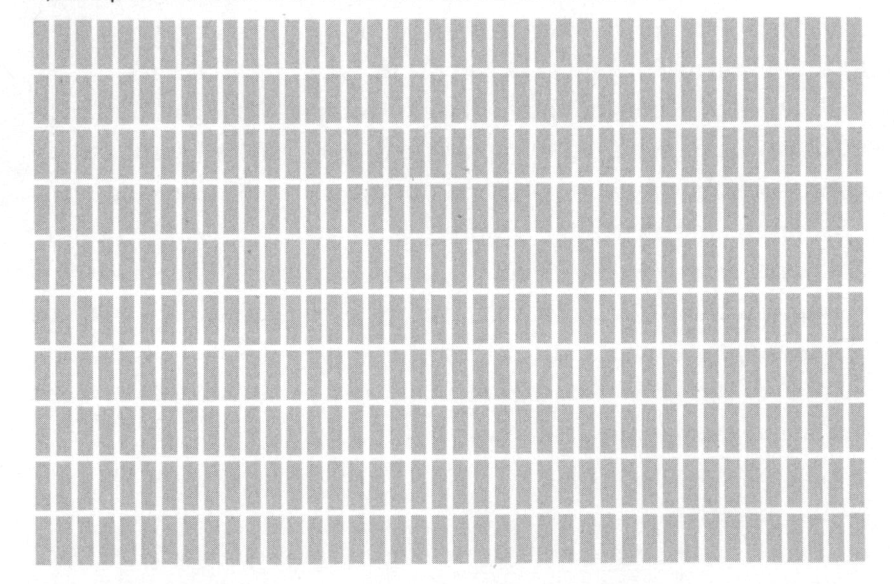

Powering New York City: Land Area Needed by Nuclear and Wind Energy to Produce 16.4 Terawatt-Hours of Electricity per Year

Sources: Natural Resources Defense Council, author calculations.

decades to come. Instead, they are being shuttered for political reasons. Antinuclear groups, including Riverkeeper and the Natural Resources Defense Council, argued for years that the plant should be closed, claiming it was harming fish in the Hudson River and posing a danger to residents in and around New York City. New York governor Andrew Cuomo, a Democrat, was convinced to join the antinuclear crusade, and in early 2017 he gleefully announced that the two reactors would be closed.

The looming closure of Indian Point is part of a rash of nuclear reactor retirements across the United States that is impeding efforts to reduce greenhouse gas emissions. Indeed, despite nuclear's essential role in reducing emissions, the US nuclear sector is in the midst of a full-blown crisis. Between 2013 and 2018, American utilities closed or announced the closure of fifteen nuclear plants. The combined output of those nuclear plants is about 133 terawatt-hours per year.[11] That's about 70 percent more zero-carbon electricity than was produced by all of the solar facilities in the United States in 2017.[12] Not only are US reactors being retired early, but aside from a pair of reactors now being built in Georgia and a small reactor slated to be built in Idaho in the mid-2020s, no new nuclear plants are even being considered by electric utilities in the United States.

The result of all these closures is obvious: the United States—which has led the world in the development and deployment of nuclear energy since the days of the Manhattan Project—is rapidly becoming a nuclear laggard. That may please antinuclear activists at Greenpeace and the Sierra Club, but it's a big loss in the effort to fight climate change. In addition, the loss of America's nuclear fleet could mean higher electricity prices, a less stable electric grid, and further escalation of the land-use battles over the siting of renewable energy projects.

•

Before going further, let me be clear about where I stand on nuclear energy: if you are anti–carbon dioxide and antinuclear, you are pro-blackout. There is simply no way to slash global carbon dioxide emissions without big increases in our use of nuclear energy.

That fact has been made clear by numerous scientists. In 2011, James Hansen, one of the world's most famous climate scientists, wrote that "suggesting that renewables will let us phase rapidly off fossil fuels in the United States, China, India, or the world as a whole is almost the equivalent of believing in the Easter Bunny and Tooth Fairy." He went on to say that politicians and environmental groups "pay homage to the Easter Bunny fantasy, because it is the easy thing to do. . . . They are reluctant to explain what is actually needed to phase out our need for fossil fuels."[13]

In late 2013, Hansen and three other climate scientists wrote an open letter to environmentalists encouraging them to support nuclear. They wrote that "continued opposition to nuclear power threatens humanity's ability to avoid dangerous climate change. . . . Renewables like wind and solar and biomass will certainly play roles in a future energy economy, but those energy sources cannot scale up fast enough to deliver cheap and reliable power at the scale the global economy requires."[14] In 2015, at the UN Climate Change Conference in Paris, Ken Caldeira, a climate scientist at the Carnegie Institution for Science who was one of the coauthors of the 2013 letter, reiterated his belief that nuclear must be part of any emissions-reduction effort. "The goal is not to make a renewable energy system. The goal is to make the most environmentally advantageous system that we can, while providing us with affordable power," Caldeira said. "And there's only one technology I know of that can provide carbon-free power when the sun's not shining and the wind's not blowing at the scale that modern civilization requires. And that's nuclear power."[15]

Also in 2015, the International Energy Agency declared that "nuclear power is a critical element in limiting greenhouse gas emissions." It went on, saying that global nuclear generation capacity, which in 2018 totaled about 375 gigawatts, must more than double by 2050 if the countries of the world are to have any hope of limiting temperature increases to the two-degree scenario that is widely agreed as the acceptable limit.[16]

In May 2019, the International Energy Agency reiterated its support for nuclear by declaring that, without more nuclear energy, global carbon dioxide emissions will surge and "efforts to transition to a cleaner energy system will become drastically harder and more costly." How costly? The agency estimated that "$1.6 trillion in additional investment would be required in the electricity sector in advanced economies from 2018 to 2040" if the use of nuclear energy continues to decline. That, in turn, will mean higher prices, as "electricity supply costs would be close to $80 billion higher per year on average for advanced economies as a whole." The report also makes it clear that solar and wind energy cannot fill the gap because of growing land-use conflicts. The Paris-based agency said that "resistance to siting wind and, to a lesser extent, solar farms is a major obstacle to scaling up renewables capacity."[17]

The importance of nuclear energy in reducing greenhouse gas emissions can be seen, again, by looking at Indian Point. In 2017, the New York Independent System Operator, the nonprofit entity that manages the grid in the Empire State, issued a report that concluded that if the two reactors at Indian Point are closed as scheduled by 2021, the electricity produced by the plant will largely be replaced by three gas-fired power plants.[18] That's no surprise. Whenever nuclear reactors are shuttered, they almost always get replaced by plants that burn hydrocarbons, and that means increased emissions of carbon dioxide. By one estimate, New York's electricity-sector emissions will increase by 29 percent when Indian Point is shuttered and its output is replaced by gas-fired power plants.[19]

In 2017, the New England Independent System Operator reported that greenhouse gas emissions increased by nearly 3 percent in the year following the 2014 closure of the 604-megawatt Vermont Yankee nuclear plant.[20] Why did emissions increase? The percentage of gas-fired electricity in New England jumped by six points after the plant shutdown, to nearly 49 percent.[21]

Similar results occurred in California after state officials negotiated the premature shutdown of the San Onofre Nuclear Generating Station in 2013. After the shutdown, Lucas Davis—a professor at the University of California, Berkeley, Haas School of Business—along with Catherine Hausman, who works at the Gerald R. Ford School of Public Policy at the University of

Michigan, published a report that found that, in the first year after San Onofre closed, California's carbon dioxide emissions jumped by about nine million tons. They point out that that quantity is "the equivalent of putting 2 million additional cars on the road."[22]

The closures of Vermont Yankee and San Onofre—along with the looming closure of Indian Point—are part of a grim outlook. By the mid-2020s, the United States could prematurely retire as much as a fifth of its installed nuclear capacity.[23] What's driving the retirements? Low-cost natural gas is a major factor. In addition, nuclear plants must compete in the wholesale market with heavily subsidized electricity produced from wind and solar. Add in aging reactors, post-Fukushima regulations, and the never-ending opposition from big environmental groups, and the US nuclear sector has been taking a beating.

The closure of these plants has been cheered by the well-funded opponents of nuclear energy. For decades, nuclear energy's foes have relied on three main criticisms to justify their opposition: radiation, waste, and cost. Let's look at those in order.

From a nuclear safety standpoint, it's difficult to imagine a scarier scenario than what happened on March 11, 2011. An earthquake measuring 9.0 on the Richter scale hit 130 kilometers off the Japanese coastline. Within minutes of the earthquake, a series of seven tsunamis slammed into the Fukushima Daiichi nuclear plant. The backup diesel generators, designed to keep the nuclear plant's cooling water pumps operating, quickly failed. A day later, a hydrogen explosion blew the roof off the Unit 1 reactor building. Over the next few days, similar explosions hit Units 2 and 3.[24] Three reactors melted down.[25] It was the worst nuclear accident since Chernobyl in 1986. In the wake of the Fukushima disaster, Greenpeace did its utmost to instill fear of radiation. In a March 22, 2011, op-ed in the *New York Times*, the head of Greenpeace International, Kumi Naidoo, declared that "nuclear energy is an expensive and deadly distraction from the real solutions." Naidoo claimed that nuclear energy is "inherently unsafe" and that the list of possible illnesses caused by "radiation is horrifying: genetic mutations, birth defects, cancer, leukemia."[26]

Despite Greenpeace's efforts to instill "radiophobia" in the minds of consumers, the reality is that nuclear energy remains the safest form of electricity production.[27] The facts show that the accident at Fukushima led to exactly two deaths. About three weeks after the tsunami hit the reactor complex, the bodies of two workers were recovered at the plant. They didn't die of radiation. They drowned.[28]

I am not minimizing the seriousness of what happened at Fukushima. Cleaning up the mess there will take decades and cost hundreds of billions of dollars. Nevertheless, despite all the hype and fearmongering, exactly zero deaths at Fukushima have been attributed to radiation. I repeat, *no one in Japan or anywhere else* has been killed by the radiation from the accident at Fukushima. You won't hear that from the world's biggest antinuclear groups, including Friends of the Earth, Greenpeace, and Sierra Club. All have made radiophobia a central tenet of their campaigns against nuclear energy. Nevertheless, the facts are clear.

In 2013, the UN Scientific Committee on the Effects of Atomic Radiation released a report on Fukushima, which found that "no radiation-related deaths have been observed among nearly 25,000 workers involved at the accident site. Given the small number of highly exposed workers, it is unlikely that excess cases of thyroid cancer due to radiation exposure would be detectable in the years to come." (Thyroid cancer is among the most common maladies caused by excessive exposure to radiation.) The UN committee was made up of eighty scientists from eighteen countries.[29]

In 2018, Gerry Thomas, a professor at Imperial College London, said that radiation fears at Fukushima are overblown. In an interview on *60 Minutes Australia*, Thomas said she had been to Fukushima many times and would have no hesitation about going back to what she called "a beautiful part of the country." Thomas, who runs the Chernobyl Tissue Bank and is an expert on the effects of radiation, also said that no more than 160 people will die from radiation poisoning due to the Chernobyl accident.[30] That's far fewer than the thousands of deaths that were predicted. What about deaths from radiation due to Fukushima? Thomas said there have been "absolutely none. No one has died from radiation poisoning."[31] In a later

60 Minutes report, Thomas told reporter Tom Steinfort, "The one thing that we have learnt from both Chernobyl and Fukushima is that it actually wasn't the radiation that's done the health damage to the people in the surrounding areas. It's their fear of radiation. There's been far more psychological damage than there have actually physical damage because of the two accidents."[32]

Antinuclear groups and others have fanned the fears of radiation by claiming there is no safe dosage level. The truth is that we are exposed to radiation all the time. Not only that, radiation can be therapeutic and is widely used in medical treatments for numerous conditions, including cancer. Despite these facts, the nuclear industry has been constrained by the policy known as ALARA, meaning it must keep radiation levels "as low as reasonably achievable." But as energy analyst James Conca pointed out in a 2018 article in *Forbes*, following ALARA means that the nuclear industry must now spend billions of dollars "protecting against what was once background levels" of radiation. Radiation is "one of the weakest mutagenic and cytogenic agents on Earth," he continued. "That's why it takes so much radiation to hurt anyone."[33]

Conca's article agrees with a 2016 analysis published by the Genetics Society of America about the radiation impacts on human health after the bombings of Hiroshima and Nagasaki. The study concluded that "public perception of the rates of cancer and birth defects among survivors and their children is greatly exaggerated when compared to the reality revealed by comprehensive follow-up studies."[34]

In addition to hyping fears about radiation, antinuclear campaigners routinely claim that the radioactive waste produced by nuclear reactors cannot be disposed of safely, and therefore no new nuclear plants should be built. Again, that's simply not true. As Michael Shellenberger, the founder of Environmental Progress and one of the world's staunchest proponents of nuclear energy, points out, nuclear energy's waste stream is actually one of its greatest virtues. While sitting in Environmental Progress's office on Telegraph Avenue in Berkeley, a few blocks from the University of California's campus, Shellenberger told me that nuclear energy is "the only way to make electricity production that contains all of its toxic waste. All of it." He continued, saying that nuclear energy prevents its waste "from going into the environment and yet people think that the waste from nuclear plants is a big problem."

Shellenberger and others have pointed out that the nuclear waste issue is not a technical problem; it's a political problem. That can be seen by looking, one more time, at Indian Point. During my visit to the plant, two Entergy employees, Jerry Nappi and Brian Vangor, showed me where the company stores the spent fuel from the reactors. On the north side of the facility, on an area that's maybe the size of two tennis courts, there were about thirty large steel-and-concrete cylinders, known in the industry as dry casks. Each cask stands about 15 feet (4.5 meters) tall and 8 feet (2.4 meters) in diameter and weighs about one hundred tons.

As I looked at the row of casks, I was struck by the fact that, throughout the entire operating history of the plant, which began producing electricity in 1962, two years after I was born, the bulk of its spent fuel could fit inside such a small area. And what if some terrorist wannabe decided to cart off one of the casks? When I asked Nappi about that possibility, he replied, "It's impossible, basically." He then pointed to the massive machine that Entergy was using to maneuver the dry casks around the site. The machine, which ran on huge metal tracks, moved at about one mile per hour—not exactly the kind of getaway vehicle that would be needed by someone hoping to pilfer a bit of radioactive material.

The dry casks at Indian Point are part of the nuclear waste that has been created by the US nuclear-energy sector. Since the 1950s, when construction on the New York facility began, the domestic sector has produced about 80,000 tons of high-level waste. That may sound like a lot. But consider this fact: if you collected all of that waste in one place and stacked it about 10 yards (9.1 meters) high, it would cover an area roughly the size of a single soccer pitch.[35]

The key to nuclear waste is proper management. France provides one of the world's best examples of proper nuclear-waste handling. France gets about 75 percent of its electricity from its

The Indian Point Energy Center, Buchanan, New York, 2018. The dry-cask storage area can be seen on the lower right.

Source: Photo by Tyson Culver.

fleet of about sixty nuclear reactors. Furthermore, according to the World Nuclear Association, France has the highest degree of reactor standardization in the world.[36] All of the high-level radioactive waste from those plants has been collected and compacted or vitrified, and is now being safely stored near the town of La Hague in a single facility.

While France provides an example of the political will needed to deal with nuclear waste, the US Congress has demonstrated decades of political cowardice. Since 1982, when Congress passed the Nuclear Waste Policy Act—a law that requires the federal government to take all of the nuclear waste off of the hands of the nuclear utilities—the San Antonio Spurs have won five NBA championships and NASA has landed robotic rovers on Mars. Despite those terrestrial and extraterrestrial feats, the United States still doesn't have a place for long-term storage and disposal of the spent fuel coming from nuclear reactors. The result is that nuclear waste continues to be stored at dozens of sites across the country, including Indian Point.[37]

In 2011, a presidential panel, the Blue Ribbon Commission on America's Nuclear Future, summarized the situation, saying that America's policy toward spent nuclear fuel is "all but completely broken down."[38] Over the past few decades, the federal government spent about $15 billion on a waste repository at Yucca Mountain, Nevada.[39] But politics have kept the facility from opening. In 2008, while campaigning in Nevada, Barack Obama, in a bow to the state's powerful Democratic senator, Harry Reid, promised to cancel federal funding for the Yucca Mountain site. After Obama was elected, he did just that. Although it's not clear if, or when, Yucca Mountain will ever be opened, the federal government has a viable option: the dry casks now sitting at Indian Point and other nuclear facilities could be put into interim storage on land already owned by the federal government. The Department of Energy has several nuclear-focused locations that are excellent candidates for interim storage, including Savannah River Site in South Carolina, Oak Ridge National Laboratory in Tennessee, and Hanford Site in Washington State. Another location, the Waste Isolation Pilot Plant (WIPP) in New Mexico, is already being used by the federal government for disposal of radioactive waste generated by the Defense Department. It, too, could be used to store nuclear waste if Congress can muster the political will to deal with the issue.

Those federal locations already have security and safety systems in place to monitor the waste. The workers at those national laboratories have decades of experience with nuclear materials, and the communities near the labs are nuclear-savvy and want to keep the jobs the sites provide. In addition, they are plenty big. For instance, WIPP covers sixteen square miles (forty-one square kilometers). Using those federally owned sites for interim storage of nuclear waste will give Congress plenty of time to either open Yucca Mountain or find another disposal site. In the meantime, most of the used fuel from America's nuclear energy sector will continue to be stored at the same locations where it was used to generate electricity.

While radiation fears and waste disposal have hampered the nuclear sector, the biggest single problem facing the future of nuclear energy is cost. This can be seen by looking at recent history in the United States. In 2012, the US Nuclear Regulatory Commission (NRC) approved the construction license for the Vogtle 3 and 4 reactors near Augusta, Georgia. The Vogtle reactors, which are primarily owned by Southern Company, will be capable of producing 2,200 megawatts of electricity. The two reactors were the first to get a construction permit in the United States since 1978. They are Westinghouse AP1000s, designed to allow passive cooling and therefore be more resistant to the meltdown accident that occurred at Fukushima. When the reactors were announced, the total cost of the project was estimated at $14 billion. The project was financed, in part, by an $8.3 billion loan guarantee from the Department of Energy.[40] Shortly after the Vogtle reactors got the nod from the NRC, the agency granted a construction license for two more AP1000 reactors—Summer 2 and 3—in South Carolina.

But in 2017, Westinghouse, a subsidiary of the Japanese company Toshiba, filed for bankruptcy, citing huge losses on the Vogtle and Summer projects. Shortly after the Westinghouse bankruptcy, the owners of the Summer plant, SCANA and Santee Cooper, announced that, due to cost overruns, they would abandon the nuclear facility even though they had spent some $9 billion on the project.[41] Cost overruns also afflicted Vogtle. By 2018, the cost of the Vogtle project had soared to some $25 billion, nearly double the original amount. Despite the mounting costs, the owners of the Vogtle project decided to continue construction.[42]

The enormous cost of building large nuclear reactors like the AP1000, which will produce about the same amount of electricity as the Unit 2 reactor at Indian Point, isn't the only expense. In addition to the sky-high construction costs, companies that are trying to commercialize new reactor designs face exorbitant permitting costs. In 2015, the Government Accountability Office concluded that obtaining certification from the NRC for a new reactor is "a multi-decade process, with costs up to $1 billion to $2 billion, to design and certify or license."[43] Venture capitalists may be interested in nuclear technologies, but with permitting costs alone measured in the billions of dollars, it appears unlikely that any new nuclear reactor designs will be brought to market—or take significant market share—unless they are backed by central governments.

In fact, when looking at the global nuclear-energy sector, it's clear that state-owned companies are the only ones building significant amounts of new nuclear capacity. The state-backed model is particularly obvious in China, which is building more nuclear plants than any other country in the world. Those efforts are being led by China National Nuclear Corporation and China General Nuclear Power Group.[44] By early 2019, China had fifteen reactors under construction and several more in the development pipeline.[45] The state-backed nuclear-energy model is also observable in South Korea, which has exported nuclear technology through the state-owned Korea Electric Power Corporation. KEPCO is building the 5,600-megawatt Barakah nuclear power plant in Abu Dhabi.[46] When completed, the plant will be the world's largest single nuclear-energy project.[47] The first reactor at Barakah is expected to begin producing electricity in 2020.[48] But South Korea's politicians are planning to phase out the country's nuclear program over the next few decades and instead rely more on renewables.[49]

The Indian government has said it, too, will construct more capacity. In early 2018, it announced plans for twelve new reactors with a combined capacity of 9,000 megawatts, meaning that India will more than double the size of its nuclear fleet over the next decade or so. Ten of the reactors will use India's own pressurized heavy-water reactor design and two will use Russia's reactor design.[50]

State ownership has allowed Russia to become the undisputed leader in nuclear-energy deployment around the world. By mid-2018, Rosatom, the state-owned nuclear firm, had contracts for nearly three dozen new nuclear plants, with about a dozen under construction, including projects in Bangladesh and India. In 2018, the company began building a $20 billion nuclear plant in

Turkey, that country's first. It is slated to come online in 2023. In all, Rosatom has contracts worth about $130 billion, and many of those construction contracts were enhanced by the Russian government's willingness to provide financing for the projects.[51]

In addition to the many reactors it is building onshore, Rosatom has deployed the world's first nuclear powership. In 2018, the state-owned Russian company began testing the powership at the port of Murmansk. The ship carries two submarine-style reactors with a total electric generation capacity of 70 megawatts. According to one report, Rosatom officials plan to "tow the vessel to coastal cities in need of power, either for short-term boosts or longer-term additions to electricity supply." The ship holds enough enriched uranium to supply the two onboard reactors for twelve years. After that time, the ship will be towed back to Russia, where the spent fuel and radioactive waste will be processed. The first location for the nuclear power ship will be Pevek, a remote port in Siberia. When Rosatom moved the powership from Saint Petersburg to Murmansk, it was tailed by a Greenpeace sailboat, which carried a banner that said "Floating Nuclear Reactor? Srsly?"[52]

Those Greenpeacers conveniently ignore the decades-long history of marine propulsion. The first nuclear-powered submarine, the USS *Nautilus*, began patrolling the world's oceans in 1955. By 1962, the US Navy was operating more than two dozen nuclear submarines and thirty more were under construction. Since then, the nuclear fleets of countries like the United States, China, India, Russia, France, and others have accumulated more than 12,000 reactor years of operation time.[53] While powerships similar to the one made by Rosatom are interesting and could provide an alternative to the oil-burning powerships like those that have been used in Lebanon and Iraq, they will only provide a small fraction of the electricity needed to boost living standards around the world.

For the nuclear industry to gain greater traction in the global electricity market, it must develop reactors that are cheaper and safer than the ones now being built. Much of the effort has been aimed at designing reactors that are inherently safe, meaning that the cooling and containment systems are designed to prevent accidents and major releases of radioactive materials. Nuclear proponents believe that much of the potential lies in small modular reactors (SMRs). Generally defined as plants that have capacities of 300 megawatts or less, SMRs could be deployed as single or multiple units. In theory, SMRs could be cheaper than the reactors now being built because many of the components could be fabricated in a factory rather than on the construction site. Having a centralized production facility could allow a dedicated workforce at one location to test, build, and ship the reactors—by barge, rail, or truck—to the final destination. Concentrating the workforce in one place should also accelerate the learning curve and allow the company (or companies) producing the reactor to streamline production, reduce costs, and therefore build more reactors faster.

NuScale Power, a US-based company that is owned by construction giant Fluor Corporation, is planning to build a smaller version of the light-water reactors commonly used around the world. The electrical output of each NuScale reactor is projected to be 60 megawatts.[54] By contrast, the Westinghouse AP1000, the reactor type now being built at Vogtle, has an electrical output of 1,110 megawatts.[55]

In theory, that smaller size gives NuScale's customers more flexibility. If a NuScale customer wants more generation at a future date, it can add capacity in 60-megawatt increments. NuScale has garnered some $226 million in grants from the Department of Energy.[56] After it gets licensing from the NRC, it plans to build its first reactor at Idaho National Laboratory and sell the electricity it produces to Utah Associated Municipal Power Systems and Energy Northwest in Washington.[57] But even though it has a financially secure parent company, federal grant money, a federally owned site for its project, and a customer for its electricity, NuScale is unlikely to begin producing electricity from its reactor until the mid-to-late 2020s.

Among the most prominent—and perhaps most promising—SMR designs are ones that use molten salt. Rather than using fuel rods like conventional reactors, this design mixes the nuclear fuel into a salt mixture. Molten-salt reactors have a proven track record.

The Department of Energy tested the design in the 1960s at Oak Ridge National Laboratory, and the molten-salt reactor there ran for six years.[58] Terrestrial Energy, a Canadian company, is developing a molten-salt reactor that can run for seven years without having to be refueled.[59] Terrestrial plans to build a 190-megawatt reactor in Ontario by 2030 and it says the power plant will be cost competitive with ones fueled by natural gas.[60]

Another company with a promising molten-salt reactor design is ThorCon, which hopes to use shipyards to build its reactor: a 250-megawatt model that will be deployed on oceangoing hulls. ThorCon wants to build vessels to compete with the Rosatom nuclear powerships and the fuel-oil-fired ships that have been deployed by Karadeniz Holding in Iraq and Lebanon. The problem is that ThorCon needs about $1 billion to build the first copy of its design and it hasn't been able to raise the money.

The problem with the designs being promoted by NuScale, Terrestrial, ThorCon, and the other nuclear start-ups can be summed up in one word: commercialization. The reactor designs may sound appealing on paper, but unless or until those reactors can be built—and by that, I mean built by the dozens or hundreds—they cannot and will not make a significant contribution to the Terawatt Challenge. Indeed, the longer it takes for those companies to get their products into commercial use, the less likely it is that nuclear energy will make a big contribution to the electricity grids of the future. Electricity producers need to make prompt decisions about the type of generators they will be deploying in the decades ahead. They can't wait for years while new nuclear reactors are developed, tested, and permitted.

Robert Hargraves, the cofounder of ThorCon, told me in a telephone interview that his company believes it can build its reactors for a cost of about $1 per watt, a price point that would allow the company's molten-salt reactors to compete with natural-gas-fired generators on initial capital costs. Hargraves said that for companies like his—which need hundreds of millions of dollars to build and deploy a first-of-its-kind reactor—"it's hard to find people willing to put that much money in and have an eight-year payback. Investors are afraid there will be a regulatory roadblock that will prevent them from getting their money back." In addition, he said that "so many people oppose nuclear energy it makes regulators reluctant to say yes to new reactor designs." And then, of course, there's the problem with financing. "The World Bank won't touch a project like ours," Hargraves told me.[61]

In addition to SMRs that use fission, some ambitious companies continue chasing the promise of fusion. Among those companies is TAE Technologies, a California-based firm that says it is on "a purposeful path to commercial fusion energy."[62] In early 2019, the company predicted it would begin commercialization of its design in 2023.[63] It has raised $600 million in investment capital and counts former energy secretary Ernest Moniz as one of its board members.[64] But fusion faces many challenges, including building containment systems that can handle the enormous amounts of heat generated by fusion.[65]

In short, deploying new nuclear reactors that utilize different chemistries than the light-water designs that dominate today's market will be difficult and extremely expensive. New nuclear technologies will also have to be able to overcome the public's long-standing distrust. Given the ongoing friction facing nuclear energy, which technologies and fuels are most likely to be used to meet the Terawatt Challenge?

20

FUTURE GRID

What is a soul? It's like electricity. We don't really know what it is, but it's a force that can light a room.

—RAY CHARLES

High in the Chouf mountains southeast of Beirut, a resort called Bkerzay has installed a United Nations of electrons.[1] The parking areas at the resort are covered with 100 kilowatts of solar panels from China. Those panels convert Lebanese sunlight into electricity that is stored in a stone building a short walk down the hill from the cars. Inside it are racks upon racks of lead-acid batteries—about 300 kilowatt-hours' worth—that were designed in Bulgaria and manufactured in India. When the batteries need replacing, they will be shipped from a port in Lebanon back to India, where they will be smelted and turned into new batteries.

"The resort has its own restaurant. They have their own pottery workshop. So, it is almost a village. And we can feed the whole thing here, no restrictions. Everything is connected to our system," said Marwan El Khoury, an engineer at E24 Solutions, the Beirut-based company that designed and installed the system at Bkerzay. As we walked around the property, I peppered him with questions. Why are they using lead-acid batteries instead of lithium ion? El Khoury replied that "these batteries are much cheaper and they do the job." And what about the future? "Every single day we are getting better," he replied. E24 is continuing to do research and development on batteries and solar systems. The result, he said, is that the company is building "more robust systems," as well as more "reliable, longer-lasting batteries and the system is getting cheaper and performing better."

The microgrid at Bkerzay shows how interconnected global supply chains in everything from solar panels to lead-acid batteries will help address the Terawatt Challenge. Global trade in all of the commodities needed to produce and store electricity—lithium, lead, uranium, natural gas, coal, solar panels—is accelerating. That trade must continue growing if we are to be able to produce the electricity the world will need over the decades ahead.

My visit to Bkerzay forced me to reconsider some of my skepticism about solar and storage. Solar prices are falling. That is leading to huge increases in solar output. Between 2012 and 2017, global solar output more than quadrupled, to some 442 terawatt-hours.[2] Battery prices are falling as well. The combination of lower-cost solar and storage will result in more options for remote resorts like Bkerzay, as well as for people who are living in rural areas. E24's founder, Antoine Saab, believes that more city dwellers will be using solar and storage as well. During an interview at his office in central Beirut, Saab told me, "Microgrids are the future." The Terawatt Challenge cannot be met by extending the traditional grid to everyone who lacks electricity. "Utilities will not be able to upgrade at the same speed as the demand in power," he said. Given those facts, Saab said, it "will become inevitable for each one of us to start generating energy from solar."

The microgrid at Bkerzay shows, yet again, how integrity, capital, and fuel determine the shape of a given grid. Thanks to its isolation in the Chouf mountains, the solar and battery system at the resort has integrity. No one is going to sneak up and cart off the panels and batteries or start stealing electricity. That integrity assures that the capital invested in the system won't disappear. Further, the system makes economic sense because it competes with diesel-fired electricity from local generators, which can cost anywhere from 30 to 50 cents per kilowatt-hour. Those factors helped determine that the grid would be fueled by the sun.

Microgrids like the one at Bkerzay will likely continue to proliferate in locations where they make economic sense. But the challenge facing widespread solar-and-storage deployment is formidable. The problem, as usual, is scale. In 2017, the combined output of every solar-energy project on the planet—including the electricity produced by the 8.5 kilowatts of solar panels on the roof of my house in Austin—totaled just two million barrels of oil equivalent per day. That may sound significant. But that same year, global energy use totaled some 271 million barrels per day.[3] In other words, all global solar-energy production amounts to slightly more than 0.5 percent of global energy demand.

Making solar and storage work at the terawatt scale will require billions of tons of material to be mined, transported, and recycled. Those materials include silica, copper, lead, zinc, and lithium, as well as enormous quantities of rare-earth elements and cobalt. Mining and smelting all that stuff will have significant impacts on people and the environment. Much of the cobalt used in lithium-ion batteries is produced in the Congo, with a major portion of that supply coming from mines that use child labor. (CNN did an excellent series on this topic in 2018, called "Dirty Energy.")[4] In addition, hybrid vehicles, wind turbines, and other green technologies require large quantities of rare earth elements. China controls an estimated 80 percent of the market in rare earths, which include elements like neodymium, dysprosium, terbium, yttrium, lanthanum, and praseodymium.[5] The production of those elements has taken a heavy environmental toll in the city of Baotou, in Inner Mongolia. In 2015, the BBC's Tim Maughan reported on the industrial pollution in Baotou, including "an artificial lake filled with a black, barely liquid toxic sludge." He went on, saying that China's dominance of the rare earth market is "less about geology and far more about the country's willingness to take an environmental hit that other nations shy away from."[6]

Lithium-ion batteries have declined dramatically in price over the past few years. But they must compete with other types of storage, including lead-acid batteries. Lead-acid battery companies have a big advantage over their lithium counterparts because they already have a global network of recyclers. About 99 percent of all lead-acid batteries are now being recycled. Meanwhile, the overwhelming majority of the lithium-ion batteries now being used are not being recycled. By 2018, only about 3 percent of lithium batteries were being recycled. The percentage is low because lithium-ion batteries are costly to recycle. Researchers are looking at new methods of recycling those batteries, but it's not clear if, or when, they will be ready to handle the millions of tons of lithium-ion batteries that are now being manufactured and deployed all over the world.[7]

From a manufacturing standpoint, lithium-ion batteries pose significant supply-chain risks. Those risks include the sourcing of the materials needed to make the batteries as well as disposal of the batteries themselves. Given those facts, it's apparent that meeting the Terawatt Challenge will require far more than solar panels and lots of big batteries. It will require fuels that are cheap, reliable, and scalable. We also need those fuels to be low or no carbon. Finally, they must have high power density; that is, they must be capable of producing lots of energy from small footprints.

Natural gas will be a fuel of the future because it fits the criteria I just laid out: It is relatively low cost and low carbon and it can be produced from a relatively small footprint. Better yet, it is abundant, and new deposits of the fuel are being discovered and produced in staggering quantities in countries all over the world. Over the past two decades, huge gas fields have been found onshore in the United States, offshore in Israel, and offshore in Africa. In fact, despite surging global consumption of natural gas, proved reserves keep growing. Between 1997 and 2017, proved global gas reserves increased by more than 50 percent.[8]

It appears counterintuitive, but the more natural gas we find, the more we find. Global gas reserves now stand at some 193 trillion cubic meters (6.8 quadrillion cubic feet). That's enough to last for fifty-two years at current rates of production. The shale revolution has made the United States into the dominant player in the global natural-gas business. Thanks to this revolution—which has been fueled by the use of horizontal drilling, hydraulic fracturing, and other technologies—the United States has enjoyed "the fastest and biggest addition to world energy supply that has ever occurred in history."[9]

The result of that massive addition to world energy supplies can easily be seen in the growth of the US liquefied natural gas business. Before going further, allow me to explain how the business works. Freezing natural gas to a temperature of about -162 degrees Celsius (-260 degrees Fahrenheit) turns it into a liquid. Liquefying the fuel reduces the volume of the gas by a factor of six hundred and allows it to be profitably loaded and shipped on large oceangoing vessels. Countries with big natural gas resources such as Nigeria, Qatar, and Russia, and more recently the United States, convert their natural gas into LNG so they can ship it to overseas customers.

Back in the mid-2000s, the consensus among the world's top energy experts was that the United States was running out of natural gas. For instance, in 2007, Lee Raymond, who was then the CEO of ExxonMobil, declared that "gas production has peaked in North America."[10] The expectation of declining natural-gas production led to huge investments in LNG import terminals, and by the late 2000s the United States had built about twenty-three billion cubic feet per day of LNG import capacity.[11]

But the shale revolution turned conventional wisdom on its head. Between 2007 and 2019, US gas production soared, going from fifty billion cubic feet per day to about ninety billion cubic feet per day.[12] That's an increase of 80 percent in just twelve years. Unable to use all of that natural gas domestically, US producers began looking overseas, and by the end of 2018, the United States was exporting about four billion cubic feet of LNG per day.[13] In mid-2020, the Energy Information Administration expects those exports will hit 10.6 billion cubic feet per day.

By going from a prospective LNG importer to a real-world LNG exporter, the United States has disrupted the global gas market. Over a span of a little more than a decade, the United States created a swing of nearly thirty-four billion cubic feet per day in the global LNG market. How much gas is that? It's equal to 1.5 times China's daily natural gas consumption.[14]

As a natural-gas superpower, the United States is now providing the methane equivalent of coal to Newcastle, or, perhaps, ice to Eskimos. In May 2019, Saudi Aramco, the largest oil company in the world, signed a twenty-year deal with Sempra Energy. The deal calls for Sempra to provide five million tons of LNG per year to Aramco. The LNG will come from a new liquefaction facility that Sempra is building in Port Arthur, Texas.[15] The deal with Aramco provides the latest example of US natural gas going to the Persian Gulf. Since 2017, LNG tankers have been delivering cargoes of US natural gas to customers in Kuwait and the United Arab Emirates.[16]

The staggering growth in global natural-gas production (which has jumped by more than 50 percent since 2000) and the corresponding growth in the global LNG trade are forcing a realignment of expectations for energy producers and consumers around the world. It is also helping decarbonize the US electricity grid. That decarbonization is occurring because low-cost gas is increasingly displacing coal as the fuel of choice for US utilities.

I have long promoted natural gas to nuclear as the logical way forward for policymakers. Increasing use of natural gas and nuclear provides the best no-regrets policy on climate change, because those two sources will have minimal negative impact on the economy and environment while providing significant decarbonization. Global natural-gas production has grown so rapidly that forecasters are predicting that, by 2040, global gas-fired generation will nearly equal the amount produced from coal-fired plants.[17]

To take advantage of the clean-burning properties of natural gas, countries all over the world are expanding their gas infrastructure. In early 2018, the Indian government announced it would build eleven new LNG import terminals that will convert the supercooled fuel back into the gaseous phase so it can be distributed via pipelines to power plants and other consumers. By announcing so much import capacity, India made clear its plans to become one of the world's biggest consumers of seaborne natural gas. The LNG will not only help India produce more electricity; it will also help cut carbon dioxide emissions and improve air quality, because it will reduce the country's reliance on coal.[18]

At this point, there's no doubt that LNG will play a bigger role in the global fuel mix over the next few decades. The graphic shows global generation capacity by fuel, and how that capacity is likely to change over the next three decades. The projections, which come from the International Energy Agency's *World Energy Outlook 2017*, show that global solar capacity may exceed global wind capacity as soon as 2025. The report also expects nuclear capacity to grow slowly through 2040, coal-fired capacity to plateau, and natural gas-fired capacity to gain significant market share.[19]

To be sure, the growth in natural-gas use will come with some costs. Anti–fossil fuel activists are opposing natural-gas extraction and the construction of natural-gas pipelines. Those activists have targeted hydraulic fracturing, the process used by producers to pump sand and water into underground rock formations so that they surrender the oil and gas trapped inside them. There's no doubt that oil and gas extraction has negative environmental effects. In Oklahoma, wells that are used to dispose of the wastewater from hydraulic fracturing have been correlated with earthquakes. In Colorado and Texas, drilling sites have encroached on suburban neighborhoods, much to the consternation of local residents irritated by the truck traffic and noise that often comes with drilling rigs that can operate twenty-four hours per day for weeks at a time.

Natural gas, like every source of energy, has its virtues and its weaknesses. But natural gas's virtues outweigh its environmental downsides. Natural gas will fuel the future because it can be used for a multitude of purposes, including transportation, power generation, space heating, and cooking, as well as fertilizer, steel, and plastic production. Natural gas emits about half as much carbon dioxide during combustion as coal. It's also cleaner than coal when looking at traditional air pollutants like sulfur dioxide and nitrogen oxide.

Another reason why natural gas will be a fuel of the future is that global gas reserves are enormous. At current rates of consumption, global gas reserves are projected to last more than fifty years. Add in the fact that gas-fired generators can be used to offset the intermittency of wind and solar, and it becomes clear that natural gas will be one of the main fuels used to meet the Terawatt Challenge.

While I'm bullish about natural gas and solar—and hopeful about nuclear energy—there are no magic bullets when it comes to meeting the Terawatt Challenge. Doubling global electricity supplies won't be easy. But a look back at history shows that it can be done. For instance, between 1990 and 2017, global electricity production more than doubled.[20]

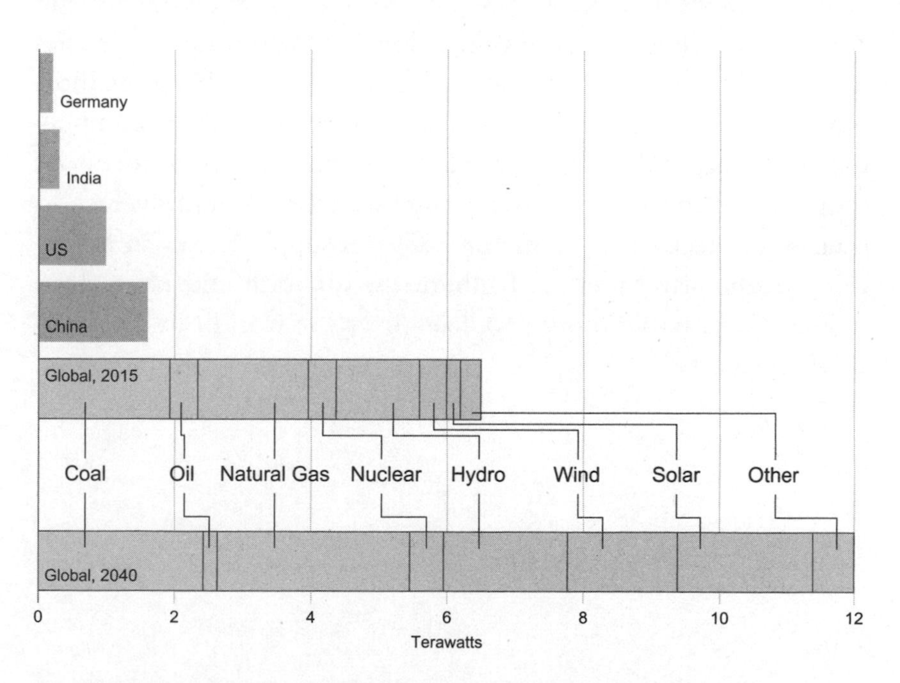

The Terawatt Challenge: Electric Generation Capacities in Various Countries and Globally in 2015, and Projected Global Capacity in 2040

Sources: IEA, CIA.

We will need all of the available fuels around the world—coal, nuclear, oil, natural gas, hydro, solar, wind, and geothermal—to meet the Terawatt Challenge. But meeting that challenge isn't solely about producing more electricity; it's also about sparing rural landscapes and viewsheds, protecting land for wildlife and people, and preserving wildlands and seascapes. The Terawatt Challenge requires us to produce more electricity, more cleanly, on smaller footprints. We will need dense, scalable fuels to power dense, skyscraper-filled cities. We cannot wreck the biosphere in our attempt to produce lower-carbon electricity. Instead, we must focus on land sparing. That means overcoming the incessant fear-mongering about nuclear energy and embracing its potential. It means moving past the feel-good rhetoric about renewables and facing the land-use conflicts that are already limiting the growth of solar and wind energy.

Electricity is among the youngest of the sciences. The innovation in nuclear technologies, and every facet of electricity and electronics, will continue. We are achieving staggering efficiencies in generating, transmitting, and storing electricity. Nanotechnology, cheaper batteries, better solar panels, more efficient computing, and artificial intelligence will help drive those advances. As progress continues, the cost of using electricity should decline, so our use of electricity can increase. Global supply chains in everything from solar panels and lithium to uranium and natural gas will help assure that more people all over the world can plug into modern society.

CONCLUSION

> I sing the body electric.
>
> —WALT WHITMAN

As I traveled around the world doing the reporting for this book, I interviewed dozens of people. I put the same question to as many of them as I could: What does electricity mean? The replies were as varied as the places I visited. Riad Chedid, the engineering professor at the American University of Beirut, replied to my question thusly: "Electricity means prosperity. It means good life. It means business."

New York University professor Carolyn Kissane said electricity "means life." In Iceland, Gísli Katrínarson, a project manager at Advania, a cloud-computing company, said, "Opportunity." Among the many replies to the question, though, one theme kept recurring: electricity has become a human right. E24's Antoine Saab replied to my question by saying: "Electricity means you are a human being."

The humanist response to the Terawatt Challenge is obvious: we cannot shrink away from it. We cannot keep hundreds of millions of people in the dark due to worries about what may happen with regard to climate change. In the twenty-first century, where education, government, business, and politics are all happening online, being unplugged—being off the grid—means being truly powerless. During an interview in Reykjavik, Birgitta Jónsdóttir, a leader of Iceland's Pirate Party, told me, "Electricity is absolutely a human right." Jónsdóttir, who served four years in the Althing, the Icelandic parliament, has been a leading proponent of transparency in government, digital privacy, and direct democracy. "Almost all modern democracy functions, be it on the federal or the city or town levels," are being conducted on the Internet, she said. That reliance, in turn, means that "it's impossible to have any democratic innovation and more forms of direct democracy without strong, stable, cheap electricity."

Electricity isn't just essential to innovation and prosperity; it is essential to *us*. Our bodies are electrical. As British poet Percy Bysshe Shelley put it, we "are no more than electrified clay."[1] In 2012, British physiology professor Frances Ashcroft published *The Spark of Life*, a book that details how electricity pulses through our bodies via ion channels. Those channels are small gated pores that sit in the membrane of every cell of every organism on the planet. Ion channels, Ashcroft explains, "act as our windows on the world," and each of our "sensory experiences—from listening to a Mozart quartet to judging where a tennis ball will fall—depends on their ability to translate sensory information into electrical signals that the brain can interpret."[2]

Not only that, our basal metabolic rate—that is, the amount of energy consumed by our bodies while at rest—is measured in a term that we normally associate with electricity: watts. Our bodies generally consume about 100 watts of power and, of that amount, about a fifth, or 20 watts, is consumed by our brains.[3]

In thinking about how electricity travels inside our bodies, and how it has changed humanity, it's almost inevitable that we think in philosophical or even religious terms. Electricity words overlap with spiritual ones. We don't just learn; we are enlightened. We don't just change; we are transformed. We don't turn away from evil; we see the light.

The concept of light as godlike pervades the world's biggest religions. The first few lines of the Bible (Genesis 1:3–4) tell of God's command: "'Let there be light,' and there was light. God saw how good the light was. God then separated the light from the darkness." When God wanted to punish the Egyptians, he told

Moses to stretch his hand to "the sky so that darkness spreads over Egypt, darkness that can be felt" (Exodus 10:21). Psalm 119:105 says, "Your word is a lamp for my feet, a light on my path." And in Isaiah 60:1, God is compared to light: "Rise up in splendor! Your light has come, the glory of the Lord shines upon you."

Jesus told his followers: "I am the light of the world. Whoever follows me will never walk in darkness, but will have the light of life" (John 8:12). Later in John, he says, "You are going to have the light just a little while longer. Walk while you have the light, before darkness overtakes you" (John 12:35–37).

The Koran declares that "Allah is the Light of the heavens and the earth. . . . Allah guides to His Light whom He wills."[4] While Allah, Jehovah, and Jesus are the creators and the Light, Satan is the prince of darkness. The occult flourishes in the dark. As one Bible-study guide puts it, "Darkness evokes everything that is anti-God."[5]

Electricity has conferred on us a bit of the creative power that God showed in Genesis. With the flip of a switch, we can kill the anti-God and banish darkness. With a touch of our mobile phones we can ensure safe passage through a strange hotel or garage at night. With quadrillions of electrons at our beck and call, we can create as much light as we want. The religious symbolism of electricity shone through in one of my discussions with Joyashree Roy. (As you may recall, Joyashree hosted us in Kolkata and introduced me to Rehena Jamadar, the woman living in Majlishpukur.) Joyashree talked about electricity in terms that verged on the evangelical. She said that with electricity, we humans can engage in "co-creation with nature." But none of that creating—none of that godlike-ness—happens without electricity. "If you are in the dark, if you are absorbed in the dark, darkness absorbs you too," Joyashree said. "So you do not see the light and you cannot bring the light to others."

The humanist response to the Terawatt Challenge is simple: it is to bring light and power to others so that those who are living in the dark can come into the bright light of modernity and progress. Making that happen won't be easy. Electrification requires societies to have integrity. It takes capital and fuel. But the trends toward greater electrification and higher living standards appear unstoppable. Electricity nourishes humans like no form of energy ever has. We need more human flourishing, not less. Sure, as we produce more electricity, we will have an effect on the environment. We will have to mine more copper and lead, refine more uranium and lithium, drill more gas wells, manufacture more solar panels, and build many more nuclear reactors. As we do all of those things and produce more electricity, we will emit more carbon dioxide, and that will have an effect on the climate. Despite the near-continual warnings about climate change, I remain optimistic that we can, and will, adapt to whatever changes are coming. We cannot stand still. Nor can we deny modernity to the billions of people who are living in Low-Watt and Unplugged places.

More people are living longer, freer, healthier lives today than at any other time in human history and they are doing so, in large part, because of electricity. My optimism about the future—and the future of electrification—is boundless. Advances in computers, batteries, generators, lights, microchips, and motors will continue the inexorable advance of civilization and modernity. Electrifying the entire world—bringing cheap, abundant, reliable electricity to every person on this planet—will take time. But it can be done.

Appendix A

SI NUMERICAL DESIGNATIONS AND POWER UNITS

We use many numerical designations—milli, mega, nano—on a regular basis without recognizing that they are part of the International System of Units. The system is commonly known as SI, the abbreviation for Système International d'Unités. (France was instrumental in the effort to harmonize units of measure.) Given that most people are only passingly familiar with these designations, see below for a review of all of the numerical designations—from yocto to yotta.

Number	Prefix	Symbol
10^{-24}	yocto-	Y
10^{-21}	zepto-	Z
10^{-18}	atto-	A
10^{-15}	femto-	F
10^{-12}	pico-	P
10^{-9}	nano-	N
10^{-6}	micro-	μ
10^{-3}	milli-	M
10^{-2}	centi-	C
10^{-1}	deci-	D
10^{1}	deka-	Da
10^{2}	hecto-	H
10^{3}	kilo-	K
10^{6}	mega-	M
10^{9}	giga-	G
10^{12}	tera-	T
10^{15}	peta-	P
10^{18}	exa-	E
10^{21}	zetta-	Z
10^{24}	yotta-	Y

1 watt (W) = 1 joule/second (J/s) or 0.00134 horsepower
1 kilowatt = 1,000 watts or 1.35 horsepower (hp)
1 megawatt = 1,000,000 watts
1 gigawatt = 1,000,000,000 watts
1 terawatt = 1,000,000,000,000 watts

Source: Math.com.

Appendix B

DOLLAR VALUE OF ELECTRICITY STORED IN COMMON BATTERIES

Batteries are ubiquitous, but the amount of energy they can store is tiny. For instance, the batteries that run our iPhones hold about 6 watt-hours of electricity. At 12 cents per kilowatt-hour (the average price of residential electricity in the United States) those 6 watt-hours cost about seven-one-hundredths of a penny.

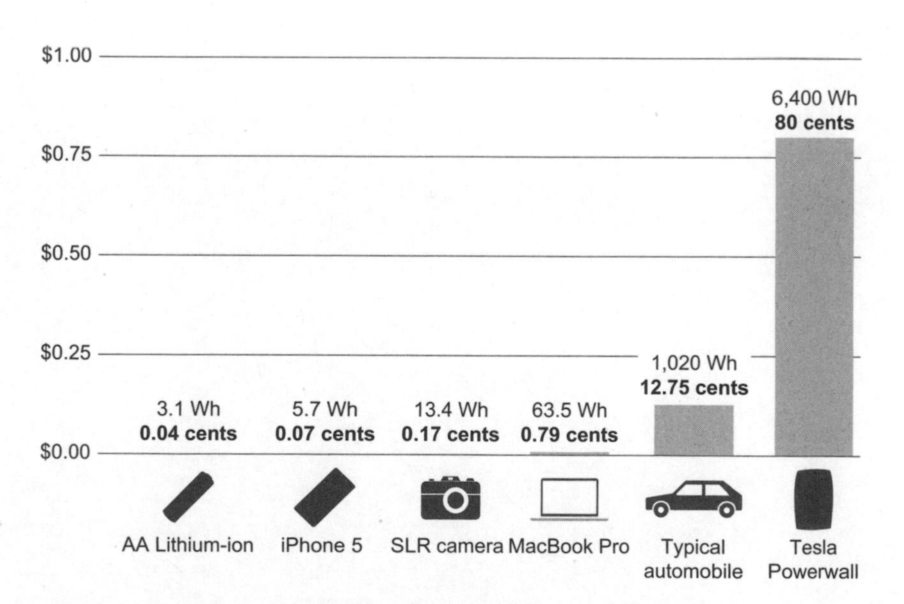

Sources: Author calculations, Allaboutbatteries.com, Apple.com, Tesla.com.

NOTES

Introduction: Barrio Antón Ruíz

1. Quoted by Ben Geman, "Nonprofit Aims to Help Africa Build Energy for Economic Development," *Axios*, February 26, 2019, www.axios.com/startup-africa-energy-manufacturing-economy-power-35e46508-3bd3-4f0c-a48c-64e534599933.html.

2. I later calculated the family's per-capita electricity consumption at about 1,300 kilowatt-hours per year. That's about the same amount of electricity as an average resident of countries like Peru or Colombia.

3. The Rhodium Group's comparison was made by counting the total number of customer-hours spent without electricity.

4. Trevor Houser and Peter Marsters, "The World's Second Largest Blackout," Rhodium Group, April 12, 2018, https://rhg.com/research/puerto-rico-hurricane-maria-worlds-second-largest-blackout/.

5. Wikipedia, s.v. "Solar Storm of 1849," last modified April 9, 2019, 06:38, https://en.wikipedia.org/wiki/Solar_storm_of_1859.

6. Deborah Byrd, "$40 Billion a Day for Solar Super-Storms," *EarthSky*, January 20, 2017, http://earthsky.org/earth/40-billion-a-day-for-solar-super-storms.

7. Rebecca Smith, "Russian Hackers Reach U.S. Utility Control Rooms, Homeland Security Officials Say," *Wall Street Journal*, July 23, 2018, www.wsj.com/articles/russian-hackers-reach-u-s-utility-control-rooms-homeland-security-officials-say-1532388110.

8. "Recognizing the Duty of the Federal Government to Create a Green New Deal," H.R. Res 109, 116th Cong. (2019), www.congress.gov/bill/116th-congress/house-resolution/109/text.

9. By the end of 2018, Amazon owned or controlled about 4,700 megawatts of electric generation capacity. For comparison, Croatia had about 4,900 megawatts and Laos had about 4,500 megawatts. See, "Country Comparison: Electricity—Installed Generating Capacity," CIA World Factbook, www.cia.gov/LIBRARY/publications/the-world-factbook/rankorder/2236rank.html.

10. *BP Statistical Review of World Energy 2018* (London: BP, 2018). Between 1985 and 2015, global electricity production increased by 145 percent. Thus, electricity generation grew more than twice as fast as oil consumption, which grew by 60 percent over that time period. It also grew faster than growth in coal (85 percent), natural gas (111 percent), and nuclear consumption (74 percent).

11. Annual global oil and gas revenues total roughly $2.9 trillion. This estimate is based on *BP Statistical Review of World Energy 2018*, which puts global oil consumption at ninety-eight million barrels per day and natural-gas consumption at about sixty-three million barrels of oil equivalent per day. Together, oil and gas use totals about fifty-nine billion barrels of oil equivalent per year. At $50 per barrel, that consumption implies annual sales of $2.9 trillion.

12. In 2017, US electricity sales totaled $390 billion. See, Energy Information Administration, "Revenue from Sales of Electricity to Ultimate Customers," www.eia.gov/electricity/annual/html/epa_02_06.html. The United States, which produces about 4,200 terawatt-hours of electricity per year, accounts for roughly 16 percent of global generation. Therefore, 6 × $400 billion = $2.4 trillion.

13. In 2016, global car sales totaled about eighty-four million units. At $25,000 per car, the auto industry had revenues of about $2.1 trillion. For car sales data, see "Global Car Sales Up by 5.6% in 2016 Due to Soaring Demand in China, India and Europe," JATO, February 9, 2017, www.jato.com/usa/global-car-sales-5-6-2016-due-soaring-demand-china-india-europe/. For average price of new cars, see "China's Got Sales, U.S. Has Riches," *Automotive News*, March 16, 2015, www.autonews.com/article/20150316/RETAIL01/303169975/chinas-got-sales-u.s.-has-riches. The global pharmaceutical business has annual revenues of about $1 trillion. See, Business Research Company, "The Growing Pharmaceuticals Market: Expert Forecasts and Analysis," *Market Research Blog*, May 16, 2018, https://blog.marketresearch.com/the-growing-pharmaceuticals-market-expert-forecasts-and-analysis.

14. Energy Information Administration, "Revenue from Sales of Electricity to Ultimate Customers."

15. In 2018, those three companies reported revenues thusly: Ford: $160.3 billion, General Electric: $121.6 billion, General Motors: $147 billion. Total: $428.9 billion.

16. "Global Greenhouse Gas Emissions Data," Environmental Protection Agency, www.epa.gov/ghgemissions/global-greenhouse-gas-emissions-data.

17. Vaclav Smil, "Moore's Curse and the Great Energy Delusion," *The American*, November 19, 2008, www.aei.org/publication/moores-curse-and-the-great-energy-delusion/.

Chapter 1. Electricity 101

1. Erin Wayman, "The Earliest Evidence of Hominid Fire," Smithsonian.com, April 4, 2012, www.smithsonianmag.com/science-nature/the-earliest-example-of-hominid-fire-171693652/.

2. A day consists of 86,400 seconds. If 400,000 years = 86,400 seconds, then 1 year = 0.216 seconds. Therefore, the 140 years of the Electric Age would last 30.24 seconds.

3. "Electric Current," SI Units Explained, www.si-units-explained.info/ElectricCurrent/#.WESLOHeZPNw.

4. Independence Hall Association, "Electricity," The Electric Ben Franklin, UShistory.org, www.ushistory.org/franklin/science/electricity.htm. "Benjamin Franklin," APS Physics, www.aps.org/programs/outreach/history/historicsites/franklin.cfm.

5. William Beaty, "How Are Watts, Ohms, Amps, and Volts Related?" April 2, 2000, http://amasci.com/elect/vwatt1.html.

6. Thanks to Jonathan Lesser and Leonardo Giacchino, for providing a similar example in their book, *Fundamentals of Energy Regulation* (Reston, VA: Public Utilities Reports Inc., 2013).

7. James Gleick, *The Information: A History, a Theory, a Flood* (New York: Vintage, 2011), 126, 170.

8. "Forms of Energy Basics," Energy Kids, EIA, www.eia.gov/kids/energy.php?page=about_forms_of_energy-basics.

9. In 2017, global electricity production was 25,500 terawatt-hours. That works out to about 70 TWh/day or 2.9 TWh per hour. There are about one billion automobiles in the world, and each auto battery can hold about 1,000 watt-hours. Thus, all the world's auto batteries can store about 1 trillion watt-hours, or 1 terawatt-hour of energy. That quantity would provide less than thirty minutes of electricity for the world.

10. In SI, units named after individuals are capitalized. Thus, the watt, named after James Watt, is noted with a capital W, and the volt, named after Volta, is noted with a capital V. The abbreviation for hour is not capitalized because it is not named for an individual. For reference, 1,000 watts is equal to 1.34 horsepower.

11. EIA data, "Table 1.2 Summary Statistics for the United States, 2007–2017," www.eia.gov/electricity/annual/html/epa_01_02.html, see table 4.2.B. In the summer of 2017, total US installed utility scale capacity was 1.084 terawatts.

Chapter 2. The Transformative Power of Electricity

1. Alex Epstein, the author of the 2014 book *The Moral Case for Fossil Fuels* (New York: Portfolio) has made a strong and sustained case for the role that energy has in human flourishing.

2. Wolfgang Schivelbusch, *Disenchanted Night: The Industrialization of Light in the Nineteenth Century* (Berkeley: University of California Press, 1995), 81–82.

3. Schivelbusch, *Disenchanted Night*, 45.

4. Schivelbusch, *Disenchanted Night*, 158.

5. Schivelbusch, *Disenchanted Night*, 35.

6. Wikipedia, s.v. "Great Chicago Fire," last modified May 10, 2019, 20:38, https://en.wikipedia.org/wiki/Great_Chicago_Fire.

7. Schivelbusch, *Disenchanted Night*, 52.

8. David E. Nye, *Electrifying America: Social Meanings of a New Technology* (Cambridge, MA: MIT Press, 1990), 5.

9. Phillip F. Schewe, *The Grid: A Journey Through the Heart of Our Electrified World* (Washington, DC: Joseph Henry Press, 2007), 38.

10. Maury Klein, *The Power Makers: Steam, Electricity, and the Men Who Invented Modern America* (New York: Bloomsbury, 2008), 171.

11. Klein, *The Power Makers*, 172.

12. Max Roser, "Light," *Our World in Data*, https://ourworldindata.org/light.

13. Department of Energy, "Energy Savings Estimates of Light Emitting Diodes in Niche Lighting Applications," October 2008, xiv, www.energy.gov/sites/prod/files/maprod/documents/Energy_Savings_Light_Emitting_Diodes_Niche_Lighting_Apps.pdf.

14. Albania consumes about 7.1 terawatt-hours per year; Latvia consumes about 6.7 terawatt-hours. "Country Comparison: Electricity—Consumption," CIA World Factbook, www.cia.gov/library/publications/the-world-factbook/rankorder/2233rank.html.

15. BBC History, s.v. "James Watt (1736–1819)," www.bbc.co.uk/history/historic_figures/watt_james.shtml.

16. Charles R. Morris, *The Dawn of Innovation: The First American Industrial Revolution* (New York: PublicAffairs, 2012), 110.

17. Craig R. Roach, *Simply Electrifying: The Technology That Transformed the World, from Benjamin Franklin to Elon Musk* (Dallas: BenBella Books, 2017), 27.

18. Not all electricity is highly ordered. Lightning and static are obvious examples of disordered electrical energy.

19. Encyclopedia of Earth, s.v. "Energy Transitions," by Arnulf Grubler, June 3, 2008, https://editors.eol.org/eoearth/wiki/Energy_transitions.

20. Vaclav Smil, "Power Density Primer," May 8, 2010, http://vaclavsmil.com/wp-content/uploads/docs/smil-article-power-density-primer.pdf.

21. Vaclav Smil, *Power Density: A Key to Understanding Energy Sources and Uses* (Cambridge, MA: MIT Press, 2015), 86–87.

22. Smil, *Power Density*, 166.

23. Nye, *Electrifying America*, 228.

24. Smil, *Power Density*, 166.

25. "Ford Rouge Timeline," The Henry Ford, www.thehenryford.org/visit/ford-rouge-factory-tour/history-and-timeline/timeline. See also, "Henry Ford River Rouge Plant," Detroit: The History and Future of Motor City, http://detroit1701.org/Ford%20Rouge%20Plant.html.

26. Nye, *Electrifying America*, 225.

27. Eliot Spitzer, *Con Edison's July 1999 Electric Service Outages: A Report to the People of the State of New York* (Albany, March 9, 2000), G-2, G-3.

28. See, Jayanth C. P. Koliyadu et al., "Optimization and Characterization of High-Harmonic Generation for Probing Solid Density Plasmas," *Photonics* 4, no. 2 (March 30, 2017), www.mdpi.com/2304-6732/4/2/25.

29. Corkum and others use what's known as chirped pulse amplification, in which they stretch short bursts of light to decrease its intensity. They then amplify and compress that same pulse, which greatly increases its intensity. The 2018 Nobel Prize in Physics was awarded for the pioneering work done on this type of laser. One of the recipients was a Canadian, Donna Strickland, who became just the third woman to win the physics Nobel. See, Emily Conover, "Dazzling Laser Feats Earn These Physicists a Nobel," *Science News*, October 2, 2018, www.sciencenews.org/article/dazzling-laser-feats-earn-these-physicists-nobel.

30. Wm. A. Wulf, "Great Achievements and Grand Challenges," *The Bridge* (Fall/Winter 2000): 5, www.nae.edu/File.aspx?id=7327.

31. Yilmaz Bayar and Hasan Alp Özel, "Electricity Consumption and Economic Growth in Emerging Economies," *Journal of Knowledge Management, Economics and Information Technology* 4, no. 2 (April 2014).

32. Roger Andrews, "Electricity and the Wealth of Nations," *Energy Matters* (blog), November 22, 2015, http://euanmearns.com/electricity-and-the-wealth-of-nations/.

33. Kashif Imran and Masood Mashkoor Siddiqui, "Energy Consumption and Economic Growth: A Case Study of Three SAARC Countries," *European Journal of Social Sciences* 16, no. 2 (2010), www.scribd.com/document/39466448/Energy-Consumption-and-Economic-Growth.

34. Keith Bradsher, "China's Unemployment Swells as Exports Falter," *New York Times*, February 5, 2009, www.nytimes.com/2009/02/06/business/worldbusiness/06yuan.html. Goldman Sachs analysts have indicated that they also use electricity data as a proxy for output in China.

35. Xi Chen and William D. Nordhaus, "The Value of Luminosity Data as a Proxy for Economic Statistics" (working paper no. 16317, National Bureau of Economic Research, Cambridge, MA, August 2010), www.nber.org/papers/w16317.pdf.

36. J. Vernon Henderson, Adam Storeygard, and David N. Weil, "Measuring Economic Growth from Outer Space," *American Economic Review* 102, no. 2 (April 2012), www.aeaweb.org/articles?id=10.1257/aer.102.2.994.

37. OECD/IEA, "Electric Power Consumption (kWh per capita)," World Bank, http://data.worldbank.org/indicator/EG.USE.ELEC.KH.PC?year_high_desc=true.

38. Cited by Marian L. Tupy in *The False Promise of Gleneagles: Misguided Priorities at the Heart of the New Push for African Development* (Washington, DC: Cato Institute, 2009), https://object.cato.org/sites/cato.org/files/pubs/pdf/dpa9.pdf.

39. Credit here to Edward Glaeser and his book *Triumph of the City: How Our Greatest Invention Makes Us Richer, Smarter, Greener, Healthier, and Happier* (New York: Penguin, 2011).

Chapter 3. The Vertical City

1. Quoted in Richard Rhodes, *Energy: A Human History* (New York: Simon & Schuster, 2017).

2. "Brooklyn Bridge," History.com, last updated August 21, 2018, www.history.com/topics/brooklyn-bridge.

3. "Population History of New York from 1790–1990," Boston University Physics, http://physics.bu.edu/~redner/projects/population/cities/newyork.html.

4. David W. Dunlap, "Consumed in Fire, Cloaked in Ice, Equitable's Headquarters Fell 100 Years Ago," *City Room* (blog), *New York Times*, January 8, 2012, https://cityroom.blogs.nytimes.com/2012/01/08/consumed-in-fire-cloaked-in-ice-equitables-headquarters-fell-100-years-ago/.

5. For a map of Edison's first grid, see Kelsey Campbell-Dollaghan, "The Forgotten Story of NYC's First Power Grid," *Gizmodo*, January 26, 2015, https://gizmodo.com/the-forgotten-story-of-nycs-first-power-grid-1681857054. For a walking map, see Google Maps, https://bit.ly/2VkeewT.

6. The transformer is to the Electricity Age as the microprocessor is to the Information Age. In 1885, Westinghouse read an article in an English engineering journal about a device that could increase, or decrease, the voltage in a system that employed alternating current. Westinghouse, a man whose personal integrity and business acumen were legendary, sensed the importance of what was then called a "secondary generator." He quickly dispatched a lieutenant to Europe to secure the rights to the patent on the transformer. Westinghouse understood that, by using transformers, he could boost the voltage

on AC transmission lines and thereby pump more electricity through them over longer distances. In 1896, Westinghouse built the world's first high-voltage electric system using alternating current. Using 11,000-volt transmission lines, he connected the hydroelectric station at Niagara Falls to the city of Buffalo, twenty miles (thirty-two kilometers) away. After Westinghouse proved the efficiency of AC transmission at Niagara Falls, generators all over the world began copying Westinghouse's high-voltage design. By 1900, electricity producers were moving electricity over 60,000-volt lines. By 1912, those voltages had more than doubled to 150 kilovolts. By 1930, transmission lines were handling 240 kilovolts. Today, electricity companies routinely use high-voltage lines that operate at 360 and 720 kilovolts. For more on this, see Jill Jonnes, *Empires of Light: Edison, Tesla, Westinghouse, and the Race to Electrify the World* (New York: Random House, 2003), 123–125. See also, Matthew H. Brown and Richard P. Sedano, *Electricity Transmission: A Primer* (Washington, DC: National Council on Electricity Policy, 2004), 2, www.energy.gov/sites/prod/files /oeprod/DocumentsandMedia/primer.pdf.

7. Frederick Dalzell, *Engineering Invention: Frank J. Sprague and the U.S. Electrical Industry* (Cambridge, MA: MIT Press, 2009).

8. Dalzell, *Engineering Invention*, 768.

9. Paul Israel, *Edison: A Life of Invention* (New York: John Wiley & Sons, 1998), 223.

10. Joseph J. Cunningham, *New York Power* (Self-published, CreateSpace, 2013), 19.

11. Frank Rowsome Jr., *The Birth of Electric Traction: The Extraordinary Life and Times of Inventor Frank Julian Sprague* (Self-published, CreateSpace, 2013).

12. Cunningham, *New York Power*, 20.

13. Robert Bradley Jr., "Horsepower Sure Beats Horses!" *MasterResource* (blog), September 29, 2009, http://masterresource.org/?p=5016.

14. Eric Morris, "From Horse Power to Horsepower," *Access* (Spring 2007): 7, www .accessmagazine.org/spring-2007/horse-power-horsepower/.

15. Gregory Curtis, "Review: A Wild Ride Through History in a 'Farewell to the Horse,'" *Wall Street Journal*, February 9, 2018, www.wsj.com/articles/review-a -wild-ride-through-history-in-a-farewell-to-the-horse-1518212178.

16. Robert C. Kennedy, "Gold at 160. Gold at 130," *On This Day* (blog), *New York Times*, http://movies2.nytimes.com/learning/general/onthisday/harp/1016.html.

17. Jim Mackin, "The Ninth Avenue El," *Bloomingdale History* (blog), September 13, 2013, https://bloomingdalehistory.com/2013/09/13/the-ninth-avenue-el/.

18. Dalzell, *Engineering Invention*.

19. Dalzell, *Engineering Invention*.

20. Rowsome's biography of Sprague was completed in the early 1960s, but Rowsome was unable to find a publisher for the manuscript. The book was finally published in 2013 by Sprague's grandson, John Sprague.

21. Rowsome, *The Birth of Electric Traction*.

22. Dalzell, *Engineering Invention*.

23. Rowsome, *The Birth of Electric Traction*.

24. Rowsome, *The Birth of Electric Traction*.

25. Rowsome, *The Birth of Electric Traction*.

26. Recently in downtown Austin, I rode an elevator twenty-one floors to the top floor of the Chase Tower, a 1980s-era building. The trip covered about 210 feet in 37 seconds, meaning I traveled at a rate of 342 feet per minute, or 3.9 miles per hour, or 6.2 kilometers per hour

27. Carol Parise, Barbara Sternfeld, Steven Samuels, and Ira B. Tager, "Brisk Walking Speed in Older Adults Who Walk for Exercise," *Journal of the American Geriatrics Society* 52, no. 3 (March 2004), www.ncbi.nlm.nih.gov/pubmed/14962157.

28. Evelyn Jutte, "Frank J. Sprague," Elevator Museum, https://theelevatormuseum.org/sprague.php.

29. Emily Badger, "Why Even the Hyperloop Probably Wouldn't Change Your Commute Time," *New York Times*, August 10, 2017, www.nytimes.com/2017/08/10/upshot/why-even-the-hyperloop-probably-wouldnt-change-your-commute-time.html.

30. NYC Buildings, "DOB Launches 'Stay Safe. Stay Put.' Campaign for Elevator Safety," news release, July 5, 2016, www1.nyc.gov/site/buildings/about/pr-elevator-safety-campaign.page.

31. "Population History of New York from 1790–1990," Boston University Physics, http://physics.bu.edu/~redner/projects/population/cities/newyork.html.

32. Matt Ridley, "17 Reasons to Be Cheerful," *Matt Ridley Online* (blog), March 1, 2012, www.rationaloptimist.com/blog/17-reasons-to-be-cheerful-1/.

33. Carly Ledbetter, "10 of the Most Expensive (and Beautiful) Apartments Around the World," *HuffPost Life*, December 9, 2014, www.huffingtonpost.com/2014/12/09/most-expensive-apartments-in-the-world_n_6295480.html.

34. Richard Florida, "The Staggering Value of Urban Land," *Citylab*, November 2, 2017, www.citylab.com/equity/2017/11/the-staggering-value-of-urban-land/544706/.

35. Celine Ge, "Living the High Life: Hong Kong Tops World Charts for Skyscrapers—and Most of Them Are Residential," *South China Morning Post*, December 31, 2015, www.scmp.com/news/hong-kong/economy/article/1896807/living-high-life-hong-kong-tops-world-charts-skyscrapers-and.

36. NYC Buildings, "DOB Launches 'Stay Safe. Stay Put.' Campaign."

Chapter 4. The New (Electric) Deal

1. Vladimir Lenin, "Communism Is Soviet Power and Electrification of the Whole Country," December 22, 1920, in *Seventeen Moments in Soviet History*, http://soviethistory.msu.edu/1921-2/electrification-campaign/communism-is-soviet-power-electrification-of-the-whole-country/.

2. Vaclav Smil, *Creating the Twentieth Century: Technical Innovations of 1867–1914 and Their Lasting Impact* (New York: Oxford University Press, 2005).

3. In 1902, the United States produced about 6 gigawatt-hours of electricity. By 1930, production exceeded 114 gigawatt-hours. See, US Census Bureau, *Historical Statistics of the United States, Colonial Times to 1970* (1975) s.v. "Chapter S: Energy," 821, www2.census.gov/library/publications/1975/compendia/hist_stats_colonial-1970/hist_stats_colonial-1970p2-chS.pdf.

4. Klein, *The Power Makers*, 397.

5. Klein, *The Power Makers*, 402.

6. Klein, *The Power Makers*, 403.

7. Division of Investment Management, Securities and Exchange Commission, *The Regulation of the Public-Utility Holding Companies* (Washington, DC: June 1995), www.sec.gov/news/studies/puhc.txt.

8. Division of Investment Management, *The Regulation of the Public-Utility Holding Companies*, 13.

9. *Public Utility Holding Company Act of 1935: 1935–1992* (Washington, DC: EIA, 1993), www.eia.gov/electricity/archive/0563.pdf.

10. "Electric Bond and Share Company," http://snaccooperative.org/ark:/99166/w6713g3m.

11. "Electric Bond and Share Company," SNAC, no. 41264488, http://socialarchive.iath.virginia.edu/ark:/99166/w6713g3m.

12. John Williams, *The Untold Story of the Lower Colorado River Authority* (College Station: Texas A&M University Press, 2016), 81.

13. "Deal Books: Texas Power & Light Company," Lehman Brothers archives, Baker Library and Bloomberg Center, Harvard Business School, Boston, www.library.hbs.edu/hc/lehman/Data-Resources/Companies-Deals/Texas-Power-Light-Company. See also, Williams, *Lower Colorado River Authority*, 81.

14. US Inflation Calculator, www.usinflationcalculator.com.

15. Data, "Average Retail Price of Electricity, United States, Annual," EIA, www.eia.gov/electricity/data/browser/#/topic/7?agg=0,1&geo=g&endsec=vg&linechart=ELEC.PRICE.US-ALL.A~ELEC.PRICE.US-RES.A~ELEC.PRICE.US-COM.A~ELEC.PRICE.US-IND.A&columnchart=ELEC.PRICE.US-ALL.A~ELEC.PRICE.US-RES.A~ELEC.PRICE.US-COM.A~ELEC.PRICE.US-IND.A&map=ELEC.PRICE.US-ALL.A&freq=A&ctype=linechart<ype=pin&rtype=s&maptype=0&rse=0&pin=.

16. John A. Garraty and Peter Gay, eds., *The Columbia History of the World* (New York: Harper & Row: 1972), 1015.

17. Franklin D. Roosevelt, "Campaign Address in Portland, Oregon on Public Utilities and Development of Hydro-Electric Power," September 21, 1932, American Presidency Project, www.presidency.ucsb.edu/ws/index.php?pid=88390.

18. "Labor Force, Employment, and Unemployment, 1929–39: Estimating Methods," Bureau of Labor Statistics, technical note, www.bls.gov/opub/mlr/1948/article/pdf/labor-force-employment-and-unemployment-1929-39-estimating-methods.pdf.

19. "United States Unemployment Rate," Infoplease, www.infoplease.com/business-finance/labor-and-employment/united-states-unemployment-rate.

20. Franklin D. Roosevelt, "Campaign Address in Portland, Oregon on Public Utilities."

21. World Atlas, s.v. "Largest Landslide Victories in US Presidential Election History," last updated April 25, 2017, www.worldatlas.com/articles/largest-landslide-victories-in-us-presidential-election-history.html.

22. 270 to Win, s.v. "1932 Presidential Election," www.270towin.com/1932_Election/.

23. Wikipedia, s.v. "1912 United States Presidential Election," last updated April 29, 2019, 01:10, https://en.wikipedia.org/wiki/United_States_presidential_election,_1912.

24. Thomas P. Hughes, *Networks of Power: Electrification in Western Society, 1880–1930* (Baltimore: Johns Hopkins University Press, 1983), 286–287.

25. David Kennedy, "Speech to Oregon Electric Cooperatives" (speech, Salem, OR, November 30, 2017).

26. Franklin D. Roosevelt, "Annual Message to Congress," January 4, 1935, American Presidency Project, www.presidency.ucsb.edu/documents/annual-message-congress-3.

27. Wikipedia, s.v. "Sam Rayburn," last updated May 17, 2019, 06:33, https://en.wikipedia.org/wiki/Sam_Rayburn.

28. Paul E. Anderson, "Sam Rayburn and Rural Electrification," *East Texas History*, http://easttexashistory.org/items/show/73.

29. Anthony Champagne, *Congressman Sam Rayburn* (New Brunswick: Rutgers University Press, 1984), 48.

30. Biographical Directory of the United States Congress, s.v. "Wheeler, Burton Kendall, 1882–1975," http://bioguide.congress.gov/scripts/biodisplay.pl?index=w000330.

31. Burton K. Wheeler, *Yankee from the West: The Candid Story of the Freewheeling U.S. Senator from Montana* (New York: Doubleday, 1962), 306.

32. Wheeler, *Yankee from the West*, 295.

33. Wheeler, *Yankee from the West*, 307.

34. Encyclopedia of the Great Plains, s.v. "Norris, George W. (1861–1944)," http://plainshumanities.unl.edu/encyclopedia/doc/egp.pg.058. See also, "History of the Unicameral," Nebraska Legislature, https://nebraskalegislature.gov/about/history_unicameral.php.

35. Wikipedia, s.v. "George W. Norris," last updated April 13, 2019, 21:22, https://en.wikipedia.org/wiki/George_W._Norris. Wikipedia, s.v. "Sam Rayburn." "George Norris," United States Senate, www.senate.gov/artandhistory/history/minute/George_Norris.htm.

36. Walt Sehnert, "Sen. George W. Norris vs. Henry Ford," *McCook Gazette*, June 9, 2008, www.mccookgazette.com/story/1435485.html.

37. *Encyclopedia Britannica Online*, s.v. "George W. Norris," www.britannica.com/biography/George-W-Norris.

38. Wheeler, *Yankee from the West*, 308.

39. Wheeler, *Yankee from the West*, 313.

40. Will Rogers is my fifth cousin. My great-grandmother on my father's side was Will Rogers's second cousin.

41. Wikipedia, s.v. "Will Rogers," last updated May 6, 2019, 15:20, https://en.wikipedia.org/wiki/Will_Rogers#Newspapers_and_magazines.

42. Wheeler, *Yankee from the West*, 310.

43. The holding companies fought the Public Utility Holding Company Act all the way to the Supreme Court. In 1938, the high court upheld the law. See, Paul W. White, "The Public Utility Holding Company Act," http://paulwwhite.com/the-public-utility-company-act-of-1935_316.html. Also in 1938, Insull was found dead of a heart attack in the Paris subway. Once one of the most influential businessmen in America, he was found with no identification or money in his pockets. His holding company had gone bankrupt in 1932, leaving Insull owing $16 million more than his net worth. He was, according to one account, "too broke to be bankrupt." In 1934, Insull was tried on fraud charges but was acquitted. See, "Samuel Insull," *They Made America*, PBS, www.pbs.org/wgbh/theymadeamerica/whomade/insull_hi.html.

44. "Public vs. Private Power: From FDR to Today," *Frontline*, PBS, www.pbs.org/wgbh/pages/frontline/shows/blackout/regulation/timeline.html.

45. The REA was first created by Roosevelt in 1935 through an executive order. The passage of the Rural Electrification Act of 1936 made the Rural Electrification Administration into a standalone agency. For the text of the act, see Rural Electrification Act of 1936, chapter 432 of the 74th Cong., approved May 20, 1936, amended through P.L. 115–334, enacted December 20, 2018, https://legcounsel.house.gov/Comps/Rural%20Electrification%20Act%20Of%201936.pdf.

46. George W. Norris, *Fighting Liberal: The Autobiography of George W. Norris*, 2nd ed. (Lincoln, NE: Bison Books, 2009), 325.

Chapter 5. Wiring the Superpower

1. Williams, *Lower Colorado River Authority*, 105.

2. Johnson joined the US House on April 10, 1937. Wikipedia, s.v. "Lyndon B. Johnson," last updated May 12, 2019, 13:58, https://en.wikipedia.org/wiki/Lyndon_B._Johnson.

3. Williams, *Lower Colorado River Authority*, 85.

4. US Census Bureau, *Historical Statistics of the United States, Colonial Times to 1970* (1975) s.v. "Chapter S: Energy," 816.

5. There are two Colorado Rivers in the United States. The one discussed here begins and terminates within the state of Texas. The other Colorado originates in the state of Colorado and terminates in the Gulf of Baja.

6. "Highland Lakes and Dams," LCRA, www.lcra.org/water/dams-and-lakes/Pages/default.aspx. Today, Buchanan has 54.9 megawatts of capacity. Mansfield has 108 megawatts of capacity.

7. Ronnie Dugger, *The Politician: The Life and Times of Lyndon Johnson* (New York: W. W. Norton, 1982), 209.

8. Robert Caro, *The Path to Power*, vol. 1 of *The Years of Lyndon Johnson* (New York: Vintage, 1983), 524–527.

9. Caro, *The Path to Power*, 527.

10. Williams, *Lower Colorado River Authority*, 86.

11. Williams, *Lower Colorado River Authority*, 86.

12. Williams, *Lower Colorado River Authority*, 84–86.

13. "Quick Facts: Texas," US Census Bureau, www.census.gov/quickfacts/TX.

14. In 1938, the co-op signed up about 3,000 families in the Texas Hill Country for electricity service. See, "The Cooperative Story," PEC, www.pec.coop/about-us/cooperative-difference/cooperative-story/.

15. "Our Service Areas," PEC, www.pec.coop/Home/Energy_Services/Service_Area_Map.aspx.

16. In 2018, Lorin and I bought property in the Hill Country and became PEC customers. For number of customers added, see *2017 Annual Report* (Johnson City, TX: Pedernales Electric Cooperative, 2018), https://2bqwe7212tygr0q3a7b9e1bv-wpengine.netdna-ssl.com/wp-content/uploads/2018/06/PEC-AR-2017.pdf. For revenue data, see *2017 Financial Report* (Johnson City, TX: Pedernales Electric Cooperative, 2018), https://2bqwe7212tygr0q3a7b9e1bv-wpengine.netdna-ssl.com/wp-content/uploads/2018/06/PEC-AR-FINANCIALS-2017.pdf.

17. "Our Members," National Rural Utilities Cooperative Finance Corporation, www.nrucfc.coop/content/cfc/about_cfc/our-members.html.

18. US Census Bureau, *Historical Statistics of the United States, Colonial Times to 1970* (1975) s.v. "Chapter S: Energy," 829–830.

19. US Census Bureau, *Historical Statistics of the United States, Colonial Times to 1970* (1975) s.v. "Chapter S: Energy," 827.

20. Carolyn Dimitri, Anne Effland, and Nielson Conklin, *The 20th Century Transformation of U.S. Agriculture and Farm Policy* (Washington, DC: US Department of Agriculture, 2005), 2, https://ageconsearch.umn.edu/bitstream/59390/2/eib3.pdf.

21. Garraty and Gay, *Columbia History*, 1015.

22. Peter Z. Grossman, *U.S. Energy Policy and the Pursuit of Failure* (Cambridge: Cambridge University Press, 2013).

23. *2015–2016 Annual Directory & Statistical Report* (Arlington, VA: American Public Power Association, 2016), 26–27. The investor-owned utilities have a total of about 88 million customers out of 135 million electric customers in the United States.

24. EIA, "Exelon-Pepco Merger Could Create Largest U.S. Electric Utility," *Today in Energy*, October 21, 2015, www.eia.gov/todayinenergy/detail.php?id=23432. Exelon has about 8.5 million customers out of 135 million US customers, or about 6.3 percent of the total.

25. "The Top 10 Biggest Power Companies of 2014," *Power Technology*, October 1, 2014, www.power-technology.com/features/featurethe-top-10-biggest-power-companies-of-2014-4385942/.

26. "State Grid: China," *Fortune* Global 500, http://fortune.com/global500/state-grid/.

27. *BP Statistical Review of World Energy 2018*.

28. "GDP per Capita, Current US$," World Bank, https://data.worldbank.org/indicator/NY.GDP.PCAP.CD?locations=CN.

29. US Census Bureau, *Historical Statistics of the United States, Colonial Times to 1970* (1975) s.v. "Chapter S: Energy," 821.

30. US Census Bureau, *Historical Statistics of the United States, Colonial Times to 1970* (1975) s.v. "Chapter S: Energy," 224.

31. US Census Bureau, *Historical Statistics of the United States, Colonial Times to 1970* (1975) s.v. "Chapter S: Energy," 225.

32. OECD/IEA, "Electric Power Consumption (kWh per Capita)," World Bank, 2014, https://data.worldbank.org/indicator/eg.use.elec.kh.pc.

33. Norris, *Fighting Liberal*, 318–319.

34. Caro, *The Path to Power*, 524–527.

Chapter 6. Women Unplugged

1. "Reddy Kilowatt," YouTube, posted September 19, 2013, www.youtube.com/watch?v=5_kW2KZ04wk.

2. For more, see Intergovernmental Panel on Climate Change, "Chapter 5: Sustainable Development, Poverty Eradication and Reducing Inequalities," 2018, www.ipcc.ch/site/assets/uploads/2018/11/sr15_chapter5.pdf.

3. OECD/IEA, "Electric Power Consumption (kWh per Capita)," World Bank, 2014, https://data.worldbank.org/indicator/eg.use.elec.kh.pc.

4. According to Mriduchhanda Chattopadhyay, who has been surveying the residents of Majlishpukur, the village is about 90 percent Muslim.

5. Michael Lipka and Conrad Hackett, "Why Muslims Are the World's Fastest-Growing Religious Group," *Fact Tank* (blog), Pew Research Center, April 6, 2017, www.pewresearch.org/fact-tank/2017/04/06/why-muslims-are-the-worlds-fastest-growing-religious-group/.

6. "The Changing Global Religious Landscape," Pew Research Center, April 5, 2017, www.pewforum.org/2017/04/05/the-changing-global-religious-landscape/.

7. "Religion and Education Around the World," Pew Research Center, December 13, 2016, www.pewforum.org/2016/12/13/religion-and-education-around-the-world/.

8. Nicholas Jackson, "Data Analyst Hans Rosling: People Vote for Washing Machines," *Atlantic*, December 8, 2010, www.theatlantic.com/technology/archive/2010/12/data-analyst-hans-rosling-people-vote-for-washing-machines/67680/.

9. Joseph D. Coppock, "Organization and Operations of the Electric Home and Farm Authority," in *Government Agencies of Consumer Instalment Credit* (Washington, DC: National Bureau of Economic Research, 1940), www.nber.org/chapters/c4943.pdf.

10. Coppock, "Organization and Operations," 97.

11. Thanks to Paul Bowers, CEO of Georgia Power, for his tutorial on the history of Reddy Kilowatt.

12. Abul Barkat, "Economic and Social Impact Evaluation Study of the Rural Electrification Program in Bangladesh," Human Development Research Center, October 2002, xxi, http://citeseerx.ist.psu.edu/viewdoc/download?doi=10.1.1.469.671&rep=rep1&type=pdf.

13. Taryn Dinkelman, "The Effects of Rural Electrification on Employment: New Evidence from South Africa" (Princeton University, August 2010), 25, https://rpds.princeton.edu/sites/rpds/files/media/dinkelman_electricity_0810.pdf.

14. Shahidur R. Khandker et al., "Who Benefits Most from Rural Electrification? Evidence in India" (paper presented at the Agricultural & Applied Economics Association's 2012 Annual Meeting, Seattle, Washington, August 12–14, 2012), http://ageconsearch.umn.edu/bitstream/125090/2/AliR.pdf.

15. *Female Genital Mutilation/Cutting: A Statistical Overview and Exploration of the Dynamics of Change* (New York: UNICEF, 2009), www.unicef.org/media/files/FGCM_Brochure_Lo_res.pdf.

16. UNESCO Institute for Statistics, "Literacy Rate, Adult Female (% of Females Ages 15 and Above)," World Bank, data.worldbank.org/indicator/SE.ADT.LITR.FE.ZS?end=2012&start=1970&view=chart&year_high_desc=false.

17. *The State of the World's Children 2015: Executive Summary* (New York: UNICEF, 2014), Table 9, 87, www.unicef.org/publications/files/SOWC_2015_Summary_and_Tables.pdf.

18. World Atlas, s.v. "Which Country First Gave Women the Right to Vote?" last updated March 18, 2019, www.worldatlas.com/articles/first-15-countries-to-grant-women-s-suffrage.html. This source notes that there were a few cases of women's suffrage that predate New Zealand's move, but that those cases were temporary.

19. Matt Novak, "How the 1920s Thought Electricity Would Transform Farms Forever," *Gizmodo*, June 3, 2013, https://paleofuture.gizmodo.com/how-the-1920s-thought-electricity-would-transform-farms-510917940.

20. "Timeline: U.S. Women's Rights, 1848–1920," Infoplease, www.infoplease.com/spot/womens-rights-movement-us.

Chapter 7. My Refrigerator Versus the World

1. Henry Schlesinger, *The Battery: How Portable Power Sparked a Technological Revolution* (New York: HarperCollins, 2010), 176.

2. Other analysts have used their refrigerator as a benchmark. In 2013, Todd Moss, who was then a senior fellow at the Center for Global Development, used his refrigerator's

consumption to compare his energy use with residents of several African countries. See, Todd Moss, "My Fridge Versus Power Africa," Center for Global Development, September 9, 2013, www.cgdev.org/blog/my-fridge-versus-power-africa. Moss's refrigerator consumed 459 kilowatt-hours per year.

3. Michael Bluejay, "How Much Electricity Does My Refrigerator Use?" *Saving Electricity* (blog), April 2014, http://michaelbluejay.com/electricity/refrigerators.html.

4. Laura Cozzi et al., "Population Without Access to Electricity Falls Below 1 Billion," IEA, October 30, 2018, www.iea.org/newsroom/news/2018/october/population-without-access-to-electricity-falls-below-1-billion.html.

5. Schewe, *The Grid*, 78.

6. Alan D. Pasternak, *Global Energy Futures and Human Development: A Framework for Analysis* (Livermore, CA: Lawrence Livermore National Laboratory, 2000), 17, https://e-reports-ext.llnl.gov/pdf/239193.pdf.

7. *Human Development Report 2015* (New York: UN Development Programme, 2015), 3, http://hdr.undp.org/sites/default/files/2015_human_development_report_1.pdf.

8. The UN describes the HDI as a "composite index focusing on three basic dimensions of human development: to lead a long and healthy life, measured by life expectancy at birth; the ability to acquire knowledge, measured by mean years of schooling and expected years of schooling; and the ability to achieve a decent standard of living, measured by gross national income per capita. The HDI has an upper limit of 1.0." That is, the closer a country is to a ranking of 1 on the HDI, the better off it is in terms of fulfilling the needs of its citizens.

9. "Alan Pasternak," obituary, *San Francisco Chronicle*, September 20, 2010, www.legacy.com/obituaries/sfgate/obituary.aspx?n=alan-pasternak&pid=145700627.

10. Nicolas Van Praet, "A Look Inside Quebec's Fort Knox of Maple Syrup," *Globe and Mail*, December 31, 2014, www.theglobeandmail.com/news/national/a-look-inside-quebecs-fort-knox-of-maple-syrup/article22262093/.

11. *BP Statistical Review of World Energy 2018*.

12. IEA, "Key World Energy Statistics," 2013.

13. "The World's Longest Power Transmission Lines," *Power Technology*, February 17, 2014, www.power-technology.com/features/featurethe-worlds-longest-power-transmission-lines-4167964/. Note that in early 2019, China commissioned a high-voltage line that spanned more than 3,000 kilometers. See, "World's Highest Voltage Transmission Project Starts Operation," UNTV, January 2, 2019, www.untvweb.com/news/worlds-highest-voltage-transmission-project-starts-operation/.

14. "251 Publicly Owned Electric & Gas Utilities (US)," Utility Connection, www.utilityconnection.com/page2e.asp#muni_util.

15. "Corporate Profile," State Grid Corporation of China, www.sgcc.com.cn/ywlm/aboutus/profile.shtml.

Chapter 8. The Power Imperatives: Integrity, Capital, and Fuel

1. "History of Electricity," Institute for Energy Research, www.instituteforenergyresearch.org/history-electricity/.

2. Black Max 3,600/4,500 Watt Portable Gas Generator, available for $299.98 from Sam's Club on May 19, 2019, www.samsclub.com/sams/blackmax-black-ma-generator/prod18650560.ip.

3. Firman W01781 2100/1700 Watt Recoil Start Gas Portable Generator, available for $663.36 from Amazon on May 19, 2019, www.amazon.com/Firman-W01781-Whisper-Inverter-Generator/dp/B01M0SIRSZ.

4. Daron Acemoglu and James A. Robinson, *Why Nations Fail: The Origins of Power, Prosperity, and Poverty* (New York: Crown Business, 2012), 3–4.

5. World Bank data. In 2017, the Netherlands GDP totaled $830 billion. See, World Bank and OECD, "GDP (Current $US)," https://data.worldbank.org/indicator/ny.gdp.mktp.cd?most_recent_value_desc=true.

6. *World Energy Investment 2019* (Paris: IEA, May 2019), 10.

7. Pers. comm. with Edison Electric Institute.

8. James Conca, "Electricity and Jobs in America," *Forbes*, August 3, 2017, www.forbes.com/sites/jamesconca/2017/08/03/electricity-and-jobs-in-america/#671c5f6267f1.

9. "Wyoming: State Profile and Energy Estimates," EIA, www.eia.gov/state/?sid=WY#tabs-4.

10. Alexander Richter, "Focusing on Energy Independence and Renewables, Djibouti Is Banking on Geothermal Energy," *Think Geoenergy*, September 17, 2018, www.thinkgeoenergy.com/focusing-on-energy-independence-and-renewables-djibouti-is-banking-on-geothermal-energy/.

Chapter 9. The American Way of War

1. "Transcript of Schwarzkopf Briefing," Associated Press, January 30, 1991, www.apnews.com/70207cbe2403c2e5743c256e1ef0f208.

2. Robert D. McFadden, "Gen. H. Norman Schwarzkopf, U.S. Commander in Gulf War, Dies at 78," *New York Times*, December 27, 2012, www.nytimes.com/2012/12/28/us/gen-h-norman-schwarzkopf-us-commander-in-gulf-war-dies-at-78.html.

3. The aerial bombing campaign began on January 16, 1991. See, *Encyclopedia Britannica Online*, s.v. "Persian Gulf War," www.britannica.com/event/Persian-Gulf-War.

4. "Transcript of Schwarzkopf Briefing," AP.

5. Eric Schmitt, "Raid on Iraq; The Day's Weapon of Choice, the Cruise Missile, Is Valued for Its Accuracy," *New York Times*, January 18, 1993, www.nytimes.com/1993/01/18/world/raid-iraq-day-s-weapon-choice-cruise-missile-valued-for-its-accuracy.html.

6. "CBU-94 'Blackout Bomb' BLU-114/B 'Soft-Bomb,'" GlobalSecurity.org, www.globalsecurity.org/military/systems/munitions/blu-114.htm.

7. Thomas E. Griffith Jr., *Strategic Attack of National Electrical Systems* (Maxwell Air Force Base, AL: Air University Press, 1994), 41, www.comw.org/pda/fulltext/griffith.pdf.

8. Other water-borne diseases include dysentery, enteric fever, and leptospirosis. For more, see Naveed Saleh, "The Spread of Waterborne Illnesses," August 2, 2018, *Verywell Health*, www.verywellhealth.com/what-you-should-know-about-waterborne-illnesses-4151851.

9. Griffith, *Strategic Attack*, 33.

10. Wikipedia, s.v. "Attack on the Sui-ho Dam," last updated January 21, 2019, 06:40, https://en.wikipedia.org/wiki/Attack_on_the_Sui-ho_Dam.

11. Blaine Harden, "The U.S. War Crime North Korea Won't Forget," *Washington Post*, March 24, 2015, www.washingtonpost.com/opinions/the-us-war-crime-north-korea-wont-forget/2015/03/20/fb525694-ce80-11e4-8c54-ffb5ba6f2f69_story.html?utm_term=.942d9ccc1848.

12. Griffith, *Strategic Attack*, 33.

13. "Operation Rolling Thunder," GlobalSecurity.org, www.globalsecurity.org/military/ops/rolling_thunder.htm.

14. Griffith, *Strategic Attack*, 33.

15. Barton Gellman, "Allied Air War Struck Broadly in Iraq," *Washington Post*, June 23, 1991, www.washingtonpost.com/archive/politics/1991/06/23/allied-air-war-struck-broadly-in-iraq/e469877b-b1c1-44a9-bfe7-084da4e38e41/?utm_term=.acd8ed68e5a8.

16. International Study Team, *Health and Welfare in Iraq After the Gulf Crisis: An In-Depth Assessment* (New York: Center for Economic and Social Rights, 1991), 4, www.cesr.org/sites/default/files/Health_and_Welfare_in_Iraq_after_the_Gulf_Crisis_1991.pdf.

17. John F. Burns, "The World; How Many People Has Hussein Killed?" *New York Times*, January 26, 2003, www.nytimes.com/2003/01/26/weekinreview/the-world-how-many-people-has-hussein-killed.html.

18. IEA/OECD, "Electric Power Consumption (kWh per Capita)," World Bank, https://data.worldbank.org/indicator/eg.use.elec.kh.pc. In 1990, per-capita electricity use was about 1,300 kilowatt-hours per year. That was a nearly 80 percent increase over the 726 kilowatt-hours per capita per year in 1979.

19. UN Inter-agency Group for Child Mortality Estimation, "Mortality Rate, Infant (per 1,000 Live Births) World Bank, https://data.worldbank.org/indicator/SP.DYN.IMRT.IN?locations=IQ. For GDP, see "GDP per Capita, Current US$," World Bank, https://data.worldbank.org/indicator/NY.GDP.PCAP.CD?locations=IQ.

20. Paul Lewis, "After the War; U.N. Survey Calls Iraq's War Damage Near-Apocalyptic," *New York Times*, March 22, 1991, www.nytimes.com/1991/03/22/world/after-the-war-un-survey-calls-iraq-s-war-damage-near-apocalyptic.html.

21. *Needless Deaths in the Gulf War: Civilian Casualties During the Air Campaign and Violations of the Laws of War* (New York: Human Rights Watch, 1991), www.hrw.org/reports/1991/gulfwar/CHAP4.htm.

22. Marc Santora, "A Nation at War: Helping Iraqis; Continued Fighting Delays Plans for Aid Distribution, Relief Workers Say," *New York Times*, March 25, 2003, www.nytimes.com/2003/03/25/world/nation-war-helping-iraqis-continued-fighting-delays-plans-for-aid-distribution.html.

23. *The Human Costs of War in Iraq* (Brooklyn, NY: Center for Economic and Social Rights, 2003), 7, www.cesr.org/sites/default/files/Human_Costs_of_War_in_Iraq.pdf.

24. Griffith, *Strategic Attack*, 42.

25. Barbara Crossette, "Iraq Sanctions Kill Children, U.N. Reports," *New York Times*, December 1, 1995, www.nytimes.com/1995/12/01/world/iraq-sanctions-kill-children-un-reports.html.

26. Edmund L. Andrews, "After the War: Energy; Thieves and Saboteurs Disrupt Electrical Services in Iraq," *New York Times*, June 21, 2003, www.nytimes.com/2003/06/21/world/after-the-war-energy-thieves-and-saboteurs-disrupt-electrical-services-in-iraq.html.

27. James Glanz, "The Reach of War: Insurgents' Strategy; Saboteurs May Be Aiming at Electrical and Water Sites as Summer Nears," *New York Times*, June 9, 2004, www.nytimes.com/2004/06/09/world/reach-war-insurgents-strategy-saboteurs-may-be-aiming-electrical-water-sites.html.

28. James Glanz, "U.S. Agency Finds New Waste and Fraud in Iraqi Rebuilding Projects," *New York Times*, February 1, 2007, www.nytimes.com/2007/02/01/world/middleeast/01reconstruction.html.

29. "Transcript for Sept. 14," *Meet the Press*, NBC, September 14, 2003, www.nbcnews.com/id/3080244/ns/meet_the_press/t/transcript-sept/#.W-s9AS_Mzvc.

30. "Iraq Index," Brookings Institution, various dates, www.brookings.edu/iraq-index/.

31. Hadeel al Sayegh, "Iraq Pays High Price for Lack of Electricity," *National*, July 13, 2012, www.thenational.ae/business/iraq-pays-high-price-for-lack-of-electricity-1.361492. See also, Griffith, *Strategic Attack*, 41.

32. Arwa Ibrahim, "Electricity Cuts Across Iraq Make Life Unbearable in Summer Heat," Al Jazeera, July 21, 2018, www.aljazeera.com/news/2018/07/electricity-cuts-iraq-life-unbearable-summer-heat-180731111220743.html.

33. Mohammed Ebraheem, "Terrorist Attack Targets Power Transmission Line in Iraq," *Iraqi News*, December 31, 2018, www.iraqinews.com/features/terrorist-attack-targets-power-transmission-line-in-iraq/.

34. Isabel Coles and Ali Nabhan, "Oil-Rich Iraq Can't Keep the Lights On," *Wall Street Journal*, July 21, 2018, www.wsj.com/articles/oil-rich-iraq-cant-keep-the-lights-on-1532174400.

35. Mina Aldroubi, "Basra Health Crisis: 17,000 Admitted to Hospitals for Water Poisoning," *National*, August 29, 2018, www.thenational.ae/world/mena/basra-health-crisis-17-000-admitted-to-hospitals-for-water-poisoning-1.764991.

36. Jane Arraf, "Months of Protests Roil Iraq's Oil Capital Basra," NPR, September 27, 2018, www.npr.org/2018/09/27/651508389/months-of-protests-roil-iraqs-oil-capital-basra.

37. Edward Wong, "Trump Pushes Iraq to Stop Buying Energy from Iran," *New York Times*, February 11, 2019, www.nytimes.com/2019/02/11/us/politics/iraq-buying-energy-iran.html.

38. "Post Saddam Era, Iran-Iraq Boost Cooperation Despite American Sanctions," *Eurasian Times*, February 12, 2019, https://eurasiantimes.com/post-saddam-era-iran-iraq-boost-cooperation-despite-american-sanctions/.

39. *Encyclopedia Britannica Online*, s.v. "Iran-Iraq War," www.britannica.com/event/Iran-Iraq-War.

40. Wong, "Trump Pushes Iraq."

41. Ammar Karim, "Electricity Crisis Leaves Iraqis Gasping for Cool Air," Phys.org, August 1, 2018, https://phys.org/news/2018-08-electricity-crisis-iraqis-gasping-cool.html.

Chapter 10. Beirut's Generator Mafia

1. Pers. comm. by author.

2. Youssef Diab, "2 Killed in Armed Clash Between 'Hezbollah' Supporters in Lebanon's Sidon," *Asharq Al-Awsat*, October 4, 2017, https://eng-archive.aawsat.com/youssef-diab/news-middle-east/2-killed-armed-clash-hezbollah-supporters-lebanons-sidon.

3. "Lebanon," in *International Religious Freedom Report for 2011* (Washington, DC: US Department of State, 2011), https://2009-2017.state.gov/documents/organization/193107.pdf.

4. Elie Bouri and Joseph El Assad, "The Lebanese Electricity Woes: An Estimation of the Economical Costs of Power Interruptions," *Energies* 9 (2016): 583. Bouri and Assad's figure is for 2014. That year, Lebanon's GDP was $45.7 billion.

5. Alex Dziadosz, "Can Green Energy Beat Lebanon's 'Generator Mafias?'" Bloomberg, February 26, 2018, www.bloomberg.com/news/features/2018-02-26/can-green-energy-beat-lebanon-s-generator-mafias.

6. Ziad K. Abdelnour, "The Corruption Behind Lebanon's Electricity Crisis," *Middle East Intelligence Bulletin* 5, no. 8–9 (August–September 2003).

7. Zeena Saifi and Sarah El Sirgany, "Hezbollah: 'Mission accomplished' Against ISIS in Lebanon," CNN, August 29, 2017, www.cnn.com/2017/08/29/middleeast/hezbollah-isis/index.html.

8. John Daniszewski, "Israel Knocks Out Beirut Electricity," *Los Angeles Times*, April 16, 1996, http://articles.latimes.com/1996-04-16/news/mn-59120_1_northern-israel. See also, *Civilian Pawns: Laws of War Violations and the Use of Weapons on the Israel-Lebanon Border* (New York: Human Rights Watch, 1996), www.hrw.org/reports/1996/Israel.htm#P231_56403.

9. Rebecca Trounson, "Israel Bombs Lebanese Bridges, Power Plants," *Los Angeles Times*, June 25, 1999, http://articles.latimes.com/1999/jun/25/news/mn-50004.

10. "No Additional U.S. Air-to-Ground Missiles to Israel," Human Rights Watch, May 23, 2000, www.hrw.org/news/2000/05/22/no-additional-us-air-ground-missiles-israel.

11. "No Additional U.S. Air-to-Ground Missiles to Israel." See also, Robert Fisk, "Lebanon's People Bear the Brunt of Israeli Wrath," *Independent*, February 9, 2000, www.independent.co.uk/news/world/middle-east/lebanons-people-bear-the-brunt-of-israeli-wrath-5372062.html.

12. Richard Black, "Environmental 'Crisis' in Lebanon," BBC News, July 31, 2006, http://news.bbc.co.uk/2/hi/science/nature/5233358.stm.

13. Lin Noueihed, "Israel Allows UN Survey of Lebanon Oil Slick," Reuters, August 21, 2006. For capacity and age of plant, see Wikipedia, s.v., "Jieh," last updated February 22, 2019, 16:16, http://en.wikipedia.org/wiki/Jieh.

14. Declan Walsh, "Life Amid the Blood and Bombs as Besieged Hospital Battles On," *Guardian*, August 14, 2006, www.theguardian.com/world/2006/aug/14/syria.israel4.

15. "Lebanon: Deliberate Destruction or 'Collateral Damage'? Israeli Attacks on Civilian Infrastructure," Amnesty International, August 22, 2006, www.amnesty.org/en/documents/MDE18/007/2006/en/.

16. Farouk Fardoun et al., "Electricity of Lebanon: Problems and Recommendations," *Energy Procedia* 19 (2012): 310–320, https://ac.els-cdn.com/S1876610212009812/1-s2.0-S1876610212009812-main.pdf?_tid=d92e7896-fcf1-4969-8a46-4a030ab4bcd8&acdnat=1545082789_8eeaa16d96916b1c3d5cd21660ab3fa8.

17. Lebanon has about 200 megawatts of hydro capacity. But actual output from that capacity is only marginal as nearly all of the country's hydro operations need maintenance and upgrades. For capacity, see "Hydropower in Lebanon," World Energy Council, www.worldenergy.org/data/resources/country/lebanon/hydropower/. For output, see "Lebanon—Electricity Production from Hydroelectric Sources," IndexMundi, www.indexmundi.com/facts/lebanon/electricity-production-from-hydroelectric-sources.

18. OECD/IEA, "Electricity Production from Oil Sources (% of Total)," World Bank, https://data.worldbank.org/indicator/EG.ELC.PETR.ZS?end=2014&start=1960&year_high_desc=true.

19. Alan Shihadeh et al., *Effect of Distributed Electric Power Generation on Household Exposure to Airborne Carcinogens in Beirut* (Beirut: American University of Beirut, 2013), 4, 9, www.aub.edu.lb/ifi/Documents/publications/research_reports/2012-2013/20130207ifi_rsr_cc_effect%20Diesel.pdf.

20. OECD/IEA, "Electricity Production from Oil Sources (% of Total)," World Bank, https://data.worldbank.org/indicator/EG.ELC.PETR.ZS.

21. Simon Tisdall, "The Turkish 'Power Ship' Keeping the Lights on in Lebanon," *Guardian*, April 11, 2013, www.theguardian.com/world/2013/apr/11/turkish-power-ship-lights-on-lebanon. See also, Soumyajit Dasgupta, "What Is Karadeniz Powership?" *Marine Insight*, July 21, 2016, www.marineinsight.com/types-of-ships/what-is-karadeniz-powership/.

Chapter 11. It's Not Possible to Keep the Lights on Without Coal

1. "Transcript of Carter's Address to the Nation About Energy Problems," *New York Times*, April 19, 1977, www.nytimes.com/1977/04/19/archives/transcript-of-carters-address-to-the-nation-about-energy-problems.html.

2. OECD/IEA, "Electricity Production from Coal Sources (% of Total)," World Bank, https://data.worldbank.org/indicator/EG.ELC.COAL.ZS?locations=IN.

3. Annie Gowen, "India's Huge Need for Electricity Is a Problem for the Planet," *Washington Post*, November 6, 2015, www.washingtonpost.com/world/asia_pacific/indias-huge-need-for-electricity-is-a-problem-for-the-planet/2015/11/06/a9e004e6-622d-11e5-8475-781cc9851652_story.html.

4. Clyde Russell, "Column—Coal Going from Winner to Loser in India's Energy Future: Russell," CNBC, February 20, 2019, www.cnbc.com/2019/02/20/reuters-america-column-coal-going-from-winner-to-loser-in-indias-energy-future-russell.html.

5. Roger Pielke Jr., "A Positive Path for Meeting the Global Climate Challenge," *E360*, October 18, 2010, http://e360.yale.edu/features/a_positive_path_for_meeting_the_global_climate_challenge.

6. Credit to Ben Heard, the founder of Bright New World, who said almost those exact words in the documentary *Juice: How Electricity Explains the World*.

7. "India's Intended National Determined Contribution: Working Towards Climate Justice," Intended Nationally Determined Contribution to the UN Framework Convention on Climate Change, submitted October 1, 2015, 5, www4.unfccc.int/submissions/INDC/Published%20Documents/India/1/INDIA%20INDC%20TO%20UNFCCC.pdf.

8. *The Emissions Gap Report 2017* (New York: UN Environmental Programme, 2017), xxi, https://wedocs.unep.org/bitstream/handle/20.500.11822/22070/EGR_2017.pdf.

9. *The Emissions Gap Report 2017*, 38.

10. IEA, *World Energy Outlook 2018*.

11. *BP Statistical Review of World Energy* (London: BP, 2018), 47, www.bp.com/content/dam/bp/business-sites/en/global/corporate/pdfs/energy-economics/statistical-review/bp-stats-review-2018-electricity.pdf.

12. Timothy Gardner, "Bloomberg's Charity Donates $64 Million to 'War on Coal,'" Reuters, October 11, 2017, www.reuters.com/article/us-usa-coal-bloomberg/bloombergs-charity-donates-64-million-to-war-on-coal-idUSKBN1CG2M5.

13. *Bloomberg Philanthropies 2018 Annual Report* (New York: Bloomberg Philanthropies, 2018), 18, 62, www.bbhub.io/dotorg/sites/34/2018/05/Bloomberg-Philanthropies-Annual-Report-2018.pdf.

14. Sebastien Malo, "Michael Bloomberg to Spend $500 Million to Close Coal Plants," Reuters, June 7, 2019, www.reuters.com/article/us-climate-change-coal-usa/michael-bloomberg-to-spend-500-million-to-close-coal-plants-idUSKCN1T82I5.

15. "Moving Beyond Coal," *ELAW Advocate* (Winter 2017), 1, www.elaw.org/system/files/attachments/advocate/fulldownload/ELAW_Winter_2017_full.pdf.

16. "Ex-Im Bank Won't Finance Vietnam Coal-Fired Power Plant," Reuters, July 18, 2013, www.reuters.com/article/2013/07/18/us-usa-vietnam-coal-idUSBRE96H15X20130718.

17. Anna Yukhananov and Valerie Volcovici, "World Bank to Limit Financing of Coal-Fired Plants," Reuters, July 17, 2013, www.reuters.com/article/2013/07/16/us-worldbank-climate-coal-idUSBRE96F19U20130716.

18. Karl Mathiesen, "World Bank Dumps Kosovo Plant, Ending Support for Coal Worldwide," *Climate Home News*, October 10, 2018, www.climatechangenews.com/2018/10/10/world-bank-dumps-support-last-coal-plant/.

19. Pers. comm. with LNG analyst at a US firm.

20. *BP Statistical Review of World Energy 2018.*

21. Frederick Kuo, "A New Coal War Frontier Emerges as China and Japan Compete for Energy Projects in Southeast Asia," *South China Morning Post*, April 1, 2018, www.scmp.com/comment/insight-opinion/article/2139667/new-coal-war-frontier-emerges-china-and-japan-compete-energy.

22. *BP Statistical Review of World Energy 2018.*

23. *BP Statistical Review of World Energy 2018.*

24. "JV Agreement Signed with Malaysia for 1,320 MW Power Plant in Maheshkhali," Energynewsbd.com, July 20, 2016, http://energynewsbd.com/details.php?id=646.

25. Christine Shearer et al., *Boom and Bust 2018: Tracking the Global Coal Plant Pipeline*, (San Francisco: CoalSwarm, 2018), https://endcoal.org/wp-content/uploads/2018/03/BoomAndBust_2018_r6.pdf.

26. Georgia Brown, "British Power Generation Achieves First Ever Coal-Free Day," *Guardian*, April 21, 2017, www.theguardian.com/environment/2017/apr/21/britain-set-for-first-coal-free-day-since-the-industrial-revolution.

27. Matt McGrath, "UK and Canada Lead Global Alliance Against Coal," BBC News, November 16, 2017, www.bbc.com/news/science-environment-42014244.

28. Brad Plumer and Nadja Popovich, "19 Countries Vowed to Phase Out Coal. But They Don't Use Much Coal," *New York Times*, November 16, 2017, www.nytimes.com/interactive/2017/11/16/climate/alliance-phase-out-coal.html. For a list of the members of the alliance, see Powering Past Coal Alliance, "Declaration," November 16, 2017, https://assets.publishing.service.gov.uk/government/uploads/system/uploads/attachment_data/file/660041/powering-past-coal-alliance.pdf.

29. "Thermal Power Plant Worth $1.7B to Be Established in Adana," *Daily Sabah*, October 26, 2017, www.dailysabah.com/energy/2017/10/27/thermal-power-plant-worth-17b-to-be-established-in-adana.

30. *BP Statistical Review of World Energy 2018.*

31. Terry Macalister, "Meet Belcha—Europe's Biggest Carbon Polluter (and It's About to Get Even Bigger)," *Guardian*, July 22, 2009, www.theguardian.com/environment/2009/jul/22/europes-biggest-carbon-polluter-coal.

32. Andrew Kureth, "Why Poland Still Clings to Coal," *Politico*, October 17, 2017, www.politico.eu/article/why-poland-still-clings-to-coal-energy-union-security-eu-commission/.

33. "Poland's PGNiG Signs Long-Term LNG Deal with Cheniere," Reuters, November 8, 2018, www.reuters.com/article/usa-energy-pgnig/polands-pgnig-signs-long-term-lng-deal-with-cheniere-idUSL8N1XJ2GW.

34. Shearer et al., *Boom and Bust 2018*, 4. It's worth noting that the amount of coal-fired capacity has stayed high for years. In 2017, CoalSwarm estimated that 273 gigawatts of coal capacity was under construction. See, Christine Shearer et al., *Boom and Bust 2017: Tracking the Global Coal Plant Pipeline* (San Francisco: CoalSwarm, 2017), 6, https://endcoal.org/wp-content/uploads/2017/03/BoomBust2017-English-Final.pdf. In 2016, that figure was 338 gigawatts. See, Christine Shearer et al., *Boom and Bust 2016: Tracking the Global Coal Plant Pipeline* (San Francisco: CoalSwarm, 2016), http://endcoal.org/wp-content/uploads/2016/06/BoomAndBust_2016.pdf. In 2015, it was 276 gigawatts. See, Brad Plumer, "There Are 2,100 New Coal Plants Being Planned Worldwide—Enough to Cook the Planet," *Vox*, July 9, 2015, www.vox.com/2015/7/9/8922901/coal-renaissance-numbers.

35. "Country Comparison: Electricity—Installed Generating Capacity," CIA World Factbook, www.cia.gov/library/publications/the-world-factbook/rankorder/2236rank.html.

36. Agora, "The Energy Transition in the Power Sector, State of Affairs in 2017" (PowerPoint presentation, January 4, 2017), 25, www.agora-energiewende.de/en/publications/the-energy-transition-in-the-power-sector-state-of-affairs-in-2017/.

37. Stanley Reed, "Germany's Shift to Green Power Stalls, Despite Huge Investments," *New York Times*, October 7, 2017, www.nytimes.com/2017/10/07/business/energy-environment/german-renewable-energy.html.

38. Sören Amelang, "How Much Does Germany's Energy Transition Cost?" Clean Energy Wire, June 1, 2018, www.cleanenergywire.org/factsheets/how-much-does-germanys-energy-transition-cost.

39. Markus Wacket, "German Coalition Negotiators Agree to Scrap 2020 Climate Target: Sources," Reuters, January 8, 2018, www.reuters.com/article/us-germany-politics/german-coalition-negotiators-agree-to-scrap-2020-climate-target-sources-idUSKBN1EX0OU.

40. Claire Stam, "European Commission Abandons Plan to Raise Climate Ambition," Euractiv, October 2, 2018, /www.euractiv.com/section/climate-environment/news/european-commission-to-abandon-plans-for-rising-climate-ambition.

41. Erik Kirschbaum, "Germany to Close All 84 of Its Coal-Fired Power Plants, Will Rely Primarily on Renewable Energy," *Los Angeles Times*, January 26, 2019, www.latimes.com/world/europe/la-fg-germany-coal-power-20190126-story.html.

42. Chisato Tanaka, "Japan Continues to Rely on Coal-Fired Plants Despite Global Criticism," *Japan Times*, October 9, 2018, www.japantimes.co.jp/news/2018/10/09/reference/japan-continues-rely-coal-eyes-coal-fired-plants-despite-global-criticism/#.XGB-96fMyik.

43. Dennis Normile, "Bucking Global Trends, Japan Again Embraces Coal Power," *Science*, May 2, 2018, www.sciencemag.org/news/2018/05/bucking-global-trends-japan-again-embraces-coal-power.

44. SourceWatch, s.v. "Coal Power Technologies," last updated September 18, 2015, 21:29, www.sourcewatch.org/index.php/Coal_power_technologies.

45. Philip Brasor, "Japan Spends Scant Energy on Renewables," *Japan Times*, January 6, 2018, www.japantimes.co.jp/news/2018/01/06/national/media-national/japan-spends-scant-energy-renewables/#.W6wJQi3MzQg.

Chapter 12. The New (Electric) Economy

1. Form 10-K, in *Amazon 2017 Annual Report* (Seattle: Amazon, 2018), 11, https://ir.aboutamazon.com/static-files/917130c5-e6bf-4790-a7bc-cc43ac7fb30a.

2. Paul R. La Monica, "Apple Reaches $1,000,000,000,000 Value," CNN, August 2, 2018, https://money.cnn.com/2018/08/02/investing/apple-one-trillion-market-value/index.html?iid=EL.

3. Lydia DePillis, "Amazon Is Now Worth $1,000,000,000,000," CNN, September 4, 2018, https://money.cnn.com/2018/09/04/technology/amazon-1-trillion/index.html.

4. Eric Jhonsa, "Amazon's Spending on the Cloud Is Growing, but Not Nearly as Fast as Facebook's," *TheStreet*, August 9, 2018, www.thestreet.com/technology/amazons-cloud-capital-spending-is-growing-but-not-as-fast-as-facebooks-14678361.

5. Metcalfe's Law, named for Robert Metcalfe, the inventor of Ethernet, says that the value of a telecommunications network grows by the square of the number of nodes on the network. A network with a single telephone is pretty much useless. But as telephones proliferate, the value of the phone network to individual users rises exponentially.

6. Saba Hamedy, "YouTube Just Hit a Huge Milestone," *Mashable*, February 27, 2017, https://mashable.com/2017/02/27/youtube-one-billion-hours-of-video-daily/#ReJAlTXFQSq4.

7. "Cisco Global Cloud Index: Forecast and Methodology, 2016–2021 White Paper," Cisco, November 19, 2018, www.cisco.com/c/en/us/solutions/collateral/service-provider/global-cloud-index-gci/white-paper-c11-738085.html. Like other major technology companies, Cisco is big electricity user. In 2017, Cisco used more than 1.6 terawatt-hours of electricity, which is nearly as much as the amount used by Apple. See, *2017 Corporate Social Responsibility Report* (San Jose, CA: Cisco, 2017), 102, www.cisco.com/c/dam/assets/csr/pdf/CSR-Report-2017.pdf.

8. Peter W. Huber and Mark P. Mills, *The Bottomless Well: The Twilight of Fuel, the Virtue of Waste, and Why We Will Never Run Out of Energy* (New York: Basic Books, 2005), 34–35.

9. Electricity data is from each company's corporate reports. Market cap data was obtained from Macrotrends and is for the last trading date of each year.

10. Farhad Manjoo, "Can Washington Stop Big Tech Companies? Don't Bet On It," *New York Times*, October 25, 2017, www.nytimes.com/2017/10/25/technology/regulating-tech-companies.html.

11. Rochelle Toplensky, "EU Fines Google €2.4bn Over Abuse of Search Dominance," *Financial Times*, June 27, 2017, www.ft.com/content/9554a8bc-5b12-11e7-b553-e2df1b0c3220.

12. Daniel Boffey, "Google Appeals Against EU's €2.4bn Fine Over Search Engine Results," *Guardian*, September 11, 2017, www.theguardian.com/technology/2017/sep/11/google-appeals-eu-fine-search-engine-results-shopping-service.

13. "Search Engine Market Share," desktop/laptop, Net Marketshare, www.netmarketshare.com/search-engine-market-share.aspx?qprid=4&qpcustomd=0.

14. Todd Haselton and Laura Feiner, "Apple's Cash Hoard Now at $225.4 Billion," CNBC, April 30, 2019, www.cnbc.com/2019/04/30/apple-now-has-225-billion-cash-on-hand.html. "GDP (Current US$)," World Bank, https://data.worldbank.org/indicator/NY.GDP.MKTP.CD?year_high_desc=true.

15. "Paradise Papers: Apple's Secret Tax Bolthole Revealed," BBC News, November 6, 2017, www.bbc.com/news/world-us-canada-41889787.

16. Rex Nutting, "Amazon Is Going to Kill More American Jobs than China Did," *MarketWatch*, March 15, 2017, www.marketwatch.com/story/amazon-is-going-to-kill-more-american-jobs-than-china-did-2017-01-19.

17. Hayley Peterson, "More than 7,000 Stores Are Closing in 2019 as the Retail Apocalypse Drags On—Here's the Full List," *Business Insider*, May 21, 2019, www.businessinsider.com/stores-closing-in-2019-list-2019-3.

18. David Ng'ang'a, "Here's Why You Can't Block Mark Zuckerberg on Facebook," *Business Insider*, September 6, 2017, www.businessinsider.com/heres-why-you-cant-block-mark-zuckerberg-on-facebook-2017-9.

19. Sebastian Huempfer, "Facebook Remains the Dominant Social Platform," *Social Media Today*, May 30, 2017, www.socialmediatoday.com/social-networks/facebook-remains-dominant-social-platform-infographic.

20. Karen Gilchrist, "Newspapers to Bid for Antitrust Exemption to Tackle Google and Facebook," CNBC, July 10, 2017, www.cnbc.com/2017/07/10/newspapers-bid-for-antitrust-exemption-to-tackle-google-and-facebook.html.

21. Roy Greenslade, "Why Facebook Is Public Enemy Number One for Newspapers, and Journalism," *Guardian*, September 20, 2016, www.theguardian.com/media/greenslade/2016/sep/20/why-facebook-is-public-enemy-number-one-for-newspapers-and-journalism.

22. Gilchrist, "Newspapers to Bid."

23. Brad Tuttle, "Amazon Just Became the Second Company to Reach $1 Trillion. Here's How Much Jeff Bezos Is Worth Now," *Time*, September 4, 2018, http://time.com/money/5386380/amazon-1-trillion-jeff-bezos-net-worth/.

24. Joel Kotkin, "Today's Tech Oligarchs Are Worse Than the Robber Barons," *Daily Beast*, August 11, 2016, www.thedailybeast.com/todays-tech-oligarchs-are-worse-than-the-robber-barons.

25. Lauren Thomas and Courtney Reagan, "Watch Out, Retailers. This Is Just How Big Amazon Is Becoming," CNBC, July 13, 2018, www.cnbc.com/2018/07/12/amazon-to-take-almost-50-percent-of-us-e-commerce-market-by-years-end.html.

26. Greg Ip, "The Antitrust Case Against Facebook, Google, and Amazon," *Wall Street Journal*, January 16, 2018, www.wsj.com/articles/the-antitrust-case-against-facebook-google-amazon-and-apple-1516121561.

27. Love the Sales, "Amazon's Prime Day Website Crash Costs Them Nearly $100 Million in Sales," news release, July 16, 2018, www.lovethesales.com/press/amazons-prime-day-website-crash-costs-them-nearly-100-million-in-sales.

28. Rich Miller, "Another Major Data Center for Prineville?" *Data Center Knowledge*, April 9, 2012, www.datacenterknowledge.com/archives/2012/04/09/another-major-data-center-for-prineville/.

29. "World's Largest Data Center: 350 E. Cermak," *Data Center Knowledge*, April 13, 2010, www.datacenterknowledge.com/special-report-the-worlds-largest-data-centers/worlds-largest-data-center-350-e-cermak/.

30. Michael Kassner, "What to Look For in a Data Center Backup Generator," *TechRepublic*, October 27, 2014, www.techrepublic.com/article/what-to-look-for-in-a-data-center-backup-generator/.

31. *Diesel Generator Set* (Deerfield, IL: Caterpillar, 2013), http://s7d2.scene7.com/is/content/Caterpillar/LEHE0341-02.

32. Amazon doesn't report its electricity use. Therefore, I benchmarked its usage to match that of Alphabet, which has the highest electricity use of the four companies that

report their electricity consumption. But Amazon has far more computing power and data centers than Alphabet. Thus, my estimate of the Giant Five's consumption numbers should be seen as a minimum.

33. In 2015, Ireland consumed about 24 terawatt-hours of electricity. See, "Country Comparison: Electricity—Consumption," CIA World Factbook, www.cia.gov/library/publications/the-world-factbook/rankorder/2233rank.html.

34. Arman Shehabi et al., *United States Data Center Energy Usage Report* (Berkeley: Lawrence Berkeley National Laboratory, 2016), https://eta.lbl.gov/publications/united-states-data-center-energy.

35. Company data, calculations by author.

36. Michael Kassner, "Microsoft's $750 Million Data Center Investment in Wyoming," *TechRepublic*, March 7, 2015, www.techrepublic.com/article/microsofts-750-million-data-center-investment-in-wyoming/.

37. Herman K. Trabish, "How Microsoft and a Wyoming Utility Designed a Data Center Tariff That Works for Everyone," *Utility Dive*, December 20, 2016, www.utilitydive.com/news/how-microsoft-and-a-wyoming-utility-designed-a-data-center-tariff-that-work/430807/.

38. John Mulligan, "Microsoft Forced to Build Dublin Power Station to Service Huge Data Centre," *Irish Independent*, September 16, 2017, www.independent.ie/business/microsoft-forced-to-build-dublin-power-station-to-service-huge-data-centre-36137561.html.

39. *Google Environmental Report 2018* (Mountain View, CA: Google, 2018), 48, https://storage.googleapis.com/gweb-sustainability.appspot.com/pdf/Google_2018-Environmental-Report.pdf.

40. "Country Comparison: Electricity—Consumption," CIA World Factbook, www.cia.gov/library/publications/the-world-factbook/rankorder/2233rank.html.

41. In 2018, WikiLeaks leaked an internal Amazon document that says that, as of October 2015, the company was operating 116 data centers around the world. See, Yevgeniy Sverdlik, "WikiLeaks Publishes What It Says Is a List of Amazon Data Centers," *Data Center Knowledge*, October 12, 2018, www.datacenterknowledge.com/amazon/wikileaks-publishes-what-it-says-list-amazon-data-centers.

42. Justine Brown, "AWS Commands Public IaaS Market with 45% of Q3 Revenue," *CIO Dive*, November 1, 2016, www.ciodive.com/news/aws-commands-public-iaas-market-with-45-of-q3-revenue/429442/.

43. Janakiram MSV, "How AWS Has Turned into an Unstoppable Juggernaut—an Analysis from re:Invent 2018," *Forbes*, December 2, 2018, www.forbes.com/sites/janakirammsv/2018/12/02/how-aws-has-turned-into-an-unstoppable-juggernaut-an-analysis-from-reinvent-2018/#2d490f3669be.

44. Benjamin Wootton, "Who's Using Amazon Web Services?" Contino, January 26, 2017, www.contino.io/insights/whos-using-aws. See also, Ron Miller, "NFL Teams with AWS on Statistics Package Driven by Machine Learning," *Techcrunch*, November29,2017,https://techcrunch.com/2017/11/29/nfl-teams-with-aws-on-statistics-package-driven-by-machine-learning/.

45. Naomi Nix, "CIA Tech Official Calls Amazon Project 'Transformational,'" Bloomberg, June 20, 2018, www.bloomberg.com/news/articles/2018-06-20/cia-tech-official-calls-amazon-cloud-project-transformational.

46. AWS, "Announcing the New AWS Secret Region," news release, November 20, 2017, https://aws.amazon.com/blogs/publicsector/announcing-the-new-aws-secret-region/.

47. Naomi Nix, "Inside the Nasty Battle to Stop Amazon from Winning the Pentagon's Cloud Contract," Bloomberg, December 20, 2018, www.bloomberg.com/news/features/2018-12-20/tech-giants-fight-over-10-billion-pentagon-cloud-contract.

48. "Amazon Wind and Solar Farms," Amazon, www.amazon.com/p/feature/e9gomtbrh5qk4yp.

49. Brad Smith, "With Our Latest Energy Deal, Microsoft's Cheyenne Datacenter Will Now Be Powered Solely by Wind Energy, Keeping Us on Course to Build a Greener, More Responsible Cloud," *Microsoft on the Issues* (blog), November 14, 2016, https://blogs.microsoft.com/on-the-issues/2016/11/14/latest-energy-deal-microsofts-cheyenne-datacenter-will-now-powered-entirely-wind-energy-keeping-us-course-build-greener-responsible-cloud/.

50. "Greening the Grid: How Google Buys Renewable Energy," Google, Environment, https://environment.google/projects/ppa/.

51. Casey Anderson, "New Apple Headquarters Sets Records in Solar and Green Building," *Renewable Energy World*, March 3, 2017, www.renewableenergyworld.com/ugc/articles/2017/02/28/new-apple-headquarters-sets-records-in-solar-and-green-building.html.

52. Hamilton estimated the power density in Apple's data center was 200 watts per square foot, which is equal to 2,160 watts per square meter. See, James Hamilton, "I Love Solar Power But . . . ," *Perspectives* (blog), March 17, 2012, https://perspectives.mvdirona.com/2012/03/i-love-solar-power-but/.

53. Mark P. Mills, "Energy and the Information Infrastructure Part 1: Bitcoins & Behemoth Datacenters," *RealClearEnergy*, September 19, 2018, www.realclearenergy.org/articles/2018/09/19/energy_and_the_information_infrastructure_part_1__bitcoins__behemoth_datacenters_110339.html.

54. Gary Cook and Elizabeth Jardim, *Clicking Clean Virginia: The Dirty Energy Powering Data Center Alley* (Washington, DC: Greenpeace, 2019), 11, www.greenpeace.org/usa/wp-content/uploads/2019/02/Greenpeace-Click-Clean-Virginia-2019.pdf. See also, Gary Cook, *Clicking Clean: Who Is Winning the Race to Build a Green Internet?* (Washington, DC: Greenpeace, 2017), 86–87, https://storage.googleapis.com/planet4-international-stateless/2017/01/35f0ac1a-clickclean2016-hires.pdf. According to the 2019 Greenpeace report, by the end of 2018, Amazon had 1,686 megawatts of capacity at its Virginia data centers alone. The 2017 report shows that in 2016 the company had about 542 megawatts of capacity at its other US data centers and another 886 megawatts of generating capacity in its overseas operations. Given AWS's rapid growth over the past few years, I estimated that Amazon's domestic data center capacity outside of Virginia doubled in size between 2016 and the end of 2018. Therefore, 1,686 + 542 + 542 + 886 = 3,656 megawatts.

55. Business Wire, "Amazon Announces Three New Renewable Energy Projects to Support AWS Global Infrastructure," news release, April 8, 2019, www.apnews.com/Business%20Wire/f69d0796a2fc4ee4838de67c0ebf7f5a.

56. This estimate counts all of the on-site generators at their data centers, as well as all of their owned or contracted renewable-energy capacity.

57. "Hoover Dam," Bureau of Reclamation, last updated August 1, 2018, www.usbr.gov/lc/hooverdam/faqs/powerfaq.html.

58. *Power System: Los Angeles' Power Generation and Transmission* (Los Angeles: Los Angeles Department of Water and Power, 2015), http://d3n8a8pro7vhmx.cloudfront.net/themes/5595dcbfebad640bf5000001/attachments/original/1430387153/LADWP_Power_System_Fact_Sheet.pdf?143038715.

59. "Power Plants," Who We Are, Austin Energy, last updated March 7, 2019, https://austinenergy.com/ae/about/company-profile/electric-system/power-plants/.

60. See, for instance, Microsoft Pay, www.microsoft.com/en-us/payments; Apple Pay, www.apple.com/apple-pay/; Google Pay, https://pay.google.com/about/#friends; Amazon Pay, https://pay.amazon.com/us.

Chapter 13. Electrified Cash

1. AnnaMaria Andriotis, "Visa to Card Customers: Lose the Signature," *Wall Street Journal*, January 12, 2018, www.wsj.com/articles/visas-new-tune-on-card-signatures-dont-sign-on-the-dotted-line-1515776401?mg=prod/accounts-wsj.

2. *Annual Report 2015* (Foster City, CA: Visa, 2015), http://s1.q4cdn.com/050606653/files/doc_financials/annual/VISA-2015-Annual-Report.pdf.

3. Michael Fitzgerald, "How Visa Protects Your Data," *Fast Company*, October 19, 2011, www.fastcompany.com/1784751/how-visa-protects-your-data.

4. Annual Report 2018 (Foster City, CA: Visa, 2018), https://s1.q4cdn.com/050606653/files/doc_financials/annual/2018/Visa-2018-Annual-Report-FINAL.pdf.

5. The data center covers about 140,000 square feet (13,000 square meters). See, Tony Kontzer, "Inside Visa's Data Center," Network Computing, May 29, 2013, www.networkcomputing.com/networking/inside-visas-data-center/1599285558. The average Walmart store covers 134,000 square feet. See, "Profile: Walmart Inc (WMT.N)," Reuters, www.reuters.com/finance/stocks/companyProfile?symbol=WMT.N.

6. Fitzgerald, "How Visa Protects Your Data." The Visa facility has seven pods, each with 8 megawatts of standby generation capacity.

7. Kontzer, "Inside Visa's Data Center."

8. *Visa Inc. at a Glance* (Foster City, CA: Visa, 2015), https://usa.visa.com/dam/VCOM/download/corporate/media/visa-fact-sheet-Jun2015.pdf.

9. Zhai Yun Tan, "Is It Time to Write Off Checks?" NPR, March 3, 2016, www.npr.org/2016/03/03/468890515/is-it-time-to-write-off-checks.

10. "Fedwire Funds Service—Annual," Board of Governors of the Federal Reserve System, last updated February 7, 2019, www.federalreserve.gov/paymentsystems/fedfunds_ann.htm.

11. Zion Market Research, "Growth of Global Mobile Wallet Market Share to Cross over USD 3,142.17 Billion by 2022: Zion Market Research," news release, March 16, 2018, https://globenewswire.com/news-release/2018/05/16/1507295/0/en/Growth-of-Global-Mobile-Wallet-Market-Share-to-Cross-Over-USD-3-142-17-billion-by-2022-Zion-Market-Research.html.

12. Thomas McGath, "M-PESA: How Kenya Revolutionized Digital Payments," *N26 Magazine*, April 9, 2018, https://mag.n26.com/m-pesa-how-kenya-revolutionized-mobile-payments-56786bc09ef.

13. Rishi Iyengar, "50 Days of Pain: What Happened When India Trashed Its Cash," CNN, January 4, 2017, https://money.cnn.com/2017/01/04/news/india/india-cash-crisis-rupee/index.html.

14. Nupur Anand, "The Narendra Modi Government Tried to Kill Cash. It Ended Up Strangling the Economy," *Quartz*, November 8, 2017, https://qz.com/india/1123463/demonetisation-the-narendra-modi-government-tried-to-kill-cash-it-ended-up-strangling-the-economy/.

15. GlobalData, "Mobile Wallet Is Gradually Displacing Cash in India, says GlobalData," news release, February 22, 2018, www.globaldata.com/mobile-wallet-gradually-displacing-cash-india-says-globaldata/.

16. Carter Graydon, "What Is Cryptocurrency?" CCN, September 16, 2014, www.cryptocoinsnews.com/cryptocurrency/.

17. Paul Vigna and Michael Casey, *The Age of Cryptocurrency: How Bitcoin and Digital Money Are Challenging the Global Economic Order* (New York: St. Martin's Press, 2015).

18. Patrick Howell O'Neill, "The Curious Case of the Missing Mt. Gox Bitcoin Fortune," *CyberScoop*, June 21, 2017, www.cyberscoop.com/bitcoin-mt-gox-chainalysis-elliptic/.

19. "Warning: Enigma Hacked; Over $470,000 in Ethereum Stolen So Far," *Hacker News*, August 20, 2017, http://thehackernews.com/2017/08/enigma-cryptocurrency-hack.html.

20. Simon Denyer, "The Bizarre World of Bitcoin 'Mining' Finds a New Home in Tibet," *Washington Post*, September 12, 2016, www.washingtonpost.com/world/asia_pacific/in-chinas-tibetan-highlands-the-bizarre-world-of-bitcoin-mining-finds-a-new-home/2016/09/12/7729cbea-657e-11e6-b4d8-33e931b5a26d_story.html.

21. "Rising Power Prices in Iceland," Askja Energy, March 17, 2017, https://askjaenergy.com/2017/03/17/rising-power-prices-in-iceland/.

22. EIA, "Table 5.3. Average Price of Electricity to Ultimate Customers," in *Data Power Monthly* (April 2019), www.eia.gov/electricity/monthly/epm_table_grapher.php?t=epmt_5_3.

23. "Electricity Prices for Industrial Consumers, Second Half 2016," Eurostat, June 28, 2017, http://ec.europa.eu/eurostat/statistics-explained/index.php/File:Electricity_prices_for_industrial_consumers,_second_half_2016_(EUR_per_kWh)_YB17.png. Euro prices were converted to dollars on November 5, 2017.

24. Invest in Iceland, http://datacenter.invest.is.

25. Wikipedia, s.v. "Economy of Iceland," last updated April 29, 2019, 23:11, https://en.wikipedia.org/wiki/Economy_of_Iceland.

26. "Electricity Generation," Orkustofnun: National Energy Authority, www.nea.is/geothermal/electricity-generation/.

27. Mark Coppock, "The World's Cryptocurrency Mining Uses More Electricity Than Iceland," *Digital Trends*, July 7, 2017, www.digitaltrends.com/computing/bitcoin-ethereum-mining-use-significant-electrical-power/.

28. Carolyn Beeler, "Bitcoin's Sky-Rocketing Energy Use Is a Viral Story. We Checked the Math," PRI, December 20, 2017, www.pri.org/stories/2017-12-20/bitcoins-sky-rocketing-energy-use-viral-story-we-checked-math. For New Zealand data, see "Country Comparison: Electricity—Consumption," CIA World Factbook, www.cia.gov/library/publications/the-world-factbook/rankorder/2233rank.html.

29. "Bitcoin (BTC)," CoinGecko, www.coingecko.com/en/price_charts/bitcoin/usd.

30. "Ethereum (ETH)," CoinGecko, www.coingecko.com/en/price_charts/ethereum/usd.

31. Kate Rooney, "There's Been a Mysterious Surge in $100 Bills in Circulation, Possibly Linked to Global Corruption," CNBC, February 27, 2019, www.cnbc.com/2019/02/27/theres-been-a-mysterious-surge-in-100-bills-in-circulation-possibly-linked-to-global-corruption.html.

Chapter 14. Watts into Weed

1. His three-bedroom house was using about 7,000 kilowatt-hours per month. An average home in Colorado consumes less than 700 kilowatt-hours per month. See, *Household Energy Use in Colorado* (Washington, DC: EIA, 2009), www.eia.gov/consumption/residential/reports/2009/state_briefs/pdf/co.pdf.

2. "3 Types of Marijuana Grow Lights. Which Will Yield the Biggest Crop?" How to Marijuana, www.how-to-marijuana.com/marijuana-grow-lights.html.

3. Some growers claim they can produce three outdoor crops per year by utilizing greenhouses for part of the grow cycle. See, Alchimia Grow Shop, "Off-Season Marijuana Crops Outdoors," *Alchimia* (blog), February 23, 2015, www.alchimiaweb.com/blogen/off-season-marijuana-crops-outdoors/.

4. Evan Mills, "Not-so-Green Greenhouses for Cannabis Hyper-Cultivation," February 26, 2018, 5, https://docs.google.com/viewer?a=v&pid=sites&srcid=ZGVmYXVsdGRvbWFpbnxtaWxsc2VuZXJneWFzc29jaWF0ZXN8Z3g6NDA4MWQyZGNlNjUwMWE5Ng.

5. Gina S. Warren, "Regulating Pot to Save the Polar Bear: Energy and Climate Impacts of the Marijuana Industry," *Columbia Journal of Environmental Law* 40, no. 3 (2015): 403, www.columbiaenvironmentallaw.org/regulating-pot-to-save-the-polar-bear-energy-and-climate-impacts-of-the-marijuana-industry/.

6. In 2016, *Cannabis Business Times* reported that growers were seeing production levels of 1.6 grams per watt. There are roughly 454 grams per pound. Therefore, they are claiming in excess of three pounds per 1,000 watts. See, "Measuring Yield," *Cannabis Business Times*, October 6, 2016, www.cannabisbusinesstimes.com/article/measuring-yield/.

7. Mary Pols, "Did You Know Marijuana Is America's Most Energy-Intensive Crop?" *Portland Press Herald*, January 15, 2017, www.pressherald.com/2017/01/15/whats-the-most-energy-intensive-crop-in-america/.

8. Evan Mills, "The Carbon Footprint of Indoor *Cannabis* Production," *Energy Policy* 46 (2012): 59, https://s3.amazonaws.com/dive_static/diveimages/cannabis-carbon-footprint.pdf.

9. Charles Fishman, *Can EVs, Pot, and Data Save U.S. Electricity Demand?* (Chicago: Morningstar, 2018), www.researchpool.com/provider/morningstar/consolidated-edison-inc-ed-can-evs-pot-and-data-save-us-electricity-demand.

10. Peter Maloney, "Data Centers, EVs and Cannabis Poised to Boost Demand," American Public Power Association, December 10, 2018, www.publicpower.org/periodical/article/data-centers-evs-and-cannabis-poised-boost-demand.

11. Ballotpedia, s.v. "Colorado Marijuana Legalization Initiative, Amendment 64 (2012)," https://ballotpedia.org/Colorado_Marijuana_Legalization_Initiative,_Amendment_64_(2012).

12. Pers. comm. with Emily Backus, May 21, 2018.

13. Grace Hood, "Nearly 4 Percent of Denver's Electricity Is Now Devoted to Marijuana," Colorado Public Radio, February 19, 2018, www.cpr.org/news/story/nearly-4-percent-of-denver-s-electricity-is-now-devoted-to-marijuana.

14. "Country Comparison: Electricity—Consumption," CIA World Factbook, www.cia.gov/library/publications/the-world-factbook/rankorder/2233rank.html.

15. "State Marijuana Laws in 2018 Map," Governing, last updated November 7, 2018, www.governing.com/gov-data/safety-justice/state-marijuana-laws-map-medical-recreational.html.

16. Chris Morris, "Colorado Was the First State to Legalize Marijuana. Now Its Governor Won't Rule Out Recriminalizing It," *Yahoo! Finance*, April 20, 2018, https://finance.yahoo.com/news/colorado-first-state-legalize-marijuana-162942435.html.

17. Monterey Bud, "Total Marijuana Sales Topped $53.3 Billion for 2016," Marijuana.com, January 18, 2017, www.marijuana.com/news/2017/01/total-marijuana-sales-topped-53-3-billion-for-2016/.

18. Chris Bennett, "Marijuana Farming Is Now for US Agriculture," *AgWeb*, January 8, 2018, www.agweb.com/mobile/article/marijuana-farming-is-now-for-us-agriculture-naa-chris-bennett/.

19. "Table 2.2. Sales and Direct Use of Electricity to Ultimate Customers," EIA, www.eia.gov/electricity/annual/html/epa_02_02.html.

20. David Ferris, "Utilities Struggle to Control Appetites in Energy-Hungry Marijuana Industry," *E&E News*, August 8, 2014, www.eenews.net/stories/1060004230.

21. In 2015, Peru's electricity use was about 41 terawatt-hours. See, "Country Comparison: Electricity—Consumption," CIA World Factbook, www.cia.gov/library/publications/the-world-factbook/rankorder/2233rank.html.

22. "Country Comparison: Electricity—Consumption," CIA World Factbook, www.cia.gov/LIBRARY/publications/the-world-factbook/rankorder/2233rank.html.

23. "Colorado," Institute for Energy Research, www.instituteforenergyresearch.org/states/colorado/.

24. "6-16-8.—Requirements Related to Operation of Recreational Marijuana Businesses," City of Boulder Municipal Code, April 17, 2019, https://library.municode.com/co/boulder/codes/municipal_code?nodeId=TIT6HESASA_CH16REMA_6-16-8REREOPREMABU.

25. Robert Walton, "Boulder Levies Extra Electric Charge on Marijuana Growers," *Utility Dive*, November 12, 2014, www.utilitydive.com/news/boulder-levies-extra-electric-charge-on-marijuana-growers/332012/. See also, "Land Use," Boulder County, www.bouldercounty.org/departments/land-use/.

26. "935 CMR 500.000 Adult Use of Marijuana," Massachusetts Cannabis Control Commission (2018), 51, https://mass-cannabis-control.com/wp-content/uploads/2018/03/SEC-OFFICIAL_935cmr500.pdf. Note that the ordinance puts the power density limits at 50 watts per square foot and 36 watts per square foot. There are 10.7 square feet per square meter. Therefore, the density limits are 535 and 385.2 watts per square meter, respectively. For date of ordinance, see "Massachusetts Cannabis Regulators: 36 Watts Will Be the Limit for Cannabis Cultivators," Margolin & Lawrence, May 8, 2018, http://blog.margolinlawrence.com/massachusetts-cannabis-update-state-imposes-strict-energy-limits-on-marijuana-cultivators.

27. Dan Adams, "Growing Concern: Marijuana Rules Could Mean Lower Quality Product," *Boston Globe*, March 8, 2018, www.bostonglobe.com/metro/2018/03/08/skeptical-led-lights-marijuana-growers-decry-lighting-efficiency-rule/AC9qm18nmfM9rgaUrl26QK/story.html.

28. Katie Leichliter et al., *SevenLeaves 2017: Indoor Horticulture Lighting Study* (Sacramento, CA: Sacramento Municipal Utility District, 2018), 5, www.smud.org/-/media/Documents/Business-Solutions-and-Rebates/Advanced-Tech-Solutions/LED-Reports/Seven-Leaves-Indoor-Horticulture-LED-Study-Final.ashx.

29. The hospital calculations are based on EIA data. See, "Table C14. Electricity Consumption and Expenditure Intensities, 2012," EIA, May 2016, www.eia.gov/consumption/commercial/data/2012/c&e/cfm/c14.php. An in-patient hospital consumes 31 kWh

$/ft^2/yr$. There are 8,760 hours per year. Thus, 31,000 $Wh/ft^2/yr \div 8760$ hrs = 3.5 W/ft^2. There are 10.7 ft^2 per m^2. Therefore, hospital power density is 37.4 W/m^2.

The average residence in the United States has 2,100 square feet or 196 square meters. See, Bridget Mallon, "How Big Is the Average House Around the World?" *Elle Decor*, August 26, 2015, www.elledecor.com/life-culture/fun-at-home/news/a7654/house-sizes-around-the-world/. The average US residence consumes about 11,000 kilowatt-hours per year. See, "How Much Electricity Does an American Home Use?" Frequently Asked Questions, EIA, last updated October 26, 2018, www.eia.gov/tools/faqs/faq.php?id=97&t=3.

Calculations: 11,000 $kWh/yr \div 196m^2$ = 11 million $Wh/yr \div 196\ m^2$

11 million Wh $\div$ 196 m^2 = 56,122 $Wh/m^2/y$

56,122 $Wh/m^2/y \div 8760\ h/y = 6.4\ W/m^2$

30. Mills, "The Carbon Footprint," 59. See also, "Energy Up in Smoke: The Carbon Footprint of Indoor *Cannabis* Production," Energy Associates, https://sites.google.com/site/millsenergyassociates/topics/energy-efficiency/energy-up-in-smoke.

31. Schneider Electric puts the power density of an average data center at 1,722 watts per square meter. See, Neil Rasmussen, *Calculating Space and Power Density Requirements for Data Centers* (South Kingston, RI: APC by Schneider Electric, 2013), www.apc.com/salestools/NRAN-8FL6LW/NRAN-8FL6LW_R0_EN.pdf.

32. Bud, "Total Marijuana Sales."

33. Brianna Calix, "Merced Police Shut Down Indoor Marijuana Grow House," *Merced Sun-Star*, April 26, 2017, www.mercedsunstar.com/news/article146912004.html.

34. AFP, "Police Bust Gang That Used Bank Warehouses in Catalonia to Grow Huge Amounts of Marijuana," *Local ES*, December 18, 2017, www.thelocal.es/20171218/spain-breaks-ring-that-used-bank-warehouses-to-grow-marijuana.

35. Ben Harvy, "Hydroponic Drug Room at Sheidow Park Goes Up in Flames," News.com.au, February 17, 2018, www.news.com.au/national/south-australia/hydroponic-drug-room-at-sheidow-park-goes-up-in-flames/news-story/c3c04225650bf55e8cf3b5f40b4f4cc5.

36. Alicia Stice, "Investigators Find Nearly 200 Pounds of Marijuana at Illegal Pot Grow," *Coloradoan*, November 3, 2017, www.coloradoan.com/story/news/2017/11/03/berthoud-man-arrested-connection-possible-illegal-pot-grow/830187001/.

37. Rob McMillan, "San Bernardino Marijuana Grow Could Be Largest Discovered in City, Police Say," ABC7, December 13, 2017, http://abc7.com/san-bernardino-marijuana-grow-could-be-largest-found-in-city/2778188/.

38. Beatriz E. Valenzuela, "Tens of Thousands of Marijuana Plants Seized in San Bernardino," *Cannifornian*, December 15, 2017, www.thecannifornian.com/cannabis-business/cultivation/tens-thousands-marijuana-plants-seized-san-bernardino-raid/.

39. Robert Walton, "Marijuana Grow Houses Trigger 7 Summer Outages for Pacific Power," *Utility Dive*, November 6, 2015, www.utilitydive.com/news/marijuana-grow-houses-trigger-7-summer-outages-for-pacific-power/408741/.

Chapter 15. The Blackout Will Not Be Televised

1. Ted Koppel, *Lights Out: A Cyberattack, a Nation Unprepared, Surviving the Aftermath* (New York: Crown, 2015), 18–19.

2. Richard A. Serrano and Evan Halper, "Sophisticated but Low-Tech Power Grid Attack Baffles Authorities," *Los Angeles Times*, February 11, 2014, www.latimes.com /nation/la-na-grid-attack-20140211-story.html.

3. Rebecca Smith, "U.S. Risks National Blackout from Small-Scale Attack," *Wall Street Journal*, March 12, 2014, www.wsj.com/articles/u-s-risks-national-blackout-from -small-scale-attack-1394664965.

4. Sherri Fink, "Class-Action Suit Filed After Katrina Hospital Deaths Settled for $25 Million," ProPublica, July 21, 2011, www.propublica.org/article/class -action-suit-filed-after-katrina-hospital-deaths-settled-for-25-millio.

5. Sheri Fink, interview by Steve Adubato, *One on One*, NJTV, July 18, 2014, YouTube, www.youtube.com/watch?v=AXgM3DTyXFY.

6. Sherri Fink, "The Deadly Choices at Memorial," ProPublica, August 27, 2009, www.propublica.org/article/the-deadly-choices-at-memorial-826.

7. "Dr. Anna Pou," Sheri Fink website, www.sherifink.net/dr-anna-pou/.

8. Andrew V. Pestano and Jamie Guirola, "Deposition Reveals Details of Hollywood Nursing Home Heat Exposure Tragedy," NBC6, January 8, 2018, www.nbcmiami.com/ news/local/Deposition-Reveals-Details-of-Hollywood-Nursing-Home-Heat-Exposure -Tragedy-468381873.html. See also, Michael Nedelman, "Husband and Wife Among 14 Dead After Florida Nursing Home Lost A/C," CNN, October 9, 2017, www.cnn .com/2017/10/09/health/florida-irma-nursing-home-deaths-wife/index.html.

9. Pestano and Guirola, "Deposition Reveals Details."

10. *Ascertainment of the Estimated Excess Mortality From Hurricane María in Puerto Rico* (Washington, DC: Milken Institute School of Public Health, 2018), 10, https://public health.gwu.edu/sites/default/files/downloads/projects/PRstudy/Acertainment %20of%20the%20Estimated%20Excess%20Mortality%20from%20Hurricane%20 Maria%20in%20Puerto%20Rico.pdf.

11. Associated Press, *Quartz*, and Puerto Rico's Center for Investigative Journalism, "Hurricane Maria's Victims," database, https://hurricanemariasdead.com/database.html.

12. David Gilbert, "It Could Happen Here: Ukraine's Power Station Hack Is a Stark Warning to Other Countries," *Vice*, January 15, 2017, https://news.vice.com/story /ukraines-power-station-hack-is-a-stark-warning-to-the-rest-of-the-world.

13. AEP official Kip Fox provided those figures during a presentation at UT Energy Week in Austin on February 4, 2019.

14. William R. Forstchen, *One Second After* (New York: Tor, 2009).

15. *Report: USSR Nuclear EMP Upper Atmosphere Kazakhstan Test 184* (Washington, DC: Electric Infrastructure Security Council, undated), www.eiscouncil.org/APP_Data /upload/a4ce4b06-1a77-44d-83eb-842bb2a56fc6.pdf.

16. John S. Foster Jr. et al., *Report of the Commission to Assess the Threat to the United States from Electromagnetic Pulse (EMP) Attack* (Washington, DC: EMP Commission, 2004), 1, 2, www.empcommission.org/docs/empc_exec_rpt.pdf.

17. *Assessing the Threat from an EMP Attack: Executive Report* (Washington, DC: EMP Commission, 2017), 4, www.dtic.mil/dtic/tr/fulltext/u2/1051492.pdf.

18. Author interview with Sheehan, August 3, 2017.

19. On May 2, 1999, and again five days later, NATO forces dropped blackout bombs on Serbia, causing an extended blackout in Belgrade. After the attack, a NATO spokesman said the attack on the grid was part of the wider effort to pressure then Serbian president Slobodan Milosevic to withdraw his military forces from Kosovo. "The fact that lights went out across 70 percent of the country shows that NATO has its finger on the

light switch now," said the spokesman. See, Michael R. Gordon, "Crisis in the Balkans: The Overview; NATO Air Attacks on Power Plants Pass a Threshold," *New York Times*, May 4, 1999, www.nytimes.com/1999/05/04/world/crisis-balkans-overview-nato-air-attacks-power-plants-pass-threshold.html.

20. Liz Sly, "The Kalashnikov Assault Rifle Changed the World. Now There's a Kalashnikov Kamikaze Drone," *Washington Post*, February 23, 2019, www.washingtonpost.com/world/2019/02/23/kalashnikov-assault-rifle-changed-world-now-theres-kalashnikov-kamikaze-drone/.

21. Eugene K. Chow, "Forget Hackers: Squirrels Are a Bigger Threat to America's Power Grid," *Week*, January 28, 2014, http://theweek.com/articles/452311/forget-hackers-squirrels-are-bigger-threat-americas-power-grid.

22. Jon Mooallem, "Squirrel Power!" *New York Times*, August 31, 2013, www.nytimes.com/2013/09/01/opinion/sunday/squirrel-power.html.

23. Kayla Webley, "Hurricane Sandy by the Numbers: A Superstorm's Statistics, One Month Later," *Time*, November 26, 2012, http://nation.time.com/2012/11/26/hurricane-sandy-one-month-later/.

24. Sten Odenwald, "What Are Solar Storms and How Do They Affect the Earth?" Ask the Space Scientist, NASA, https://image.gsfc.nasa.gov/poetry/ask/a10624.html.

25. John Hruska, "A Massive Solar Storm Could Cost the US Economy $40 Billion per Day," *ExtremeTech*, January 23, 2017, www.extremetech.com/extreme/243255-massive-solar-storm-cost-us-economy-40-billion-per-day.

26. Deborah Byrd, "$40 Billion a Day for Solar Super-Storms," *EarthSky*, January 20, 2017, http://earthsky.org/earth/40-billion-a-day-for-solar-super-storms.

27. Electric Power Research Institute, *Considerations for a Power Transformer Emergency Spare Strategy for the Electric Utility Industry* (Washington, DC: US Department of Homeland Security Science and Technology Directorate, 2014), 13, www.dhs.gov/sites/default/files/publications/RecX%20-%20Emergency%20Spare%20Transformer%20Strategy-508.pdf.

28. Department of Energy, Office of Electricity Delivery and Energy Reliability, www.oe.netl.doe.gov/OE417_annual_summary.aspx.

29. "KeyBanc Industrial Conference" (Generac investor presentation, May 31, 2017). For the most recent presentation, see "Investor Presentations," Generac, http://investors.generac.com/phoenix.zhtml?c=232690&p=irol-presentations.

Chapter 16. The Terawatt Challenge

1. "Richard E. Smalley: Facts," Nobel Prize, www.nobelprize.org/prizes/chemistry/1996/smalley/facts/.

2. Richard E. Smalley, "Future Global Energy Prosperity: The Terawatt Challenge," *MRS Bulletin* 30 (June 2005), https://cohesion.rice.edu/NaturalSciences/Smalley/emplibrary/120204%20MRS%20Boston.pdf.

3. *Global Energy and CO2 Status Report* (Paris: IEA, 2019), www.iea.org/geco/.

4. In 2016, global installed capacity was about 6,700 gigawatts. See, *World Energy Outlook 2017* (Paris: IEA, 2017), 650.

5. Alison Stewart, "Is Density Our Destiny?" June 11, 2012, in *TED Radio Hour*, podcast, www.npr.org/templates/transcript/transcript.php?storyId=154801764.

6. Smil, *Power Density*, 163.

7. UN Department of Economic and Social Affairs, "World Population Projected to Reach 9.7 Billion by 2050," news release, July 29, 2015, www.un.org/en/development/desa/news/population/2015-report.html.

8. Michael Greenstone, "India's Air-Conditioning and Climate Change Quandary," *New York Times*, October 26, 2016, www.nytimes.com/2016/10/27/upshot/indias-air-conditioning-and-climate-change-quandary.html.

9. *Consumer Durable Sector—Air Conditioners* (Mumbai: HDFC Bank Investment Advisory Group, 2017), www.hdfcbank.com/assets/pdf/privatebanking/Sector_Update_Consumer_Durable_Air_Conditioners_April_2017.pdf.

10. *The Future of Cooling* (Paris: IEA, 2018), 11, https://webstore.iea.org/the-future-of-cooling.

11. IEA, "Air Conditioning Use Emerges as One of the Key Drivers of Global Electricity-Demand Growth," news release, May 15, 2018, www.iea.org/newsroom/news/2018/may/air-conditioning-use-emerges-as-one-of-the-key-drivers-of-global-electricity-dema.html.

12. *The Future of Cooling.*

13. *BP Statistical Review of World Energy 2018.* In 2018, China used about 6,500 TWh.

14. Peter Green, "Everything You Need to Know About Water in 500 Words or Less," *Quartz*, February 5, 2019, https://qz.com/1538507/everything-you-need-to-know-about-water-in-500-words-or-less/.

15. "Energy Efficiency for Water Utilities," Sustainable Water Infrastructure, Environmental Protection Agency, www.epa.gov/sustainable-water-infrastructure/energy-efficiency-water-utilities.

16. Hexa Research, "Water Desalination Market Size Worth USD 26.81 Billion by 2025: Hexa Research," news release, August 29, 2017, www.prnewswire.com/news-releases/water-desalination-market-size-worth-usd-2681-billion-by-2025-hexa-research-642089153.html. Hexa estimated the global desalination market was $13.3 billion in 2016. See, *Water Desalination Market Size and Forecast, by Technology (Reverse Osmosis, Multi-Stage Filtration, Multi-Effect Distillation), by Source (Seawater, Brackish Water, Wastewater), and Trend Analysis, 2014–2025* (Felton, CA: Hexa Research, 2017), www.hexaresearch.com/research-report/water-desalination-market.

17. *Global Water Desalination Market and Its Related Technologies (Reverse Osmosis, Multi-stage Flash Distillation, Multi Effect Distillation, Hybrid Electrodialysis), Its Source (Seawater, Brackish Water), Regional Trends and Forecast 2018 to 2025* (Dallas: Adroit Market Research, 2018), www.adroitmarketresearch.com/press-release/global-water-desalination-market-size-and-forecast-2018-2025.

18. Data from Carlsbad Desalination Plant, www.carlsbaddesal.com.

19. Paul Rogers, "Nation's Largest Ocean Desalination Plant Goes Up Near San Diego; Future of the California Coast?" *Mercury News*, May 29, 2014, www.mercurynews.com/2014/05/29/nations-largest-ocean-desalination-plant-goes-up-near-san-diego-future-of-the-california-coast/.

20. *BP Energy Outlook: 2018 Edition* (London: BP, 2018), data pack, 112.

21. Sammy Roth, "Unprecedented Heat in the Southwest Shatters Energy Use Records," *USA Today*, June 30, 2017, www.usatoday.com/story/news/nation-now/2017/06/30/unprecedented-heat-southwest-shatters-energy-use-records/440423001/.

22. Christopher Helman, "Natural Gas Demand Hits Record as Cold Bomb Targets Northeast," *Forbes*, January 3, 2018, www.forbes.com/sites/christopherhelman/2018/01/03/natural-gas-demand-hits-record-as-cold-bomb-targets-northeast/#64823318aacd.

23. "Country Comparison: Electricity—Installed Generating Capacity," CIA World Factbook, www.cia.gov/library/publications/the-world-factbook/rankorder/2236rank.html.

24. "Corruption Perceptions Index, 2017," Transparency International, February 21, 2018, www.transparency.org/news/feature/corruption_perceptions_index_2017.

25. *World Energy Scenarios 2016* (London: World Energy Council, 2016), 9, www.worldenergy.org/wp-content/uploads/2016/10/World-Energy-Scenarios-2016_Full-report.pdf.

Chapter 17. The All-Renewable Delusion

1. "Stop Climate Change," Greenpeace, www.coolplanet2009.org/climate-change-links-and-partners/ngos-for-sustainable-development/2-organisations-foundations-and-ngos/11-greenpeace.html.

2. Matthew McKinzie, "NRDC Analysis: Nuclear Energy and a Safer Climate Future," NRDC, September 29, 2017, www.nrdc.org/experts/matthew-mckinzie/nrdc-analysis-nuclear-energy-and-safer-climate-future.

3. PG&E Corporation, "After CPUC Decision, PG&E to Confer with Diablo Canyon Joint Proposal Parties on Path Forward," news release, January 11, 2018, http://investor.pgecorp.com/news-events/press-releases/press-release-details/2018/After-CPUC-Decision-PGE-to-Confer-With-Diablo-Canyon-Joint-Proposal-Parties-on-Path-Forward/default.aspx. See also, Hudson Sangree, "Diablo Canyon Shutdown Bill Goes to Brown," *RTO Insider*, August 22, 2018, www.rtoinsider.com/diablo-canyon-jerry-brown-98544/.

4. "Environmental Statement on Nuclear Energy and Global Warming," June 2005, 313 signatories, www.citizen.org/documents/GroupNuclearStmt.pdf.

5. Brian Kennedy, "Americans Strongly Favor Expanding Solar Power to Help Address Costs and Environmental Concerns," *Fact Tank* (blog), Pew Research Center, October 5, 2016, www.pewresearch.org/fact-tank/2016/10/05/americans-strongly-favor-expanding-solar-power-to-help-address-costs-and-environmental-concerns/.

6. *2016 Democratic Party Platform* (Orlando, FL: Democratic Platform Committee, 2016), https://democrats.org/wp-content/uploads/2018/10/2016_DNC_Platform.pdf.

7. "1972 Democratic Party Platform," July 10, 1972, American Presidency Project, www.presidency.ucsb.edu/documents/1972-democratic-party-platform.

8. See, Rebecca Riffkin, "U.S. Support for Nuclear Energy at 51%," Gallup, March 30, 2015, www.gallup.com/poll/182180/support-nuclear-energy.aspx.

9. Jeff Merkley, "Merkley, Sanders, Markey, Booker Introduce Landmark Legislation to Transition United States to 100% Clean and Renewable Energy," news release, April 27, 2017, www.merkley.senate.gov/news/press-releases/merkley-sanders-markey-booker-introduce-landmark-legislation-to-transition-united-states-to-100-clean-and-renewable-energy.

10. Jeff Merkley, *100 by '50 Act: Transitioning America to 100% Clean, Renewable Energy for All by 2050* (Washington, DC: Office of Senator Jeff Merkley, 2017), www.merkley.senate.gov/imo/media/doc/17.04.26%20100%20by%2050%201%20pager.pdf.

11. Merkley, "Merkley, Sanders, Markey, Booker."

12. Andrew Cuomo, "ICYMI: Governor Cuomo Proposes Sweeping Set of Environmental Initiatives to Lead the Nation in the Fight to Combat Climate Change," news

release, January 11, 2017, www.governor.ny.gov/news/icymi-governor-cuomo-proposes-sweeping-set-environmental-initiatives-lead-nation-fight-combat.

13. Robert Bryce, "The Appalling Delusion of 100 Percent Renewables, Exposed," *National Review Online*, June 24, 2017, www.nationalreview.com/2017/06/renewable-energy-national-academy-sciences-christopher-t-m-clack-refutes-mark-jacobson/.

14. "100% Commitments in Cities, Counties, & States," Sierra Club, www.sierraclub.org/ready-for-100/commitments.

15. "Companies," RE100, http://there100.org/companies.

16. Video of her remarks is available here: John Parkinson, "Alexandria Ocasio-Cortez Opens Freshman Orientation by Leading Protest at Nancy Pelosi's Office," ABC News, November 13, 2018, https://abcnews.go.com/Politics/alexandria-ocasio-cortez-opens-freshman-orientation-leading-protest/story?id=59165661.

17. Joseph A. Wulfsohn, "AOC Defends Green New Deal, Says Narrative Being 'Manipulated' by Trump, Other Critics," Fox News, March 22, 2019, www.foxnews.com/entertainment/aoc-insists-green-new-deal-is-not-about-what-we-have-to-cut-back-on-but-its-about-being-more-expansive.

18. "Re: Legislation to Address the Urgent Threat of Climate Change," letter, January 10, 2019, http://foe.org/wp-content/uploads/2019/01/Progressive-Climate-Leg-Sign-On-Letter-2.pdf.

19. David Roberts, "Utilities Have a Problem: The Public Wants 100% Renewable Energy, and Quick," *Vox*, October 11, 2018, www.vox.com/energy-and-environment/2018/9/14/17853884/utilities-renewable-energy-100-percent-public-opinion.

20. Robert Bryce, "The Anti-Science, Anti-Nuclear Left," *National Review Online*, December 11, 2015, http://robertbryce.com/the-anti-science-anti-nuclear-left/. In 2011, journalist Bill Tucker asked climate activist Bill McKibben why he wasn't advocating for the use of nuclear energy when rallying other activists. McKibben replied that "if I came out in favor of nuclear, I would split this movement in half."

21. Patrick Graichen, Alice Sakhel, and Christoph Podewils, *The Energy Transition in the Power Sector: State of Affairs in 2017* (Berlin: Agora Energiewende, 2017), 37, www.agora-energiewende.de/fileadmin2/Projekte/2018/Jahresauswertung_2017/Energiewende_2017_-_State_of_Affairs.pdf.

22. William Wilkes and Brian Parkin, "Germany's Economic Backbone Suffers from Soaring Power Prices," Bloomberg, September 23, 2018, www.bloomberg.com/news/articles/2018-09-24/electricity-power-prices-surge-for-german-mittelstand-merkel?utm_source=CCNet+Newsletter&utm_campaign=c857e8c585-EMAIL_CAMPAIGN_2018_09_24_12_40&utm_medium=email&utm_term=0_fe4b2f45ef-c857e8c585-20172653.

23. Elmira Aliakbari et al., eds., *Understanding the Changes in Ontario's Electricity Markets and Their Effects* (Vancouver: Fraser Institute, 2018), ii, www.fraserinstitute.org/sites/default/files/understanding-the-changes-in-ontarios-electricity-markets-web-final_0.pdf.

24. Sarah Sacheli, "Hydro Rates Wreak Havoc on Municipal Budgets," *Windsor Star*, February 13, 2017, https://windsorstar.com/news/local-news/double-whammy-hydro-rates-wreak-havoc-on-municipal-budgets.

25. Aliakbari et al., eds., *Understanding the Changes*, ii.

26. Shawn Jeffords, "Doug Ford Promises 12 Percent Cut to Hydro Rates If Elected," CTV News Toronto, April 27, 2018, https://toronto.ctvnews.ca/ontario-election-2018/doug-ford-promises-12-per-cent-cut-to-hydro-rates-if-elected-1.3905227.

27. Ashifa Kassam, "Ontario Election Results: Populist Doug Ford to Become Premier," *Guardian*, June 8, 2018, www.theguardian.com/world/2018/jun/08/ontario-election-results-populist-doug-ford-to-become-premier.

28. Elizabeth McSheffrey, "Ontario Cancelling 758 'Unnecessary and Wasteful' Renewable Energy Contracts," *National Observer*, July 13, 2018, www.nationalobserver.com/2018/07/13/news/ontario-cancelling-758-unnecessary-and-wasteful-renewable-energy-contracts.

29. Ezra Levant, "'I Never Thought I'd Live to See the Day': Doug Ford Repeals Ontario's Hated Green Energy Act," *Rebel Media*, December 7, 2018, www.therebel.media/ontario-news-doug-ford-green-energy-ezra-levant-show-december-07-2018?safari_redirect.

30. A. Odysseus Patrick, "Scott Morrison Vows Stability as Australia Ousts Yet Another Prime Minister," *Washington Post*, August 24, 2018, www.washingtonpost.com/world/australian-prime-minister-is-ousted-in-dispute-over-greenhouse-gas-emissions/2018/08/24/55f4bcce-a757-11e8-a656-943eefab5daf_story.html.

31. Angus Taylor, "Cut the Energy Virtue Signalling and Start Serving Customers," *Australian Financial Review*, September 18, 2018, www.angustaylor.com.au/media/media-releases/opinion-cut-energy-virtue-signalling-and-start-serving-customers.

32. "Angus Taylor Confirms Government 'Won't Be Replacing' Renewable Energy Target,'" *Sydney Morning Herald*, September 18, 2018, www.smh.com.au/politics/federal/angus-taylor-confirms-government-won-t-be-replacing-renewable-energy-target-20180918-p504j1.html.

33. Damien Cave, "Australia Election Results: Prime Minister Scott Morrison Seizes a Stunning Win," *New York Times*, May 18, 2019, www.nytimes.com/2019/05/18/world/australia/election-results-scott-morrison.html.

34. "Protecting the Environment and Promoting Clean Energy," Arnold Schwarzenegger, www.schwarzenegger.com/issues/milestone/protecting-the-environment-and-promoting-clean-energy.

35. David R. Baker, "California May Reach 50% Renewable Power Goal by 2020—10 Years Early," *San Francisco Chronicle*, November 13, 2017, www.sfchronicle.com/business/article/California-may-reach-50-renewable-power-goal-by-12354313.php.

36. Mark Nelson and Michael Shellenberger, "Electricity Prices in California Rose Three Times More in 2017 Than They Did in the Rest of the United States," Environmental Progress, February 12, 2018, http://environmentalprogress.org/big-news/2018/2/12/electricity-prices-rose-three-times-more-in-california-than-in-rest-of-us-in-2017.

37. The Two Hundred v. California Air Resources Board, Superior Court of the State of California, County of Fresno, filed April 27, 2018, www.thetwohundred.org/wp-content/uploads/2018/07/Complaint_signed.pdf.

38. *The Two Hundred*, Superior Court of the State of California, County of Fresno.

39. Alexei Koseff, "California Approves Goal for 100% Carbon-Free Electricity by 2045," *Sacramento Bee*, September 10, 2018, www.sacbee.com/news/politics-government/capitol-alert/article218128485.html.

40. James Bushnell, "Breaking News! California Electricity Prices Are High," Energy Institute at Haas, February 21, 2017, https://energyathaas.wordpress.com/2017/02/21/breaking-news-california-electricity-prices-are-high/.

41. "Delivering Renewable Power Affordably Throughout Southern California," Tehachapi Renewable Transmission Project, Southern California Edison, https://on.sce.com/2PO8XKC.

42. Doug Karpa, "Exploding Transmission Costs Are the Missing Story in California's Regionalization Debate," *Utility Dive*, July 5, 2018, www.utilitydive.com/news/exploding-transmission-costs-are-the-missing-story-in-californias-regional/526894/.

43. James Temple, "The $2.5 Trillion Reason We Can't Rely on Batteries to Clean Up the Grid," *MIT Technology Review*, July 27, 2018, www.technologyreview.com/s/611683/the-25-trillion-reason-we-cant-rely-on-batteries-to-clean-up-the-grid/.

44. "How Long Will Powerwall Last in an Outage?" Tesla, www.tesla.com/support/powerwall/how-long-will-powerwall-last-in-an-outage.

45. Temple, "The $2.5 Trillion Reason."

46. "Tesla Powerwall: The Complete Battery Review," EnergySage, last updated May 23, 2019, www.energysage.com/solar/solar-energy-storage/tesla-powerwall-home-battery/.

47. Matthew R. Shaner et al., "Geophysical Constraints on the Reliability of Solar and Wind Power in the United States," *Energy and Environmental Science* 4 (2018), http://pubs.rsc.org/en/content/articlelanding/2018/ee/c7ee03029k#!divAbstract.

48. James Temple, "Relying on Renewables Alone Significantly Inflates the Cost of Overhauling Energy," *MIT Technology Review*, February 26, 2018, www.technologyreview.com/s/610366/relying-on-renewables-alone-would-significantly-raise-the-cost-of-overhauling-the-energy/.

49. University of California, Irvine, "Wind and Solar Power Could Meet Four-Fifths of US Electricity Demand, Study Finds," *Science Daily*, February 27, 2018, www.sciencedaily.com/releases/2018/02/180227111639.htm.

50. "Tesla Powerwall: The Complete Battery Review."

51. Mark P. Mills, *The "New Energy Economy": An Exercise in Magical Thinking* (New York: Manhattan Institute, 2019), www.manhattan-institute.org/green-energy-revolution-near-impossible.

52. B. P. Heard et al., "Burden of Proof: A Comprehensive Review of the Feasibility of 100% Renewable-Electricity Systems," *Renewable and Sustainable Energy Reviews* 76 (September 2017), www.sciencedirect.com/science/article/pii/S1364032117304495.

53. *BP Statistical Review of World Energy 2018.*

54. *Global Trends in Renewable Energy Investment 2017* (Frankfurt: Frankfurt School of Finance & Management, 2017), 22, www.greengrowthknowledge.org/sites/default/files/downloads/resource/Global%20Trends%20in%20Renewable%20Energy%20Investment%202017_0.pdf.

55. *BP Statistical Review of World Energy 2018.*

56. *BP Statistical Review of World Energy 2018.*

57. *BP Statistical Review of World Energy 2018.*

58. *BP Statistical Review of World Energy 2018.*

59. *BP Statistical Review of World Energy 2018.*

60. Christopher T. M. Clack et al., "Evaluation of a Proposal for Reliable Low-Cost Grid Power with 100% Wind, Water, and Solar," *Proceedings of the National Academy of Sciences* 114, no. 26 (June 2017), www.pnas.org/content/114/26/6722.

61. Available on YouTube: Mark Jacobson, interview by David Letterman, *Late Show*, October 9, 2013, www.youtube.com/watch?v=AqIu2J3vRJc.

62. Stanford Energy, "Stanford-Led Study on Grid Reliability to Receive Cozzarelli Prize," news release, March 1, 2016, https://energy.stanford.edu/news/stanford-led-study-grid-reliability-receive-2015-cozzarelli-prize.

63. See a list of Jacobson's publications, "100% Wind, Water, and Solar (WWS) All-Sector Energy Roadmaps for Countries, States, Cities, and Town," Mark Z. Jacobson, Stanford University, http://web.stanford.edu/group/efmh/jacobson/Articles/I/WWS-50-USState-plans.html. The spreadsheet discussed in the 2015 report with details by state is also posted online by Stanford. On the spreadsheet, look for the tab labeled "Intermediate details by state." See, http://web.stanford.edu/group/efmh/jacobson/Articles/I/USStates.xlsx.

64. Clack et al., "Evaluation of a Proposal." It must be noted that three months after Clack's paper was published, Jacobson sued Clack—but none of his coauthors—in federal court for $10 million, claiming defamation. Jacobson also sued the National Academy of Sciences. Jacobson complained that Clack's paper damaged his reputation and made him and his coauthors "look like poor, sloppy, incompetent, and clueless researchers." See, Robert Bryce, "An Environmentalist Sues Over an Academic Disagreement," *National Review Online*, November 10, 2017, www.nationalreview.com/2017/11/environmentalist-who-claimed-us-could-run-renewables-sues-over-academic-disagreement/. In February 2019, shortly after a hearing in the case was held in federal court in Washington, DC, Jacobson suddenly withdrew the lawsuit while claiming, according to the *Stanford Daily*, that his lawsuit was "never intended to stifle academic debate." See, Alex Tsai, "Stanford Professor Retracts $10 Million Libel Suit Against Scientific Critic, Academic Journal," *Stanford Daily*, March 2, 2018, www.stanforddaily.com/2018/03/02/stanford-professor-retracts-10-million-libel-suit-against-scientific-critic-academic-journal/.

65. California covers 423,970 square kilometers. See, "US States (plus Washington D.C.) Area and Ranking," Enchanted Learning, www.enchantedlearning.com/usa/states/area.shtml.

66. *BP Statistical Review of World Energy 2018.*

67. Lee M. Miller and David W. Keith, "Observation-Based Solar and Wind Power Capacity Factors and Power Densities," *Environmental Research Letters*, October 4, 2018, http://iopscience.iop.org/article/10.1088/1748-9326/aae102/meta. The paper puts the mean power density of wind at 0.5 W/m^2 and the mean for solar at 5.4 W/m^2.

68. Leah Burrows, "The Down Side of Wind Power," *Harvard Gazette*, October 4, 2018, https://news.harvard.edu/gazette/story/2018/10/large-scale-wind-power-has-its-down-side/.

69. Wikipedia, s.v. "Contiguous United States," last updated May 16, 2019, 20:36, https://en.wikipedia.org/wiki/Contiguous_United_States.

70. Vaclav Smil, *Energy Myths and Realities: Bringing Science to the Energy Policy Debate* (Washington, DC: AEI Press, 2010), 125.

71. David Roberts, "These Huge New Wind Turbines Are a Marvel. They're Also the Future," *Vox*, October 23, 2018, www.vox.com/energy-and-environment/2018/3/8/17084158/wind-turbine-power-energy-blades.

72. In 1919, Betz conjectured that the maximum efficiency for a wind turbine is about 59 percent. However, modern turbines haven't come close to that figure, and instead have efficiencies of about 35 to 45 percent. See, "Betz Limit," Energy Education, last updated July 21, 2018, https://energyeducation.ca/encyclopedia/Betz_limit.

73. Michael Kuser, "New York Plans or Wind Energy, Related Jobs," *RTO Insider*, November 28, 2018, www.rtoinsider.com/new-york-nyserda-wind-power-106837/.

74. Leo Hickman, "Power to the People," *Guardian*, April 29, 2009, www.theguardian.com/environment/2009/apr/30/david-mckay-sustainable-energy.

75. Emily Gosden, "Wind Farm 'Needs 700 Times More Land' Than Fracking Site to Produce Same Energy," *Telegraph*, August 14, 2014, www.telegraph.co.uk/news/earth/energy/fracking/11034270/Wind-farm-needs-700-times-more-land-than-fracking-site.html.

76. David MacKay, "A Reality Check on Renewables," filmed March 2012 in Warwick, UK, TED video, www.ted.com/talks/david_mackay_a_reality_check_on_renewables?language=en.

77. Emily Gosden, "Wind and Solar a Waste of Money for UK, Prof Sir David MacKay Said in Final Interview," *Telegraph*, May 3, 2016, www.telegraph.co.uk/news/2016/05/03/wind-and-solar-a-waste-of-money-for-uk-prof-sir-david-mackay-sai/.

Chapter 18. This Land Is My Land

1. Interviewed by author. Cited in my article, "Wind Power Is an Attack on Rural America," *Los Angeles Times*, February 27, 2017, www.latimes.com/opinion/op-ed/la-oe-bryce-backlash-against-wind-energy-20170227-story.html.

2. Associated Press, "Jill Stein Reaches Plea Deal Over Arrest at Standing Rock Protest," *Guardian*, August 9, 2017, www.theguardian.com/us-news/2017/aug/09/dakota-access-pipeline-jill-stein-arrest-green-party.

3. Amy Sisk, "Dakota Access Pipeline Upends Oil Transport," *Inside Energy*, August 14, 2017, http://insideenergy.org/2017/08/14/dakota-access-pipeline-upends-oil-transport/.

4. "Proposition 112 Endorsed By," Colorado Rising, https://corising.org/endorsements/. See also, Colorado Chamber of Commerce, "Initiative 97, Oil-and-Gas Well Setback Measure, Qualifies for November Ballot," *Colorado Capital Report*, http://cochamber.com/2018/09/04/initiative-97-oil-and-gas-well-setback-measure-qualifies-for-november-ballot/.

5. "Editorial: Vote No on Proposition 112 Because It's a Ban on Oil and Gas," *Denver Post*, October 10, 2018, www.denverpost.com/2018/10/10/proposition-112-is-ban-on-oil-and-gas/. For gas production data, see "Rankings: Natural Gas Marketed Production, 2017 (million cu ft)," EIA, www.eia.gov/state/rankings/#/series/47.

6. AFP, "Thousands of Anti-coal Protesters Celebrate German Forest's Reprieve," *Guardian*, October 6, 2018, www.theguardian.com/environment/2018/oct/06/thousands-of-anti-coal-protesters-celebrate-german-forests-reprieve.

7. Robert M. Bryce, "Oil Waste Pits Trap Unwary Birds," *Christian Science Monitor*, March 19, 1990, www.csmonitor.com/1990/0319/apit.html.

8. *Encyclopedia Britannica Online*, s.v. "Deepwater Horizon Oil Spill of 2010," by Richard Pallady, last updated April 30, 2019, www.britannica.com/event/Deepwater-Horizon-oil-spill-of-2010.

9. Drew Michanowicz, "The Aliso Canyon Gas Leak Was a Disaster. There Are 10,000 More Storage Wells Just Like It," *Los Angeles Times*, May 14, 2018, www.latimes.com/opinion/op-ed/la-oe-michanowicz-aliso-canyon-gas-leak-20180514-story.html.

10. Michael Stroup, "Fracking Bans Continue to Proliferate," *Energy and Environment: Clearing the Air* (blog), National Center for Policy Analysis, Mary 19, 2014, http://environmentblog.ncpathinktank.org/fracking-bans-continue-to-proliferate/#sthash.Wjvyt1wY.dpbs.

11. Jon Hurdle, "With Governor's Signature, Maryland Becomes Third State to Ban Fracking," *StateImpact Pennsylvania*, April 4, 2017, https://stateimpact.npr.org

/pennsylvania/2017/04/04/with-governors-signature-maryland-becomes-third-state-to-ban-fracking/.

12. *2015 Annual Report* (Washington, DC: Food & Water Watch, 2016), www.foodandwaterwatch.org/sites/default/files/rpt_1703_fww2015annualreport-c2.pdf.

13. *The Urgent Case For a Ban on Fracking* (Washington, DC: Food & Water Watch, 2015), www.foodandwaterwatch.org/sites/default/files/urgent_case_for_ban_on_fracking.pdf.

14. Natural Resources Defense Council 2017 Form 990, GuideStar, https://pdf.guidestar.org/PDF_Images/2017/132/654/2017-132654926-0f6cfa7f-9.pdf.

15. Andrew Postman, "How to Block Big Polluters," NRDC, January 27, 2016, www.nrdc.org/land/fracking-community-defense/.

16. Earthjustice 2016 Form 990, GuideStar, https://pdf.guidestar.org/PDF_Images/2017/941/730/2017-941730465-0fab25cb-9.pdf.

17. "Contact Us," Earthjustice, https://earthjustice.org/about/contact.

18. Jude Clemente, "Do Wind Turbines Lower Property Values?" *Forbes*, September 23, 2016, www.forbes.com/sites/judeclemente/2015/09/23/do-wind-turbines-lower-property-values/print/.

19. Jeremy P. Jacobs, "Wisconsin 'Health Hazard' Ruling Could Shock Wind Industry," *Energy News Network*, September 17, 2015, http://midwestenergynews.com/2015/09/17/wisconsin-health-hazard-ruling-could-shock-wind-industry/.

20. Tux Turkel, "Protesters Arrested at Lincoln Windfarm," *Portland Press Herald*, November 8, 2010, www.pressherald.com/news/Protesters-arrested-at-Lincoln-maine-windfarm.html.

21. Robert Bryce, "Backlash Against Big Wind Continues," *National Review Online*, November 27, 2012, www.nationalreview.com/2012/11/backlash-against-big-wind-continues-robert-bryce/.

22. Sarah Favot, "L.A. County Supervisors to Ban Large Wind Turbines in Unincorporated Areas," *Los Angeles Daily News*, July 14, 2015, www.dailynews.com/government-and-politics/20150714/la-county-supervisors-to-ban-large-wind-turbines-in-unincorporated-areas.

23. "Board of Supervisors Ban Utility-Scale Wind Turbines in Los Angeles County," North American Windpower, July 16, 2015, www.nawindpower.com/e107_plugins/content/content.php?content.14428.

24. Halie Cook, "Wind Turbines Banned in Unincorporated Los Angeles County," KHTS, July 23, 2015, www.hometownstation.com/santa-clarita-news/los-angeles-county-news/wind-turbines-banned-in-unincorporated-los-angeles-county-157288.

25. Ros Davidson, "California Sets 50% Renewable Energy Target," *Windpower Monthly*, September 15, 2015, www.windpowermonthly.com/article/1363977/california-sets-50-renewable-energy-target.

26. Erin Mansfield, "Blittersdorf Proposes Two 500-Foot Wind Turbines in Irasburg," *VTDigger*, August 7, 2015, http://vtdigger.org/2015/08/07/blittersdorf-proposes-two-500-foot-wind-turbines-in-irasburg/.

27. Bruce Parker, "Revolt: Vermont Town Votes 274–9 Against Giant Wind Turbines," *Vermont Watchdog*, October 2, 2015, www.windaction.org/posts/43511-revolt-vermont-town-votes-274-9-against-giant-wind-turbines.

28. Kalsey Stults, "Wind Farm Application Rejected by Billings County Commission," *Bismarck Tribune*, November 15, 2016, https://bismarcktribune.com/wind-farm-application-rejected-by-billings-county-commission/article_14c9d7d6-24ad-53cc-97b9-b02b6d9404e7.html.

29. "Iowa: State Profile and Energy Estimates," EIA, last updated April 18, 2019, www.eia.gov/state/?sid=IA. See also, Associated Press, "Company Proposes 2nd Black Hawk County Spot for Wind Farm," *Washington Times*, September 17, 2015, www.washingtontimes.com/news/2015/sep/17/company-proposes-2nd-black-hawk-county-spot-for-wi/.

30. Tim Jamison, "Black Hawk County Wind Project Withdrawn," *Courier*, October 21, 2015, https://wcfcourier.com/news/local/govt-and-politics/black-hawk-county-wind-project-withdrawn/article_10fdce35-c698-54ad-9ae9-3f3b77983f02.html.

31. Kit Kennedy, "New York Adopts Historic '50 by '30' Renewables Goal," NRDC, August 1, 2016, www.nrdc.org/experts/kit-kennedy/new-york-adopts-historic-50-30-renewables-goal.

32. Robert Bryce, "Sucking Wind in the Fight for Renewable Energy," *New York Post*, March 28, 2016, https://nypost.com/2016/03/28/sucking-wind-in-the-fight-for-renewable-energy/.

33. Marcus Wolf, "Lighthouse Owners Appeal for Wind Farm Review Funds Rejected," *Watertown Daily Times*, December 11, 2018, www.watertowndailytimes.com/news03/lighthouse-owners-appeal-for-wind-farm-review-funds-rejected-20181211.

34. Marcus Wolf, "Another Motion for Wind Farm Dismissal Receives Varying Responses," *Watertown Daily Times*, December 9, 2018, www.watertowndailytimes.com/news03/another-motion-for-wind-farm-dismissal-receives-varying-responses-20181209.

35. Marcus Wolf, "Apex Withdraws Article 10 Application for Galloo Island Wind," *Watertown Daily Times*, February 9, 2019, www.watertowndailytimes.com/news03/apex-withdraws-article-10-application-for-galloo-island-wind-20190209.

36. Thomas J. Prohaska, "Lake Ontario Wind Project Is Dead, Somerset Supervisor Says," *Buffalo News*, April 11, 2019, https://buffalonews.com/2019/04/11/town-supervisors-think-somerset-yates-wind-project-is-dead-as-company-announces-indefinite-delay/.

37. I have repeatedly checked the archives of the *Times* for stories on the Lighthouse Wind project. By late May 2019, the newspaper still had not covered the story.

38. Wikipedia, s.v. "Hinton, Oklahoma," last updated February 20, 2019, 00:42, https://en.wikipedia.org/wiki/Hinton,_Oklahoma

39. Robert Bryce, "Big 'Green' and Mean: A Wind-Energy Giant Attacks Small-Town America," *National Review Online*, May 2, 2017, www.nationalreview.com/2017/05/wind-turbine-company-sues-small-towns-get-tax-credits/.

40. Bryce, "Big 'Green' and Mean."

41. John Schneider, "Lawsuit Filed Against Juniata Township," *Advertiser*, April 3, 2019, www.tuscolatoday.com/index.php/2019/04/03/lawsuit-filed-against-juniata-township/.

42. Robert Bryce, "NextEra Won't Drop Its Lawsuit Against Esther Wrightman," *National Review Online*, July 13, 2015, www.nationalreview.com/2015/07/big-wind-still-slapping-canadian-woman/.

43. NextEra Energy, "NextEra Energy Named One of the World's Most Ethical Companies for the 12th Time," news release, February 26, 2019, www.investor.nexteraenergy.com/news-and-events/news-releases/2019/02-26-2019-110955805.

44. Matthew Gardner, Robert S. McIntyre, and Richard Phillips, *The 35 Percent Corporate Tax Myth* (Washington, DC: Institute on Taxation and Economic Policy, 2017), www.itep.org/pdf/35percentfullreport.pdf.

45. For more, see Good Jobs First, www.goodjobsfirst.org/.

46. "NextEra Energy," Subsidy Tracker Parent Company Summary, Good Jobs First, https://subsidytracker.goodjobsfirst.org/parent/nextera-energy.

47. *Annual Report 2018* (Juno Beach, FL: NextEra Energy, 2018), 97, www.investor.nexteraenergy.com/~/media/Files/N/NEE-IR/reports-and-fillings/annual-reports/NextEra%20Energy_Annual_Report_2018.pdf. Of those tax credit carryforwards, $2.9 billion were federal and $344 million were state.

48. Darrel Radford, "County Towns Putting Up Walls Against Wind," *Courier-Times*, November 1, 2018, www.thecouriertimes.com/common/story.php?ID=7040&hl=County-towns-putting-up-walls-against-wind.

49. Will Lewis, "Wind Turbine Application Denied by Penn Forest Township," WFMZ-TV, December 17, 2018, www.wfmz.com/news/poconos-coal/wind-turbine-application-denied-by-penn-forest-township/938471853.

50. Robert Bradley Jr., "Enough! Martis Responds to Sinclair Re Industrial Wind (Face-to-Face Debate Urged)," *MasterResource* (blog), March 5, 2018, www.masterresource.org/wind-power-grassroots-debate/martis-sinclair-wind/.

51. "Electric Generation Capacity & Energy," California Energy Commission, last updated May 6, 2019, www.energy.ca.gov/almanac/electricity_data/electric_generation_capacity.html.

52. Rob Nikolewski, "Wind Energy in California: The Good News and Bad News," *San Diego Union-Tribune*, August 28, 2017, www.sandiegouniontribune.com/business/sd-fi-california-wind-20170825-story.html.

53. "NOT a Willing Host," Ontario Wind Resistance, http://ontario-wind-resistance.org/not-a-willing-host/.

54. For more, see European Platform Against Windfarms, www.epaw.org.

55. The Bavarian rule was put in place in 2014. See, Julian Wettengel, "Shake-Up in Bavaria's Election May Impact German Energy Policy," Clean Energy Wire, October 12, 2018, www.cleanenergywire.org/news/shake-bavarias-election-may-impact-german-energy-policy.

56. "Poland Adopts Limits on Where Wind Farms Can Be Built," Reuters, May 23, 2016, www.reuters.com/article/us-energy-poland-windfarm-idUSKCN0YE17V.

57. "Carwyn Jones Says Wales Must Grasp 'Energy Decade,'" BBC News, May 26, 2011, www.bbc.co.uk/news/uk-wales-13563168.

58. "Mid and West Wales Power Protesters at Senedd," BBC News, May 24, 2011, www.bbc.com/news/uk-wales-13498707.

59. Joshua S. Hill, "UK Government Refuses 970 MW Navitus Bay Offshore Wind Farm," *CleanTechnica*, September 11, 2015, http://cleantechnica.com/2015/09/11/uk-government-refuse-970-mw-navitus-bay-offshore-wind-farm/.

60. OffshoreWind.biz, "UK Thumbs Down Navitus Bay Project," September 11, 2015, www.offshorewind.biz/2015/09/11/uk-thumbs-down-navitus-bay-project/.

61. Among the projects rejected were Bhein Mhor and Limekiln. Alistair Munro, "Controversial Highlands Wind Farm Plan Rejected," *Scotsman*, July 13, 2015, www.scotsman.com/news/environment/controversial-highlands-wind-farm-plan-rejected-1-3829360.

62. "Sallachy and Glencassley Wind Farms Refused Consent," BBC News, November 17, 2015, www.bbc.com/news/uk-scotland-highlands-islands-34842315.

63. Iain Ramage, "'Enough Is Enough' as Loch Ness Turbines Are Rejected," *Press and Journal,* April 13, 2016, www.pressandjournal.co.uk/fp/news/highlands/887713/ness-side-turbines-rejected/.

64. "Wind Energy FAQ," Scotland Against Spin, https://scotlandagainstspin.org/wind-energy-faqs/.

65. Anca Gurzu, "Going Electric, but Not in My Backyard," *Politico*, October 9, 2018, www.politico.eu/article/going-electric-but-not-in-my-back-yard-germany-wind-coal-nuclear-power/.

66. Clemente, "Wind Turbines."

67. Michael S. McCann, "Testimony of Michael McCann on Property Value Impacts in Adams County IL," June 8, 2010, *Wind Action*, www.windaction.org/posts/26696-testimony-of-michael-mccann-on-property-value-impacts-in-adams-county-il#.W8X72S_Mzvd.

68. Sanchez Manning, "Proof Wind Turbines Take Thousands Off Your Home: Value of Houses Within 1.2 Miles of Large Wind Farms Slashed by 11%, Study Finds," *Daily Mail*, January 25, 2014, www.dailymail.co.uk/news/article-2546042/Proof-wind-turbines-thousands-home-value-homes-1-2-miles-wind-farms-slashed-11-cent-study-finds.html.

69. Yasin Sunak and Reinhard Madlener, "The Impact of Wind Farm Visibility on Property Values: A Spatial Difference-in-Differences Analysis," *Energy Economics* 55 (March 2016), www.sciencedirect.com/science/article/pii/S014098831600044X.

70. Benjamin Wehrmann, "Wind Turbines Hurt Property Prices, Study Finds," Clean Energy Wire, January 21, 2019, www.cleanenergywire.org/news/wind-turbines-hurt-property-prices-study-finds. For the full study, see Manuel Frondel et al., *Local Cost for Global Benefit: The Case of Wind Turbines* (Essen, Germany: RWI, January 2019), www.rwi-essen.de/media/content/pages/publikationen/ruhr-economic-papers/rep_18_791.pdf.

71. Robert Bryce, *Smaller Faster Lighter Denser Cheaper: How Innovation Keeps Proving the Catastrophists Wrong* (New York: PublicAffairs, 2014), Appendix E.

72. J. Mikolajczak et al., "Preliminary Studies on the Reaction of Growing Geese (*Anser anser f. domestica*) to the Proximity of Wind Turbines," *Polish Journal of Veterinary Sciences* 16, no. 4 (2013): 679–686, www.researchgate.net/publication/260561143_Preliminary_studies_on_the_reaction_of_growing_geese_Anser_anser_f_domestica_to_the_proximity_of_wind_turbines.

73. Malgorzata Karwowska et al., "The Effect of Varying Distances from the Wind Turbine on Meat Quality of Growing-Finishing Pigs," *Annals of Animal Science* 15, no. 4 (2015), https://content.sciendo.com/view/journals/aoas/15/4/article-p1043.xml.

74. R. C. N. Agnew, V. J. Smith, and R. C. Fowkes, "Wind Turbines Cause Chronic Stress in Badgers" (unpublished manuscript, 2016), https://research-repository.st-andrews.ac.uk/bitstream/handle/10023/9208/Agnew_WindTurbines_JWD_AAM.pdf?sequence=1&isAllowed=y.

75. Michael Nissenbaum, "Wind Turbines, Health, Ridgelines, and Valleys," National Wind Watch, May 9, 2010, www.wind-watch.org/documents/wind-turbines-health-ridgelines-and-valleys/.

76. Pers. comm. by author with Nissenbaum, February 12, 2010. Nissenbaum also submitted an affidavit in a hearing on a wind farm that is available here: "Dr. Michael Nissenbaum Affidavit, Record Hill Wind Appeal," *Wind Action*, September 17, 2009, www.windaction.org/documents/23332.

77. "Doctor Says Wind Turbine Noise Can Harm Health," *VTDigger*, June 2, 2010, https://vtdigger.org/2010/06/02/doctor-says-wind-turbine-noise-can-harm-health/.

78. M. A. Nissenbaum, J. J. Aramini, and C. D. Hanning, "Effects of Industrial Wind Turbine Noise on Sleep and Health," *Noise Health* (September–October 2012), www.ncbi.nlm.nih.gov/pubmed/23117539.

79. Carl V. Phillips, "Properly Interpreting the Epidemiological Evidence About the Health Effects of Industrial Wind Turbines on Nearby Residents," *Bulletin of Science, Technology & Society* 31, no. 4 (August 2011), https://eric.ed.gov/?id=EJ932840.

80. Alec N. Salt, "Wind Turbines Can Be Hazardous to Human Health," Alec Salt's Lab, Washington University School of Medicine in St. Louis, oto.wustl.edu/saltlab/Wind-Turbines.

81. Alec N. Salt and Jeffery T. Licthenhan, "Perception-Based Protection from Low-Frequency Sounds May Not Be Enough" (paper presented at Inter.noise, New York, August 19–22, 2012), https://bit.ly/2Qz0vEp.

82. Hsuan-Hsiu Annie Chen and Peter Narins, "Wind Turbines and Ghost Stories: The Effects of Infrasound on the Human Auditory System," *Acoustics Today* (April 2012), https://acousticstoday.org/wp-content/uploads/2017/09/Article_7of7_from_ATCODK_8_2.pdf. Note that this study, like Salt's work, underscores the fact that "inter-individual differences in hearing sensitivity allow some people to detect the 'inaudible.'" They continue, saying that "auditory cortical responses and cochlear modulations due to infrasound exposure have been observed, despite the subjects' lack of tonal perception. These studies provide strong evidence for infrasound impact on human peripheral and central auditory responses." In other words, just because you don't hear it doesn't mean your body isn't reacting to it.

83. Jesper Hvass Schmidt and Mads Klokker, "Health Effects Related to Wind Turbine Noise Exposure: A Systematic Review," *PLOSOne*, December 4, 2014, https://journals.plos.org/plosone/article?id=10.1371/journal.pone.0114183.

84. Milad Abbasi et al., "Impact of Wind Turbine Sound on General Health, Sleep Disturbance and Annoyance of Workers: A Pilot-Study in Manjil Wind Farm, Iran," *Journal of Environmental Health Science and Engineering* (2015), https://jehse.biomedcentral.com/articles/10.1186/s40201-015-0225-8.

85. Christian Krekel and Alexander Zerrahn, "Does the Presence of Wind Turbines Have Negative Externalities for People in Their Surroundings? Evidence from Well-Being Data," *Journal of Environmental Economics and Management* 82 (March 2017), www.sciencedirect.com/science/article/pii/S0095069616304624#!.

86. Anabela Botelho et al., "Effect of Wind Farm Noise on Local Residents' Decision to Adopt Mitigation Measures," *International Journal of Environmental Research and Public Health* (July 2017), www.ncbi.nlm.nih.gov/pmc/articles/PMC5551191/.

87. Jeremy Deaton and Owen Agnew, "With Wind Farms, Bias Is in the Eye of the Beholder," *Popular Science*, March 6, 2018, www.popsci.com/aesthetics-wind-energy.

88. Mike Hughlett, "Administrative Law Judge Says PUC Should Reject Freeborn County Wind Project," *Star Tribune*, May 17, 2018, http://m.startribune.com/administrative-law-judge-says-puc-should-reject-freeborn-county-wind-project/482980081/.

89. Mike Hughlett, "Wind Project in Southern Minnesota Gets Pushback," *Star Tribune*, November 18, 2017, www.startribune.com/wind-project-in-southern-minnesota-gets-pushback/458079653/.

90. Christine Legere, "What's Next for Falmouth Turbines?" *Cape Cod Times*, October 14, 2018, www.capecodtimes.com/news/20181014/whats-next-for-falmouth-turbines.

91. Christine Legere, "Falmouth Official Orders Plan to Dismantle Wind Turbine," *Cape Cod Times*, December 20, 2017, www.capecodtimes.com/news/20171220/falmouth-official-orders-plan-to-dismantle-wind-turbine.

92. Steven Withrow, "Falmouth Resident Tells of Lasting Turbine Troubles," *Falmouth Enterprise*, August 28, 2018, www.capenews.net/falmouth/news/falmouth-resident-shares-lasting-turbine-troubles/article_7211f529-fd23-5dbb-87ab-d6cbdaf3fd47.html.

93. Donnelle Eller, "Neighbors in Eastern Iowa Fight to Bring Down Turbines—and Win," *Des Moines Register*, November 21, 2018, www.desmoinesregister.com/story/money/business/2018/11/21/iowa-first-wind-developers-ordered-tear-down-turbines-land-use-lawsuit-supreme-court/1922334002/.

94. National Toxicology Program, *Infrasound: Brief Review of Toxicological Literature* (Bethesda, MD: National Institutes of Health, 2001), 5, https://ntp.niehs.nih.gov/ntp/htdocs/chem_background/exsumpdf/infrasound_508.pdf.

95. Email exchanges with Warren by author, February 17 and 19, 2010.

96. Author interview with Moyer, via phone, February 17, 2010.

97. The Keanes were featured in a 2014 BBC story. See, Diarmaid Fleming, "Ireland's Rural Protests Over Wind Energy," BBC News, February 2, 2014, www.bbc.com/news/world-europe-25966198.

98. Email from Keane, September 14, 2016.

99. Email from Keane, September 21, 2016.

100. Other residents in Wisconsin have also complained about the noise from wind turbines. See, "Shirley Wind Project," YouTube video, December 22, 2011, www.youtube.com/watch?v=71DxuicwCXw.

101. Paul Srubas, "Health Officials Weigh Next Step in Wind Turbine Battle," *Green Bay Press Gazette*, October 26, 2014, www.greenbaypressgazette.com/story/news/local/2014/10/26/health-officials-weigh-next-step-wind-turbine-battle/17967875/.

102. Doug Schneider, "Does Shirley Wind Farm Make Some People Sick? Depends on Which Expert You Ask," *Green Bay Press Gazette*, September 13, 2017, www.greenbaypressgazette.com/story/news/2017/09/13/does-shirley-wind-farm-make-some-people-sick-depends-which-expert-you-ask/637984001/. For more on the situation in Brown County, see Brown County Citizens for Responsible Wind Energy, www.bccrwe.com.

103. Doug Schneider, "Health Chief Got 'Such Migraines' at Wind Farm," *Green Bay Press Gazette,* March 20, 2016, www.greenbaypressgazette.com/story/news/2016/03/20/health-chief-got-such-migraines-wind-farm/82059968/.

104. "Saving the Desert Tortoise," Center for Biological Diversity, www.biologicaldiversity.org/species/reptiles/desert_tortoise/index.html.

105. Frank Eltman, "Solar Projects Can't Save the Forest for the Trees?" *San Diego Union-Tribune*, July 23, 2016, www.sandiegouniontribune.com/sdut-solar-project-cant-save-the-forest-for-the-trees-2016jul23-story.html.

106. Scott Shenk, "Spotsylvania Solar Farm Public Hearing Postponed," *Free-Lance Star*, January 25, 2019, www.fredericksburg.com/news/local/spotsylvania/spotsylvania-solar-farm-public-hearing-postponed/article_518d7fd5-eaf7-52d6-9c93-1ac8b931599b.html.

107. Scott Dance, "Go Solar, or Save the Trees? Georgetown University Solar Farm Would Clear 240-Acre Forest in Charles County," *Baltimore Sun*, January 31, 2019, www.baltimoresun.com/news/maryland/environment/bs-md-georgetown-solar-trees-20190131-story.html.

108. Robert Bryce, "New York's Energy Policy Depends on an Impossible Fantasy," *New York Post*, May 20, 2019, https://nypost.com/2019/05/20/new-yorks-energy-policy-depends-on-an-impossible-fantasy/.

109. K. Shawn Smallwood, "Comparing Bird and Bat Fatality-Rate Estimates Among North American Wind-Energy Projects," *Wildlife Society Bulletin* 37, no. 1 (March 2013): 19–33, http://onlinelibrary.wiley.com/doi/10.1002/wsb.260/abstract.

110. Joel Pagel et al., "Bald Eagle and Golden Eagle Mortalities at Wind Energy Facilities in the Contiguous United States," *Journal of Raptor Research* 47, no. 3 (September 2013): 311–315, www.researchgate.net/publication/271250740_Bald_Eagle_and_Golden_Eagle_Mortalities_at_Wind_Energy_Facilities_in_the_Contiguous_United_States.

111. Author interview with Pagel by phone, September 18, 2013.

112. *BP Statistical Review of World Energy 2018.*

113. *Eagle Conservation Plan Guidance: Module 1—Land-Based Wind Energy, version 2* (Washington, DC: US Fish and Wildlife Service Division of Migratory Bird Management, 2013), iv, http://digitalmedia.fws.gov/utils/getdownloaditem/collection/document/id/1802/filename/1803.pdf/mapsto/pdf/type/singleitem.

114. Yale Environment 360, "Wind Farms Can Act Like Apex Predators in Ecosystems, Study Finds," E360 Digest, November 5, 2018, https://e360.yale.edu/digest/wind-farms-can-act-like-apex-predators-in-ecosystems-study-finds.

115. Maria Thaker, Amod Zambre, and Harshal Bhosale, "Wind Farms Have Cascading Impacts on Ecosystems Across Trophic Levels," *Nature Ecology & Evolution* 2 (2018): 1854–1858, www.nature.com/articles/s41559-018-0707-z#author-information.

116. Thomas J. O'Shea et al., "Multiple Mortality Events in Bats: A Global Review," *Mammal Review* 46, no. 3 (January 2016): 175–190, http://onlinelibrary.wiley.com/doi/10.1111/mam.12064/abstract.

117. Amy Matthews Amos, "Bat Killings by Wind Energy Turbines Continue," *Scientific American*, June 7, 2016, www.scientificamerican.com/article/bat-killings-by-wind-energy-turbines-continue/.

118. Robert Bryce, "Stop Subsidizing the Big Wind Bullies," *New York Post*, November 9, 2017, https://nypost.com/2017/11/09/stop-subsidizing-thc-big-wind-bullies/.

119. J. Miner, "Wind Turbines Killing Tens of Thousands of Bats, Including Many on the Endangered Species List," *London Free Press*, July 20, 2016, www.lfpress.com/2016/07/20/wind-turbines-killing-tens-of-thousands-of-bats-including-many-on-the-endangered-species-list.

120. Joanna Klein, "A Summer Evening in Texas Isn't Complete Without a Bat Show," *New York Times*, August 28, 2016, www.nytimes.com/2016/08/29/science/texas-bats-show.html.

121. Author interview with Tuttle by phone, February 17, 2014.

122. Jeffrey Tomich, "Iowa Landowners Claim Win Over Clean Line Project," *E&E News*, May 19, 2017, www.eenews.net/stories/1060054786.

123. Kyle Massey, "Clean Line Shelves Arkansas Plans; Delegation Steps Up Attack," *Arkansas Business*, January 23, 2018, www.arkansasbusiness.com/article/120518/clean-line-shelves-plans-for-arkansas-delegation-steps-up-attack.

124. Kevin Randolph, "Arkansas Congressional Delegation Cheers Termination of DOE Partnerships with Clean Line Energy Partners," *DailyEnergyInsider*, March 27, 2018, https://dailyenergyinsider.com/news/11477-arkansas-congressional-delegation-cheers-termination-doe-partnerships-clean-line-energy-partners/.

125. Julian Spector, "New Hampshire Rejects Northern Pass Transmission Line Permit," Greentech Media, February 1, 2018, www.greentechmedia.com/articles/read/new-hampshire-rejects-northern-pass#gs.ezn4l_8.

126. Jeffrey Tomich, "Grain Belt Express' New Owner Makes Fresh Push for Approval," *E&E News*, November 16, 2018, www.eenews.net/energywire/2018/11/16/stories/1060106395.

127. Associated Press, "Missouri Utility Group Joins Controversial Transmission Line," Fox 2, June 3, 2016, http://fox2now.com/2016/06/03/missouri-utility-group-joins-controversial-transmission-line-2/.

128. "Citizens Tell Governor 'Grain Belt Express Not a Public Utility,'" *Caldwell County News*, January 22, 2019, www.mycaldwellcounty.com/news/citizens-tell-governor-grain-belt-express-not-public-utility.

129. Edward McKinley, "Missouri Supreme Court Clears Obstacle Blocking Grain Belt Express," *Kansas City Star*, July 17, 2018, www.kansascity.com/news/business/article215051730.html.

130. M. M. Hand et al., eds., *Renewable Electricity Futures Study* (Lakewood, CO: National Renewable Energy Laboratory, 2012), 26, www.nrel.gov/docs/fy13osti/52409-ES.pdf. Figure ES-8 shows that in a 90 percent renewable energy scenario, the United States would need about 200 million megawatt-miles of new transmission capacity. The caption for the graph says that "existing total transmission capacity in the contiguous United States is estimated at 150-200 million megawatt-miles."

131. "Transmission," Edison Electric Institute, www.eei.org/issuesandpolicy/transmission/Pages/default.aspx.

Chapter 19. The Nuclear Necessity

1. Maria Popova, "Our Friend the Atom: Disney's 1956 Illustrated Propaganda for Nuclear Energy," *Brain Pickings*, February 18, 2013, www.brainpickings.org/2013/02/18/our-friend-the-atom-disney/.

2. "Indian Point Nuclear Generating Unit 2," US Nuclear Regulatory Commission, last updated April 30, 2018, www.nrc.gov/info-finder/reactors/ip2.html.

3. Christine Macy, "Dams Across America," *Places* (January 2010), https://placesjournal.org/article/dams-across-america/.

4. Richard F. Weingroff, "The Greatest Decade 1956–1966," Highway History, Federal Highway Administration, www.fhwa.dot.gov/infrastructure/50interstate.cfm. "At Indian Pt., a History of Nuclear Power, Problems and Controversy," *New York Times*, May 6, 1983, www.nytimes.com/1983/05/06/nyregion/at-indian-pt-a-history-of-nuclear-power-problems-and-controversy.html.

5. The combined output of the two reactors—Unit 2 and Unit 3—is 2,069 megawatts. That's roughly the same output as Hoover Dam. But Indian Point's footprint—just one square kilometer—is a tiny fraction of the territory covered by Lake Mead, the body of water created by Hoover Dam. Lake Mead's surface area is about 247 square miles (640 square kilometers).

6. *Nuclear Accident at Indian Point: Consequences and Costs* (New York: NRDC, 2011), 8, www.nrdc.org/sites/default/files/NRDC-1336_Indian_Point_FSr8medium.pdf.

7. Melanie Grayce West, "Indian Point Closure Won't Leave New York in the Dark," *Wall Street Journal*, January 9, 2017, www.wsj.com/articles/indian-point-closure-wont-leave-new-york-in-the-dark-1484000729.

8. This estimate is based on the proposed South Fork wind project, a 90-megawatt facility that is expected to produce 370 gigawatt-hours per year. (For output figures, see Mark Harrington, "Comptroller: Offshore Wind Farm to Cost Ratepayers $1.62B," *Newsday*, www.newsday.com/long-island/offshore-wind-farm-to-cost-ratepayers-1-62b-comptroller-finds-1.13351693.) That means that 1 megawatt of offshore capacity will produce about 4.1 gigawatt-hours per year. Thus, to match the energy output of Indian Point, at 16,400 gigawatt-hours per year, will require 4,005 megawatts of offshore wind capacity.

9. The footprint, or capacity density, of wind energy is 3 watts per square meter. Thus, 4 billion watts divided by 3 watts per square meter = 1.333 billion square meters or 1,333 square kilometers. That's equal to about 515 square miles.

10. Note that the projected output of the offshore wind project of 4.1 gigawatt-hours per megawatt of installed capacity is significantly higher than the recorded output of onshore wind projects. For instance, in 2017, Texas had 22,637 megawatts of installed wind capacity, which produced 67,061 gigawatt-hours of electricity. Thus, in one of America's best states for wind, 1 megawatt of wind capacity produces about 3 gigawatt-hours of electricity per year. For 2017 wind capacity in Texas, see "U.S. Installed and Potential Wind Power Capacity and Generation," WindExchange, https://windexchange.energy.gov/maps-data/321. For 2017 wind output in Texas, see "Electricity Data Browser," EIA, https://bit.ly/2EwmlRU. Therefore, to replace Indian Point with onshore wind-energy produced in Texas would require about 5,533 megawatts of wind capacity. That much capacity would require 1.844 billion square meters (1,844 square kilometers) or about 711 square miles of territory.

11. Author calculations based on Nuclear Energy Institute data.

12. *BP Statistical Review of World Energy 2018*. In 2017, US solar production totaled about 78 terawatt-hours.

13. James Hansen, "Baby Lauren and the Kool-Aid," July 29, 2011, www.columbia.edu/~jeh1/mailings/2011/20110729_BabyLauren.pdf.

14. Andrew C. Revkin, "To Those Influencing Environmental Policy but Opposed to Nuclear Power," *Dot Earth* (blog), *New York Times*, November 3, 2013, https://dotearth.blogs.nytimes.com/2013/11/03/to-those-influencing-environmental-policy-but-opposed-to-nuclear-power/?_r=0.

15. Quote is from the 2017 documentary *The New Fire*, directed by David Schumacher.

16. "Nuclear Power in the World Today," World Nuclear Association, last updated February 2019, www.world-nuclear.org/info/current-and-future-generation/nuclear-power-in-the-world-today/. "Taking a Fresh Look at the Future of Nuclear Power," January 29, 2015, IEA, www.iea.org/newsroomandevents/news/2015/january/taking-a-fresh-look-at-the-future-of-nuclear-power.html.

17. *Nuclear Power in a Clean Energy System* (Paris: IEA, 2019), https://webstore.iea.org/nuclear-power-in-a-clean-energy-system.

18. "N.Y. Power Seen Sufficient After Indian Point Nuclear Retirement: Report," Reuters, December 13, 2017, www.reuters.com/article/us-new-york-entergy-indian-point/n-y-power-seen-sufficient-after-indian-point-nuclear-retirement-report-idUSKBN1E72DU.

19. "Breaking: Closure of Indian Point Would Spike Power Emissions 29%, Reversing 14 Years of Declines," Environmental Progress, January 8, 2017, http://environmentalprogress.org/big-news/2017/1/8/breaking-closure-of-indian-point-would-spike-power-emissions-29-reversing-14-years-of-declines.

20. Mary C. Serreze, "Closure of Vermont Yankee Nuclear Plant Boosted Greenhouse Gas Emissions in New England," *MassLive*, February 18, 2017, www.masslive.com/news/index.ssf/2017/02/report_closure_of_vermont_yank.html.

21. "New England Using More Natural Gas Following Vermont Yankee Closure," Institute for Energy Research, January 20, 2016, http://instituteforenergyresearch.org/analysis/new-england-using-more-natural-gas-following-vermont-yankee-closure/.

22. Lucas Davis and Catherine Hausman, "Market Impacts of a Nuclear Power Plant Closure," *American Economic Journal: Applied Economics* 8, no. 2 (2016): 92–122, http://faculty.haas.berkeley.edu/ldavis/Davis%20and%20Hausman%20AEJ%202016.pdf.

23. Rebecca Smith, "Exelon Moves to Close Two Illinois Nuclear Plants," *Wall Street Journal*, June 2, 2016, www.wsj.com/articles/exelon-moves-to-close-two-illinois-nuclear-plants-1464873850.

24. Sarah Fecht, "1 Year Later: A Fukushima Nuclear Disaster Timeline," *Scientific American*, March 8, 2012, www.scientificamerican.com/article.cfm?id=one-year-later-fukushima-nuclear-disaster.

25. Mari Saito, Kiyoshi Takenaka, and James Topham, "Insight: Japan's 'Long War' to Shut Down Fukushima," Reuters, March 8, 2013, www.reuters.com/article/2013/03/08/us-japan-fukushima-idUSBRE92417Y20130308.

26. Kumi Naidoo, "Nuclear Energy Isn't Needed," *New York Times*, March 22, 2011, www.nytimes.com/2011/03/23/opinion/23iht-ednaidoo23.html.

27. David Brown, "Nuclear Power Is Safest Way to Make Electricity, According to Study," *Washington Post*, April 2, 2011, www.washingtonpost.com/national/nuclear-power-is-safest-way-to-make-electricity-according-to-2007-study/2011/03/22/AFQUbyQC_story.html.

28. Mark Holt, Richard J. Campbell, Mary Beth Nikitin, *Fukushima Nuclear Disaster* (Washington, DC: Congressional Research Service, 2012), 1, www.fas.org/sgp/crs/nuke/R41694.pdf. See also, "Two Fukushima Daiichi Workers Drowned by Tsunami Had Been Ordered to Inspect Basement," Beyond Nuclear, August 2, 2011, www.beyondnuclear.org/home/2011/8/2/two-fukushima-daiichi-workers-drowned-by-tsunami-had-been-or.html.

29. "No Fukushima Radiation Problems: Report," News.com.au, June 1, 2013, www.news.com.au/world/breaking-news/no-radiation-problems-from-fukushima-rep/news-story/3e8c5247f469482254cec4776ddfca02.

30. For more on Thomas, see "Professor Gerry Thomas," Imperial College London, www.imperial.ac.uk/people/geraldine.thomas.

31. "Inside the Fukushima Reactor," *60 Minutes Australia*, October 22, 2018, www.youtube.com/watch?v=bSNn7fdaPVs. For Thomas's quote, see her at about the 11:15 mark.

32. "Reporter's Grave Radiation Discovery On Board a Plane," *60 Minutes Australia*, October 23, 2018, www.youtube.com/watch?v=tIphDdosaJg.

33. James Conca, "Radiation and the Value of a Human Life," *Forbes*, July 23, 2018, www.forbes.com/sites/jamesconca/2018/07/23/radiation-and-the-value-of-a-human-life/amp/.

34. Genetic Society of America, "Long-Term Health Effects of Hiroshima and Nagasaki Atomic Bombs Not as Dire as Perceived," *Science Daily*, August 11, 2016, www.sciencedaily.com/releases/2016/08/160811120353.htm.

35. "Nuclear Waste," Fundamentals, NEI, www.nei.org/fundamentals/nuclear-waste.

36. "Nuclear Power in France," World Nuclear Association, November 2018, www.world-nuclear.org/information-library/country-profiles/countries-a-f/france.aspx.

37. *Nuclear Waste Management: Key Attributes, Challenges, and Costs for the Yucca Mountain Repository and Two Potential Alternatives* (Washington, DC: Government Accountability Office, 2009), www.gao.gov/new.items/d1048.pdf.

38. Ryan Tracy, "Panel Scorns Nuclear-Waste Policy," *Wall Street Journal*, July 29, 2011, http://online.wsj.com/article/SB10001424053111904888304576476112957361004.html.

39. Rebecca Worby, "Is Yucca Mountain Back from the Dead?" *High Country News*, May 8, 2017, www.hcn.org/articles/is-yucca-mountain-back-from-the-dead.

40. EIA, "Nuclear Regulatory Commission Approves Construction of First Nuclear Units in 30 Years," *Today in Energy*, March 5, 2012, www.eia.gov/todayinenergy/detail.cfm?id=5250.

41. Thad Moore, "Santee Cooper, SCE&G Pull Plug on Roughly $25 Billion Nuclear Plants in South Carolina," *Post and Courier*, July 31, 2017, www.postandcourier.com/business/santee-cooper-sce-g-pull-plug-on-roughly-billion-nuclear/article_c173c0fa-75fb-11e7-a086-cfcd325f82e7.html.

42. Molly Samuel and Emma Hurt, "Companies Reach Deal to Keep Construction Going at Plant Vogtle," WABE, September 26, 2018, www.wabe.org/companies-reach-deal-to-keep-construction-going-at-plant-vogtle/.

43. *Technology Assessment: Nuclear Reactors* (Washington, DC: Government Accountability Office, 2015), 31, www.gao.gov/assets/680/671686.pdf.

44. "Plans for New Reactors Worldwide," World Nuclear Association, last updated April 2019, www.world-nuclear.org/information-library/current-and-future-generation/plans-for-new-reactors-worldwide.aspx.

45. "Nuclear Power in China," World Nuclear Association, last updated April 2019, www.world-nuclear.org/information-library/country-profiles/countries-a-f/china-nuclear-power.aspx.

46. "Barakah Nuclear Power Plant, Abu Dhabi," *Power Technology*, www.powertechnology.com/projects/barakah-nuclear-power-plant-abu-dhabi/.

47. Stanley Carvalho, "UAE's First Nuclear Reactor to Operate in 2018: Minister," Reuters, September 25, 2017, www.reuters.com/article/us-emirates-nuclear/uaes-first-nuclear-reactor-to-operate-in-2018-minister-idUSKCN1C0126.

48. "US NRC Set to Certify APR-1400 Reactor Design," *World Nuclear News*, May 1, 2019, http://world-nuclear-news.org/Articles/US-NRC-set-to-certify-APR-1400-reactor-design.

49. Kim Jaewon, "Company in Focus: Kepco Struggles to Go Green," *Nikkei Asia Review*, June 1, 2017, https://asia.nikkei.com/Business/Company-in-focus-Kepco-struggles-to-go-green.

50. Brian Wang, "India Approves Construction of 12 More Nuclear Reactors," *Nextbigfuture* (blog), February 16, 2018, www.nextbigfuture.com/2018/02/india-approves-construction-of-12-more-nuclear-reactors.html.

51. "Atoms for Peace," *Economist*, August 2, 2018, www.economist.com/europe/2018/08/02/the-world-relies-on-russia-to-build-its-nuclear-power-plants.

52. Andrew E. Kramer, "The Nuclear Power Plant of the Future May Be Floating Near Russia," *New York Times*, August 26, 2018, www.nytimes.com/2018/08/26/business/energy-environment/russia-floating-nuclear-power.html.

53. "Nuclear-Powered Ships," World Nuclear Association, last updated November 2018, www.world-nuclear.org/information-library/non-power-nuclear-applications/transport/nuclear-powered-ships.aspx.

54. "Small Modular Reactors (SMRs)," NuScale, www.nuscalepower.com/benefits/smallest-reactor.

55. "AP1000 Nuclear Power Plant Design," Westinghouse, www.westinghousenuclear.com/New-Plants/AP1000-PWR/Overview.

56. "NuScale Wins U.S. DOE Funding for Its SMR Technology," About Us, NuScale, www.nuscalepower.com/about-us/doe-partnership.

57. "NuScale Begins Program WIN with Governors and Utilities in Western U.S. States," Projects, NuScale, www.nuscalepower.com/our-technology/technology-validation/program-win.

58. Kevin Bullis, "Safer Nuclear Power, at Half the Price," *MIT Technology Review*, March 12, 2013, www.technologyreview.com/news/512321/safer-nuclear-power-at-half-the-price/.

59. Matt Wald, "Investigating Terrestrial Energy's Molten Salt Reactor Design," *NEI Nuclear Notes* (blog), May 5, 2015, http://neinuclearnotes.blogspot.com/2015/05/investigating-terrestrial-energys.html.

60. Leigh Phillips, "The New, Safer Nuclear Reactors That Might Help Stop Climate Change," *MIT Technology Review*, February 27, 2019, www.technologyreview.com/s/612940/the-new-safer-nuclear-reactors-that-might-help-stop-climate-change/.

61. Author interview with Hargraves, December 13, 2018.

62. "About Us," Company, TAE Technologies, https://tae.com/company/.

63. Phillips, "New, Safer Nuclear Reactors."

64. Katherine Bourzac, "Fusion Start-Ups Hope to Revolutionize Energy in the Coming Decades," *Chemical & Engineering News*, August 6, 2018, https://cen.acs.org/energy/nuclear-power/Fusion-start-ups-hope-revolutionize/96/i32.

65. Rachel Feltman, "Why Don't We Have Fusion Power?" *Popular Mechanics*, May 16, 2013, www.popularmechanics.com/science/energy/a8914/why-dont-we-have-fusion-power-15480435/.

Chapter 20. Future Grid

1. For more, see Bkerzay Guest Houses, https://bkerzay.com.

2. *BP Statistical Review of World Energy 2018.*

3. *BP Statistical Review of World Energy 2018.*

4. Nima Elbagir, Dominique van Heerden, and Eliza Mackintosh, "Dirty Energy," CNN, https://edition.cnn.com/interactive/2018/05/africa/congo-cobalt-dirty-energy-intl/.

5. James Vincent, "Rare Earth Elements Aren't the Secret Weapon China Thinks They Are," *Verge*, May 23, 2019, www.theverge.com/2019/5/23/18637071/rare-earth-china-production-america-demand-trade-war-tariffs.

6. Tim Maughan, "The Dystopian Lake Filled by the World's Tech Lust," BBC, April 2, 2015, www.bbc.com/future/story/20150402-the-worst-place-on-earth.

7. Rob Nikolewski, "New Way to Recycle Lithium-Ion Batteries Could Be a Lifeline for Electric Cars and the Environment," *Los Angeles Times*, March 16, 2018, www.latimes.com/business/technology/la-fi-lithium-ion-battery-recycling-20180316-story.html.

8. *BP Statistical Review of World Energy 2018.*

9. Mark P. Mills, "Geopolitical Implications of the 'Invisible' Digital Oil Revolution," testimony before US Senate Committee on Energy and Natural Resources,

July 18, 2017, Washington, DC, www.energy.senate.gov/public/index.cfm/files/serve?File_id=C18F97E3-A1E8-4F31-BB41-89548216C432.

10. "Exxon Says N. America Gas Production Has Peaked," Reuters, January 19, 2007, www.reuters.com/article/Utilities/idUSN2163310420050621.

11. Chris Pedersen, *US LNG: A Benchmark for the Future* (London: S&P Global Platts, 2017), 2, www.platts.com/IM.Platts.Content/InsightAnalysis/IndustrySolutionPapers/SR-us-lng-benchmark-for-the-future-052017.pdf.

12. 2007 data from *BP Statistical Review of World Energy 2018*. In March 2019, EIA predicted that gas production for 2019 would average just over ninety billion cubic feet per day. See, "Natural Gas," in *Short-Term Energy Outlook* (Washington, DC: EIA, 2019), www.eia.gov/outlooks/steo/report/natgas.php.

13. "U.S. Natural Gas Exports and Re-Exports by Point of Exit," EIA, www.eia.gov/dnav/ng/ng_move_poe2_a_EPG0_ENG_Mmcf_a.htm.

14. In 2017, China was using about twenty-three billion cubic feet per day. See, *BP Statistical Review of World Energy 2018*.

15. Natasha Turak and Tom DiChristopher, "Saudi Oil Giant Aramco Strikes Deal to Buy Natural Gas from Sempra Energy," CNBC, May 22, 2019, www.cnbc.com/2019/05/22/saudi-oil-giant-aramco-strikes-deal-to-buy-us-natural-gas-from-sempra.html.

16. "U.S. Natural Gas Exports and Re-Exports by Point of Exit," EIA, www.eia.gov/dnav/ng/NG_MOVE_POE2_A_EPG0_ENG_MMCF_A.htm.

17. *International Energy Outlook 2017* (Washington, DC: EIA, 2017), 79, www.eia.gov/outlooks/ieo/pdf/0484(2017).pdf.

18. "India Wants Eleven More LNG Import Terminals," Society of Master Mariners' Bangladesh, February 12, 2018, www.smmbd.org/india-wants-eleven-more-lng-import-terminals/.

19. *World Energy Outlook 2017*, 650. These projections come from the "New Policies Scenario."

20. *BP Statistical Review of World Energy 2018*. In 1990, electricity production totaled less than 12,000 terawatt-hours. By 2017, production exceeded 25,000 terawatt-hours.

Conclusion

1. Alok Jha, "Frances Ashcroft: We Are Controlled by Electrical Impulses," *Guardian*, June 23, 2012, www.theguardian.com/technology/2012/jun/24/frances-ashcroft-ion-channel-physiology.

2. Frances Ashcroft, *The Spark of Life: Electricity in the Human Body* (New York: W. W. Norton, 2012), 6.

3. Bryce, *Smaller Faster Lighter Denser Cheaper*, 9.

4. "Allah Guides to His Light . . . " Awaisia, www.awaisiah.com/silsila-e-awaisiah/hazrat-ghulam-muhammad/sayings/item/326-allah-guides-to-his-light.

5. Michael J. Wilkins, "Light," Bible Study Tools, www.biblestudytools.com/dictionaries/bakers-evangelical-dictionary/light.html.

BIBLIOGRAPHY

Acemoglu, Daren, and James A. Robinson. *Why Nations Fail: The Origins of Power, Prosperity, and Poverty*. New York: Crown Business, 2012.

Ashcroft, Frances. *The Spark of Life: Electricity in the Human Body*. New York: W. W. Norton, 2012.

Bryce, Robert. *Smaller Faster Lighter Denser Cheaper: How Innovation Keeps Proving the Catastrophists Wrong*. New York: PublicAffairs, 2014.

Caro, Robert A. *The Path to Power*. Vol. 1 of *The Years of Lyndon Johnson*. New York: Vintage, 1983.

Champagne, Anthony. *Congressman Sam Rayburn*. New Brunswick, NJ: Rutgers University Press, 1984.

Cunningham, Joseph J. *New York Power*. Self-published, CreateSpace, 2013.

Dalzell, Frederick. *Engineering Invention: Frank J. Sprague and the U.S. Electrical Industry*. Cambridge, MA: MIT Press, 2009.

Dugger, Ronnie. *The Politician: The Life and Times of Lyndon Johnson*. New York: W. W. Norton, 1982.

Forstchen, William R. *One Second After*. New York: Tor, 2009.

Freund, Paul, and Olav Kaarstad. *Keeping the Lights On: Fossil Fuels in the Century of Climate Change*. Oslo: Universitetsforlaget, 2007.

Garraty, John A., and Peter Gay, eds. *The Columbia History of the World*. New York: Harper & Row, 1972.

Glaeser, Edward. *Triumph of the City: How Our Greatest Invention Makes Us Richer, Smarter, Greener, Healthier, and Happier*. New York: Penguin, 2011.

Gleick, James. *The Information: A History, a Theory, a Flood*. New York: Vintage, 2011.

Gordon, Robert J. *The Rise and Fall of American Growth: The U.S. Standard of Living Since the Civil War*. Princeton, NJ: Princeton University Press, 2016.

Grossman, Peter Z. *U.S. Energy Policy and the Pursuit of Failure*. Cambridge: Cambridge University Press, 2013.

Haass, Richard N. *Intervention: The Use of American Military Force in the Post–Cold War World*. Rev. ed. Washington, DC: Brookings Institution Press, 1999.

Huber, Peter W., and Mark P. Mills. *The Bottomless Well: The Twilight of Fuel, the Virtue of Waste, and Why We Will Never Run Out of Energy*. New York: Basic Books, 2005.

Hughes, Thomas P. *Networks of Power: Electrification in Western Society, 1880–1930*. Baltimore: Johns Hopkins University Press, 1983.

Israel, Paul. *Edison: A Life of Invention*. New York: John Wiley & Sons, 1998.

Jonnes, Jill. *Empires of Light: Edison, Tesla, Westinghouse, and the Race to Electrify the World*. New York: Random House, 2003.

Klein, Maury. *The Power Makers: Steam, Electricity, and the Men Who Invented Modern America*. New York: Bloomsbury, 2008.

Koppel, Ted. *Lights Out: A Cyberattack, a Nation Unprepared, Surviving the Aftermath*. New York: Crown, 2015.

Kotkin, Joel. *The City: A Global History*. New York: Modern Library, 2005.

Landau, Sarah Bradford, and Carl W. Condit. *Rise of the New York Skyscraper, 1865–1913*. New Haven, CT: Yale University Press, 1996.

Lesser, Jonathan, and Leonardo Giacchino. *Fundamentals of Energy Regulation*. Reston, VA: Public Utilities Reports Inc., 2013.

MacKay, David J. C. *Sustainable Energy—Without the Hot Air*. Cambridge: UIT Cambridge Ltd., 2009.

Morris, Charles R. *The Dawn of Innovation: The First American Industrial Revolution*. New York: PublicAffairs, 2012.

Morris, Charles R. *A Rabble of Dead Money: The Great Crash and the Global Depression: 1929–1939*. New York: PublicAffairs, 2017.

Norris, George W. *Fighting Liberal: The Autobiography of George W. Norris*. 2nd ed. Lincoln, NE: Bison Books, 2009.

Nye, David E. *Electrifying America: Social Meanings of a New Technology*. Cambridge, MA: MIT Press, 1990.

Rapley, Bruce, and Huub Bakker, eds. *Sound, Noise, Flicker and the Human Perception of Wind Farm Activity*. Palmerston North, New Zealand: Atkinson & Rapley Consulting Ltd., 2010.

Rhodes, Richard. *Energy: A Human History*. New York: Simon & Schuster, 2017.

Ridley, Matt. *The Rational Optimist: How Prosperity Evolves*. New York: HarperCollins, 2010.

Roach, Craig R. *Simply Electrifying: The Technology That Transformed the World, from Benjamin Franklin to Elon Musk*. Dallas: BenBella Books, 2017.

Rowsome, Frank, Jr. *The Birth of Electric Traction: The Extraordinary Life and Times of Inventor Frank Julian Sprague*. Self-published, CreateSpace, 2013.

Schewe, Phillip F. *The Grid: A Journey Through the Heart of Our Electrified World*. Washington, DC: Joseph Henry Press, 2007.

Schivelbusch, Wolfgang. *Disenchanted Night: The Industrialization of Light in the Nineteenth Century*. Berkeley: University of California Press, 1988.

Schlesinger, Henry. *The Battery: How Portable Power Sparked a Technological Revolution*. New York: HarperCollins, 2010.

Smil, Vaclav. *Creating the Twentieth Century: Technical Innovations of 1867–1914 and Their Lasting Impact*. New York: Oxford University Press, 2005.

Smil, Vaclav. *Energy Myths and Realities: Bringing Science to the Energy Policy Debate*. Washington, DC: AEI Press, 2010.

Smil, Vaclav. *Power Density: A Key to Understanding Energy Sources and Uses*. Cambridge, MA: MIT Press, 2015.

Smil, Vaclav. *Prime Movers of Globalization: The History and Impact of Diesel Engines and Gas Turbines*. Cambridge, MA: MIT Press, 2010.

Tucker, William. *Terrestrial Energy: How Nuclear Power Will Lead the Green Revolution and End America's Energy Odyssey*. Savage, MD: Bartleby Press, 2008.

Vigna, Paul, and Michael J. Casey. *The Age of Cryptocurrency: How Bitcoin and Digital Money Are Challenging the Global Economic Order*. New York: St. Martin's Press, 2015.

Wheeler, Burton K. *Yankee from the West: The Candid Story of the Freewheeling U.S. Senator from Montana*. New York: Doubleday, 1962.

Williams, John. *The Untold Story of the Lower Colorado River Authority*. College Station: Texas A&M University Press, 2016

INDEX

Note: Pages followed by *fig* indicate figures; those followed by *t* indicate tables.

Abadi, Haider al-, 95
Acemoglu, Daron, 83–84
Adroit Market Research, 168
Africa, 182
Age of Cryptocurrency, The (Vigna and Casey), 138
Agora Energiewende, 174–175
air conditioning, 167–168
air pollution, 102–103, 107
ALARA (as low as reasonably achievable), 226
Al-Barouk, Khalid, 103
Alipay, 135
Aliso Canyon, 191
Alliance for Clean Energy New York, 187–188
Alphabet, 119, 122, 125, 128, 130
alternating-current motors/systems, 23, 35, 69
Amazon, xxii, 119, 123, 125, 126, 127, 128–131
Amazon Web Services (AWS), 129
American Electric Power (AEP), 156
American Geophysical Union, xx, 159
American Wind Energy Association, 202
Americans Against Fracking, 191
Ammar, Maya, 100
Amper, Dick, 211
amperes, explanation of, 5–6
Andersen, Neil P. and Elizabeth L., 206
Andrews, Roger, 19
animals, effects of wind turbines on, 202–203, 212–214
antitrust rules, 122
Antonovich, Michael D., 193
Apex Clean Energy, 194
Apple, 119, 122–123, 126, 130
Apple Pay, 135
appliances, 65–67
arc lights, 12
ArcView Group, 146
Arraf, Jane, 94
Arthaud, Jim, 193–194
Ashcroft, Frances, 246
Atlantic Wind, 198
Australia, 111, 176–177
Ausubel, Jesse, 24, 34
AWS Secret Region, 129

Baalbaki, Mustafa, 99
Backus, Emily, 146
Bangladesh, coal use in, 111

Barakah nuclear power plant (Abu Dhabi), 231
Barkat, Abul, 67
Bat Conservation International, 214
bats, wind turbines and, 212, 213–214
Battaglini, Antonella, 200
batteries, 8, 179–181, 236–237, 239, 251*fig*
Bayar, Yilmaz, 19
Bear Ridge solar project, 212
Beirut Electricity (iPhone app), 99
Bell, Alexander Graham, 7
Bent Tree wind project, 206
Betz, Albert, 187
Betz limit, 187
Beyond Coal campaign, 108–109, 171
Beyond Natural Gas campaign, 171
Beyond Oil campaign, 171
Bezos, Jeff, 124
Bing, 122
Bird Studies Canada, 213–214
birds, wind turbines and, 212–213
Birth of Electric Traction, The (Rowsome), 29
Bitcoin, 134, 136, 138–140
Bkerzay, 236–237
Black Hills Energy, 128
blackout bomb, 88, 91
blackouts
  costs of, 99, 151
  effects of, xx
  EMPs and, 156–158
  frequency of, 159–160
  medical facilities and, 153–156
  natural causes of, 158–159
  sabotage and, 152–153, 160–161
Bloomberg, Michael, 108, 109
Bloomberg Philanthropies, 108–109
BLU-114/B, 88
Blue Ribbon Commission on America's Nuclear Future, 229
Boeve, May, 173
Booker, Cory, 173
Bottom Billion, The (Collier), 21
Bouri, Elie, 99
Brand, Stewart, 166–167
Brick, Stephen, 180
Brin, Sergey, 124
British thermal units (Btus), 8
Brooklyn Bridge, 22, 25
Brown, Jerry, 177, 178
Brune, Michael, 172, 173
Buchanan Dam, 49–52
building height, 23, 24, 30
Bush, George W., 93
Bushnell, James, 179

Calcutta Electric Supply Corporation, 105–106
Caldeira, Ken, 222
California
  renewable energy and, 177–180, 177*fig*
  wind-energy backlash in, 199
California Air Resources Board (CARB), 177–178
California Independent System Operator, 153
California Wind Energy Association, 199
Calix, Brianna, 149–150
Cameron, Mary, 184
Canada
  coal use in, 112
  renewable energy and, 175–176
  wind-energy backlash in, 199
cannabis industry, 142–151, 149*fig*
capital, 80–81, 84–86
carbon dioxide emissions. *See* emissions
carbon duties, 173
carbon offsets, 147–148
carcinogen exposure, 102
Carmody, John, 51–52
Carnegie, Andrew, 14, 37
Caro, Robert, 51, 59
Carrington Event, 159
Casey, Michael J., 138
cash, 135–136, 140–141
Center for Biological Diversity, 110
Center for Social Responsibility, 92
Central Intelligence Agency (CIA), 129
Chattopadhyay, Mriduchhanda, 61

Chedid, Riad, 101–102, 245
Chen, Hsuan-Hsiu Annie, 204–205
Cheney, Dick, 93
Cheniere Energy, 113
Chernobyl, 225–226
child and infant mortality rates, 75, 92
child marriage, 68–69, 68*t*
China
  coal use in, 111
  nuclear energy in, 231
  renewable energy and, 182, 183–184
China General Nuclear Power Group, 231
China National Nuclear Corporation, 231
cholera, 92
Chowdhury, Sanjay Kar, 105–106
Cisco, 120
cities
  migration to, 17
  population growth in, 33–34
  transportation in, 27–30
  verticality of, 22–24, 30–33, 34–35
Clack, Christopher, 184–185
Clean Line Energy Partners, 214–215
Clear Air Task Force, 180
Clemente, Jude, 201
climate change
  Democratic Party on, 172
  electricity production and, xxiii
  increased demand due to, 169
cloud computing and storage, 129
cloud miners, 136–138
coal
  reliance on, 105–115
  use of in Colorado, 147–148
CoalSwarm, 113
cobalt, 238
Collier, Paul, 21
Collins, Ash, 65
communications, early, 7
Community Fracking Defense campaign, 192
computer chips, 17
Conca, James, 226
consumption estimates
  health and welfare correlates and, 76–77
  inaccurate, 73
  per capita, 74–75, 76*fig*
Cook, Ewing, 153–155
Corcoran, Tommy, 51
Corkum, Paul, 17–18
coronal mass ejections, 159
corruption, 82–83, 83*fig*
cortisol increase due to wind turbines, 203
Cozzarelli Prize, 184
credit cards, 133–134
cross-border electricity trade, 78
Cryan, Paul M., 213
cryptocurrency, 134, 136–141
Culver, Tyson, 60
Cunningham, Joseph J., 25
Cuomo, Andrew, 173, 194, 221
cyber sabotage/cyberattacks, 156, 160–161

Dakota Access Pipeline, 189, 193
dams, 41–42, 44, 49–52, 219
Dance, Scott, 211
darkness, 10–11
data-center capacity, 120
Dávila, Eladia, 156
Davis, Lucas, 223–224
dead-man's control, 31
Deepwater Horizon, 191
defamation lawsuit, 196
Delucchi, Mark, 184
Democratic Party, 171, 172
demonetization, 135–136
density. *See* power density
desalination plants, 168
Diablo Canyon, 172
direct-current (DC) electric motors, 27
disempowered people, numbers of, xxii
Disenchanted Night (Schivelbusch), 11
drones, weaponization of, 158
dry casks, 227, 229
Dugger, Ronnie, 47, 51
Duke Energy, 209

E24 Solutions, 236–237, 245
eagles, wind turbines and, 212–213
Earth First! 192
Earthjustice, 192
Eckert, J. Presper, 120
economic growth, electricity's impact on, 19–21, 21*fig*, 57, 58
economy, digital, 119–132
Edison, Thomas, 12–13, 22, 23–24, 25–27, 35, 37
Edison Electric Institute, 161
education
  disparities in, 64–65
  electrification and, 62, 67
  women and girls and, 64–65
Eisenhower, Dwight, 219
El Assad, Joseph, 99
El Khoury, Marwan, 236–237
Electric Bond and Share, 38–39, 44
electric cars, xxi
electric grid, in United States, 56, 56*t*
electric grids
  attacks on, 87–90
  control of, 78–79
  for data centers, 126–127
  Giant Five and, 120
  integrity, capital, and fuel for, 80–86
  microgrids, 236–238
  size of, 82
Electric Home and Farm Authority, 65
Electric Infrastructure Security Council, 157
Electric Power Research Institute, 159
electric rail systems, 28–30
electric vehicles, 168
Électricité de France (EDF), 57
Electricité du Liban (EdL), 96–98, 99, 102
electricity
  basics of, 3–9
  blackouts and, 152–161
  cannabis industry and, 142–151
  cities and, 22–35
  coal use for, 105–115
  consumption estimates for, 73, 74–75, 76–77, 76*fig*
  contemporary economy's reliance on, 119–132
  cost of, by country, 139
  decreasing costs of, 53
  effects of, xvii–xviii
  expansion throughout United States of, 48–59
  future of, 236–244
  generation capacity for, 243*fig*
  global spending on, xxiii, 85
  human body and, 246
  as human right, 245–246
  increased demand for, 166–170
  increasing production of, 36
  in India, 105–108
  integrity, capital, and fuel and, 80–86
  in Iraq, 87–89, 90–95
  land-use impacts and, 189–216
  in Lebanon, 96–104
  money and payment systems and, 133–141
  New Deal and, 36–47
  nuclear sources for, 217–235
  poverty and, xxii–xxiii, 20–21, 60–63
  religious symbolism of, 246–247
  renewable sources and, 171–188
  storage of, 8, 251*fig*
  Terawatt Challenge and, 165–170
  theft of, 84, 94, 149–150
  transformative power of, 10–21
  transmission of, 78
  women and girls and, 60–70
electromagnetic pulse (EMP), 156–158
electrons, controlling, 4
Elektrownia Belchatow (Poland), 113
elevators, 23, 30–33, 35
Ellis, Wright, 212
emissions
  from cannabis industry, 145
  nuclear plant shutdowns and, 223–224
  reduction efforts for, 113–114, 115
EMP Commission, 157–158
employment
  electrification and, 67–68
  women and, 67

Endangered Species Act, 210
Energiewende, 113–114
Energy Information Administration (EIA), 147, 240
Energy Myths and Realities (Smil), 186–187
energy versus power, 8–9
England, coal use in, 112
ENIAC (Electrical Numerical Integrator and Computer), 120–121
Enigma, 139
Entergy, 227
Environmental Law Alliance Worldwide (ELAW), 110
Environmental Progress, 177, 181, 226
Environmental Research Letters, 186
Environmental Working Group, 174
Enz, Dave and Rose, 208–209, 210, 211*fig*
Equitable Life Assurance Building, 23
Erblich, Timothy, 197
Ethereum, 137, 139, 140
Ethisphere Institute, 196–197
European Commission, 122
European Network of Transmission System Operators for Electricity, 200–201
European Platform Against Windfarms, 199
Ewing, Fergus, 200
excess mortality, 155
Exelon Corporation, 57

Facebook, 119, 123–124, 125, 126
Fair Employment and Housing Act, 178
Fair Housing Act, 178
Faraday cage, 160
farm production, improvements in, 53–54
Fatmagül Sultan, 102–103, 104*fig*
Federal Energy Regulatory Commission, 46, 153
Federal Power Act (1935), 46
Federal Power Commission, 46, 55
Federation of Germany Industries, 114
Fedwire Funds Service, 135
Fighting Liberal (Norris), 46, 58–59
Fink, Sheri, 154
fire, danger of, 12
First Iraq War, 91
Fisk, Jim, 28
Five Days at Memorial (Fink), 154
Florida Power and Light, 155
Fluor Corporation, 233
Fogal, Paul, 198
Food & Water Watch, 174, 191
Ford, Doug, 176
Ford, Henry, 44
Ford Motor Company, 16–17
Forstchen, William R., 156–157
fracking, 191–192, 242
France
  coal use in, 112
  nuclear-waste handling in, 227–228
  wind-energy projects in, 200
Franklin, Benjamin, 4–5
Fraser Institute, 175, 176
Frew, Bethany, 184
Friends of the Earth, 110, 174
From the Ashes (film), 109
Frondel, Manuel, 202
fuel
  availability of, 80–81, 86
  in India, 105–108
  in Lebanon, 101–104, 103*fig*
  share of global electricity output by, 109*fig*
  See also individual fuels
Fukushima Daiichi power plant (Japan), 113, 114, 224–226
Fusen hydroelectric plant (North Korea), 89
fusion energy, 235
Future of Cooling, The (IEA), 167

Gamboa, John, 178
gas lighting, 11–12, 23
gasometers, 12
Gates, Bill, 124
GDP, 75, 91

Gellman, Barton, 90–91
Generac Power Systems, 160
General Electric, 38
generator mafia, 96–98, 99–100, 102, 103
generators
for data centers, 126–127
in Iraq, 95
in North Vietnam, 90
reliance on, xiv–xx, xxii
residential backup, 160
Genesis Mining, 136–138, 140
Genetics Society of America, 226
geothermal plants, 86
Germany
anti-coal protests in, 190
coal use in, 113–114
emission-reduction efforts in, 113–114
energy production in, 182
renewable energy and, 174–175, 183
wind-energy backlash in, 199, 205
Giant Five
dominance of, 124–126, 125*fig*
electricity consumption of, 119–121, 122*fig*, 126–132, 127*fig*
generation capacity of, 132*fig*
market capitalization increases of, 122*fig*
renewable energy and, 129–131
Gibbons, Steve, 201
girls. *See* women and girls
Glass-Steagall Act, 42
Gleick, James, 7
global trade, 237
globalization, 77–78
Good Jobs First, 197
Google, 122, 125, 126
Google Pay, 135
Gould, Jay, 28, 30
Government Accountability Office, 230
Grain Belt Express, 215
Great Chicago Fire of 1871, 12
Great Depression, 39, 40
Green, Kenneth, 176
Green Energy Act (Canada), 175–176
Green New Deal, xxi, 173–174
greenhouse gas emissions. *See* emissions
Greenpeace, 113, 171, 172, 190, 224–225, 232
Greenpeace USA, 110
Greenslade, Roy, 124
Griffith, Thomas E., 90
Grossman, Peter Z., 55

Hagen, Bernie and Cheryl, 206
Hambach Forest, 190
Hamilton, James, 130
Hanford Site, 229
Hansen, James, 222
Hardscrabble Wind Power Project, 193
Hargraves, Robert, 234–235
Hausman, Catherine, 223–224
Heard, Ben, 181
Helman, Chris, 169
Hershey, Cheyney, 207
Hexa Research, 168
Hezbollah, 100–101
high-efficiency, low-emission (HELE) projects, 115
high-voltage transformers, 159
high-voltage transmission lines, 214–216
Hiroshima, 226
holding companies, 38–46, 50–51
Hoover, Herbert, 40, 41
Hoover Dam, 41
horsepower, 8
horses, 27–28
Hughlett, Mike, 206
human body, electricity and, 246
Human Rights Watch, 92, 101
Hurricane Irma, 155
Hurricane Katrina, 153–155
Hurricane Maria, xiii–xx, 80, 155–156
Hurricane Sandy, 159
Hussein, Ahmed, 94
Hussein, Saddam, 91
hydraulic fracturing, 191–192, 242

Iceland, 136–137, 139
Idaho National Laboratory, 233
Imran, Kashif, 19
incandescent lamps, 12–13
India
  air conditioning and, 167
  coal use in, 105–108
  mobile-based payment systems and, 135–136
  natural gas and, 241–242
  nuclear energy in, 231
  renewable energy and, 182
  women and girls in, 60–63
Indian Institute of Science, 212–213
Indian Point Energy Center, 217–221, 223, 227, 228*fig*
Indonesia, 111, 112
Industrial Revolution, 14, 16
infant/child mortality rates, 75, 91, 92
Information, The (Gleick), 7
Information Age, xxi–xxii, 119–121
infrasound effects, 204, 207
infrastructure as a service (IaaS), 129
Institute on Taxation and Economic Policy, 197
Insull, Samuel, 37–38, 40, 55
integrity
  importance of, 80–83, 85–86, 237
  lack of, 99, 169
  societal, 83–84
Intergovernmental Panel on Climate Change, 61
International Energy Agency (IEA), 20–21, 74, 167, 222–223, 242
International Ladies' Garment Workers' Union, 70
International System of Units (SI), 9, 249–250
Interstate Informed Citizens Coalition, 198
investor-owned utilities, 56–57, 161
ion channels, 246
Ip, Greg, 124–126
Iran, Iraq and, 94–95
Iraq, 87–89, 90–95
Iraq Index, 93
"iron law" of electricity, 106–107
Israel, Lebanon and, 100–101
Israel Defense Forces, 100–101
Ivanpah solar complex, 210

Jacobson, Mark Z., 184–185
Jamadar, Rehena, 60–63, 61*fig*
Jamhour power plant (Lebanon), 101
Japan
  coal use in, 114–115
  emission-reduction efforts in, 115
Jiyeh power plant (Lebanon), 101, 102–103
Jobs, Steve, 124
Johnson, Edward Hibberd, 27, 28
Johnson, Lyndon, 46–52, 50*fig*, 59, 90
Jónsdóttir, Birgitta, 245–246
joules (J), 8
Jounieh power plant (Lebanon), 102–103, 104*fig*

Kalashnikov Group, 158
Kammen, Dan, 185
Karadeniz Holding, 102, 104*fig*, 234
Karpa, Doug, 179
Katrínarson, Gísli, 245
Keane, Michael, 208
Keith, David, 186–187
Kennedy, David, 41–42
Kenya, 135
kerosene, 12
Kim, Jim Yong, 110
Kissane, Carolyn, 245
kite experiment, 4–5
KMGT (kilo, mega, giga, tera), 9
Knight, Rick, 133
Kopacz, Ewa, 113
Korea Electric Power Corporation (KEPCO), 231
Korean War, 89
Kosova e Re plant (Kosovo), 110
Kotkin, Joel, 124
KUB-UAV (drone), 158
Kyoto Protocol, 114

land-use impacts of renewable energy
  battles regarding, 189–190, 192–196, 210–211, 214–216
  calculations of, 184–188
Lang Chi hydropower plant (Vietnam), 90
Langrud, Dave and Birgitt, 206
lasers, 17–18
lead-acid batteries, 236–237, 239
leakage, 81–82
Lebanon
  generator mafia in, 96–98, 99–100, 102, 103
  powerships in, 102–104, 104*fig*
  refugees in, 98–99
  war in, 100–102
LED lights, 148
Leyden jar, 4
Lichtenhan, Jeffery, 204
life expectancy, 75
Lighthouse Wind, 194–195
lighting
  decreasing costs of, 13
  gas, 11–12, 23
  industrialization of, 11–12
  transformative power of, 10–11
  types of, 13
liquefied natural gas (LNG), 113, 240–242
liquefied petroleum gas (LPG), 61
literacy rates, 67, 68
lithium-ion batteries, 239
living standards, improvements in, 57–58
localism, 78
Long, Jane, 185
low-frequency noise effects, 207
Lynas, Mark, 188

MacKay, David J. C., 188
Madden, Mike, 151
Mahdi, Adel Abdul, 95
Malaysia, coal use in, 111
Manhattan Elevated Railway Company, 28
Manjoo, Farhad, 121
Mansfield Dam, 49–52
Marchetti, Cesare, 33
Marchetti's constant, 33
marijuana industry, 142–151, 149*fig*
marketing, 65–67
Markey, Edward J., 173
Mars Hill wind project, 203–204
Marshall Ford Dam, 49–52
Martis, Kevon, 198–199
Massachusetts Cannabis Control Commission, 148
Mauchly, John W., 120
Maughan, Tim, 238
maximum uptime, 126
McCann, Michael, 201
McCarthy Tétrault, 196
McKibben, Bill, 174
media coverage of land-use fights, 194–195
medical facilities, blackouts and, 153–156
Memorial Medical Center (New Orleans), 153–155
Merkel, Angela, 114
Merkley, Jeff, 173
Merrimack Manufacturing Company, 16
metal-halide lights, 143–145
Metcalf Transmission Substation (Santa Clara County, California), 152–153
microgrids, 236–238
Microsoft, 119, 122, 128, 130
Middle East, 182
Middle West Utilities Company, 38
migration patterns, 17
Migratory Bird Treaty Act, 191
Miller, Lee, 185–187
Mills, Evan, 147, 148
Mills, Mark P., 181
mobile-based payment systems, 135–136
Modi, Narendra, 135–136
molten-salt reactors, 233–234
money, electrification of, 132–141
Moniz, Ernest, 235

monopolies, 37, 42, 125*fig*
Morgan, Granger, 185
Morgan, J. P., 37
Morningstar, 145–146, 147
Morrison, Scott, 176–177
Morse, Samuel, 7
motor, electric, 24, 25–30
Mousl, Hussein, 96–97, 97*fig*
Moyer, Tony, 208
M-PESA, 135
Mt. Gox, 138–139
Mumford, Lewis, 219
Murphy, Audrey, 209–210
Muslim population, growth of, 63–65
Myers, Seth, 74, 75–76

Nagasaki, 226
Naidoo, Kumi, 224
Nakhle, Khaled, 99
Nappi, Jerry, 227
Narins, Peter, 204–205
Nasrallah, Hassan, 100
National Academy of Engineering, 18
National Center for Policy Analysis, 191
National Institutes of Health, 207
National Pact (Lebanon), 98
National Renewable Energy Laboratory, 215–216
natural gas, 239–243
Natural Resources Defense Council (NRDC), 171–172, 191–192, 221
Nautilus, USS, 232
Naval Academy, 25
Navitus Bay offshore wind project, 200
Nelson, Mark, 177
networks, 6
New Deal, 36–47, 65
New England Independent System Operator, 223
New York City
  population of, 33–34
  transportation in, 22–23
  vertical nature of, 34–35
New York Independent System Operator, 223
New York Power (Cunningham), 25
New York Public Service Commission, 40
Newcomen, Thomas, 14
newspaper industry, 124
Newton, Shelly, 195
NextEra Energy, 195–198
nighttime luminosity, 20
Nikolewski, Rob, 199
Nissenbaum, Michael, 203–204
Nixon, Richard, 90
nocebo effect, claims of, 205
noise from wind turbines, 192–193, 202–210
Nordhaus, William, 20
Norris, George, 42, 43–44, 45*fig*, 46, 55, 57–58
Norris-Rayburn Act (1936), 42, 46, 49
Northern Pass Transmission, 215
nuclear energy
  advances in, 232–235
  costs of, 230–231
  land requirements of, 219–220, 220*fig*
  necessity of, 217–235
  opposition to, 171–173
  radiation concerns and, 224–226
  waste from, 226–229
nuclear powerships, 232, 234
nuclear submarines, 232
Nuclear Waste Policy Act, 228
numerical designations, 249–250
NuScale Power, 233

Oak Ridge National Laboratory, 229, 234
Obama, Barack, 110, 229
Ocasio-Cortez, Alexandria, 173–174
O'Hanlon, Michael, 93
oil and gas industry, environmental impact of, 190–192
oil-fired generators, 102–104, 103*fig*
One Second After (Forstchen), 156–157
100 by '50 Act, 173
Operation Rolling Thunder, 89–90

Operations Center East (Visa), 133–134
Optimum Renewables, 194
Orhan Bey, 102–103
Ortiz, Iris, xiii–xx, 80–81
O'Shea, Thomas J., 213
Otis, Elisha, 30
Our World in Data (Roser), 13
Özel, Hasan Alp, 19

Pacific Gas & Electric, 153
Pacific Power, 151
Pagel, Joel, 212
Pakistan, coal use in, 111–112
paper money, 135–136, 140–141
Pasternak, Alan D., 76–77
Path to Power, The (Caro), 51, 59
Paytm, 135
Pedernales Electric Cooperative, 48, 51–53, 59
penthouses, 34
Pew Research Center, 172
Philadelphia Electrical Exposition, 27
Philippines, coal use in, 112
Phillips, Carl V., 204
phone-based payment systems, 135–136
Pielke, Roger, Jr., 106–107
Plains and Eastern Clean Line, 214–215
Poe, Edgar Allen, 11
Poland
  coal use in, 112–113
  wind-energy backlash in, 199
Polish Wind Energy Association, 199
political power
  integrity and, 83–84
  women and, 69–70
Politician, The (Dugger), 47
population growth
  in cities, 167
  in Muslim communities, 63–64, 64*fig*
Porter, Charlie, 207
Postal Telegraph Cable Company, 30–32, 32*fig*
Pou, Anna, 155
poverty
  electricity and, xxii–xxiii, 20–21, 68
  electricity rates and, 178
  levels of electrification needed for alleviation of, 77
  women and girls and, 60–63
power
  versus energy, 8–9
  instant, 10
  transformative power of, 14–15
power density
  cannabis industry and, 148, 149*fig*
  explanation of, 15–16
  importance of, 10, 16–17
  of nuclear facilities, 219
  renewable energy and, 186–187
Power Density (Smil), 16
Powering Past Coal Alliance, 112
powerships, 102–103, 232, 234
Proceedings of the National Academy of Sciences, 184–185
Progressive Conservative Party, 176
property values, renewable energy systems and, 201–202, 208, 212
Proposition 112 (Colorado), 189–190
Public Citizen, 172
public utility holding companies, 38–46, 50–51
Public Utility Holding Company Act (Rayburn-Wheeler Act; 1935), 44–46
Public Works Administration, 42
Puerto Rico, xviii–xx, 80–81, 155–156
Puerto Rico Electric Power Authority (PREPA), xiv

Radford, Darrel, 198
radiation, 224–226
Ramesh, Jairam, 106
rare earth elements, 238
Rauth, Helmut, 136–137, 140
Rayburn, Sam, 42–43, 44–45, 46, 50, 50*fig*, 55
Rayburn-Wheeler Act (1935), 42, 44–46
Raymond, Lee, 240

real estate, urban, 34–35
recycling of batteries, 239
Reddy Kilowatt, 65–67, 66*fig*
refrigerators, as benchmarks, 73–74
refugees, in Lebanon, 98
Rehabilitation Center at Hollywood Hills (Florida), 155
Reid, Harry, 229
reliability, xxi–xxii
religious affiliation
  disparities in electrification and, 75
  in Lebanon, 98
  population growth and, 63–65
renewable energy
  backlash against, 190
  cannabis industry and, 147–148
  challenges to relying on, xxi
  costs of, 174–179, 177*fig*
  in Germany, 114
  Giant Five and, 129–131, 132*fig*
  global spending on, 182, 183*fig*
  impact on property values of, 201–202
  insufficiency of, 171–188, 222
  land requirements of, 184–188, 223
  scale challenge of, 182–184
  storage of, 179–181
Renewables Grid Initiative, 200
Republican Party, 173
residential backup generators, 160
retail stores, closures of, 123
Reynolds, Anne, 187–188
Rhodium Group, xx
Richmond Union Passenger Railway, 29–30
River Rouge plant (Michigan), 16–17
Riverkeeper, 221
Robinson, James A., 83–84
Rock Island Clean Line, 214
Rockefeller, John D., 37
Roebling, John, 22
Rogers, Will, 45–46
Rollins Mountain, 192
Roosevelt, Franklin, 40–42, 43, 48–49, 51–52
Roque, Wilfredo, xiii–xx, 80–81
Rosatom, 231–232
Roser, Max, 13
Rosling, Hans, 65
Roubini, Nouriel, 140
Rowsome, Frank, Jr., 29–30
Roy, Joyashree, 60–61, 61*fig*, 69, 247
Ruffalo, Mark, 173
rural communities
  electrification of, 46–55, 54*fig*, 58–59
  in India, 60–63, 67
  price of electricity in, 39
  refusal of service to, 39
Rural Electrification Act (Norris-Rayburn Act; 1936), 46, 49
Rural Electrification Administration (REA), 46, 48–49, 51–52
Rusk, Dean, 89
Russia, nuclear energy in, 231–232
RWI, 202

Saab, Antoine, 237, 245
sabotage, xx–xxi, 92–94, 152–153, 156–158, 160–161
Salt, Alec, 204
San Onofre Nuclear Generating Station, 223–224
Sanders, Bernie, 173
Santee Cooper, 230
Santos, Nelson, 175
Saudi Aramco, 241
Savannah River Site, 229
SCANA, 230
Schivelbusch, Wolfgang, 11
Schwarzenegger, Arnold, 177
Schwarzkopf, Norman, 87–88
Scotland, 200
Scotland Against Spin, 200
Scott, Rick, 155
Second Iraq War, 92–93
secondary energy sources, 7
Sempra Energy, 241
shadow flicker, 206
shale revolution, 240
Sheehan, Michael A., 158
Shellenberger, Michael, 177, 226–227
Shelley, Percy Bysshe, 246

Shirley Wind Farm, 209
Shorten, Bill, 177
Siddiqui, Masood Mashkoor, 19
Sidon power plant (Lebanon), 101
Sierra Club, 108, 113, 171, 172, 190, 212
single-fluid theory, 5–6
skyscrapers, 23, 30–32, 34–35
SLAPP suits, 196
sleep deprivation from wind turbine noise, 203–205, 207
small modular reactors (SMRs), 233–234
Smalley, Richard, 165–166
Smil, Vaclav, 15–16, 186–187
Smith, Rebecca, 153
social media, 123–124
societal integrity, 83–84
solar energy
  capacity of, 186
  decreasing costs of, 237
  favorability of, 172
  land requirements of, 223
  land-use battles and, 210–211
  spending on, 182
  storage of, 238
solar storms, 159
solar-energy projects, 86
Solis, Hilda, 193
South Africa, 111
South Korea, 231
Southern Company, 230
Spark of Life, The (Ashcroft), 246
Sprague, Frank Julian, 23–32, 26*fig*, 35
Sprague Electric Elevator Company, 30–32
Sprague Electric Railway and Motor Company, 28
squirrels, blackouts due to, 158
Standing Rock Sioux tribe, 189
State Grid Corporation of China, 57
State of the World's Children (UNICEF), 68
steam engine, 14–15
steam locomotives, 28
Steinfort, Tom, 226
subsidy mining, 197
suffrage movement, 69–70
Sui-ho dam, 89
Summer 2 and 3 reactors, 230
Sustainable Energy—Without the Hot Air (MacKay), 188
Synergy Research Group, 129

TAE Technologies, 235
Taft, William Howard, 41
Tavanir, 94
tax avoidance, 123, 135–136, 197
Taylor, Angus, 176
tech giants. *See* Giant Five
Tehachapi Renewable Transmission Project, 179
telegraph, 7
telephone, 7
10H rule, 199
Tennessee Valley Authority, 42, 55
Tennessee Valley Authority Act (1933), 44
Terawatt Challenge, 165–170, 174, 237, 243–244, 243*fig*, 245
Terrestrial Energy, 234
Tesla, xxi
Tesla, Nikola, 23–24, 35
Tesla Powerwall, 180–181
Texas Power & Light, 39
Thai Binh 2 power plant (Vietnam), 110
Thomas, Gerry, 225–226
ThorCon, 234–235
350.org, 173, 190
time, travel and, 33
Toshiba, 230
town gas, 11–12
transformers, 35
transmission lines, land-use battles and, 214–216
transmission projects and spending, 179, 200–201
Transparency International's Corruption Perceptions Index, 169
transportation, 27–30, 32–33

trespass zoning, 198–199
Triangle Shirtwaist Factory fire, 70
Trump, Donald, 94, 189
Turkey, coal use in, 112
Turnbull, Malcolm, 176
Tuscola III, 195
Tuttle, Merlin, 214
Two Hundred, 177–178

Ukraine, cyberattack on power grid of, 156
ultrasupercritical combustion, 115
UN Climate Change Conference, 222
UN Food and Agriculture Organization, 92
UN Scientific Committee on the Effects of Atomic Radiation, 225
United Kingdom, 112
United Nations' Human Development Index (HDI), 76–77, 92
United States
  coal use in, 111
  electrification of, 48–59
  energy production in, 182
  global gas market and, 240–241
  New Deal and, 36–47
  nuclear-waste handling in, 228–229
  renewable energy and, 172–174, 177–180
  water treatment in, 168
urbanization, 166–167
US Department of Energy, 160
US Department of Homeland Security, xx
US electric grid, 56, 56*t*
U.S. Energy Policy and the Pursuit of Failure (Grossman), 55
US Export-Import Bank, 110
US Nuclear Regulatory Commission (NRC), 230, 233
Utah Associated Municipal Power Systems and Energy Northwest, 233

vacant-land myth, 190
Vangor, Brian, 227
Venmo, 135
Vermont Yankee nuclear plant, 223–224
video, electricity needs of, 120
Vietnam, coal use in, 112
Vietnam War, 89–90
Vigna, Paul, 138
Visa Inc., 133–134
Vogtle 3 and 4 reactors, 230, 233
voltage, explanation of, 5–6
voting rights, 69–70

Wales, 200
Warren, Janet, 207–208
washing machines, 65
Waste Isolation Pilot Plant (WIPP), 229
water
  contamination of, 92, 94
  need for clean, 168
waterwheels, 16
Watt, James, 14
watt-hours (Wh), 8–9
watts
  explanation of, 5–6
  as measurement of power, 8–9
wealth, electricity consumption and, 19–21, 21*fig*
weather
  blackouts due to, 159
  increased demand due to, 169
western expansion, 41–42
Western Union, 30
Westinghouse, George, 23–24, 35, 69
Westinghouse AP1000, 230, 233
Westinghouse bankruptcy, 230
whale oil, 12
Wheeler, Burton, 42, 43, 44–45, 44*fig*, 46, 55
Whole Earth Catalog, 167
Why Nations Fail (Acemoglu and Robinson), 83–84
Wilson, Woodrow, 41
Wilson dam, 41
wind shadow, 186

wind-energy projects
  backlash against, 192–201
  capital required for, 86
  effects of on animals, 212–214
  impact on property values of, 201–202, 208
  insufficiency of, 183–184, 186–188
  land requirements of, 219–220, 220*fig*, 223
  noise from, 192–193, 202–210
  public opinion on, 172
wire transfers, 135
Wisconsin Power and Light, 205–206
women and girls
  electrification and, 58–59, 60–70
  employment and, 67
  marketing to, 65–67
  political power and, 69–70
wood, burning of, 107–108
World Bank, 110
World Energy Council, 170
World Energy Outlook 2017 (IEA), 242
World Nuclear Association, 228
Wrightman, Esther, 196
Wynne, Kathleen, 176

Xiong, Chua, 210

Yankee from the West (Wheeler), 43
YouTube, 120
Yucca Mountain waste repository site, 229

Zuckerberg, Mark, 123, 124

LORIN BRYCE

Robert Bryce is the acclaimed author of five previous books, including *Smaller, Faster, Lighter, Denser, Cheaper: How Innovation Keeps Proving the Catastrophists Wrong* and *Power Hungry: The Myths of "Green" Energy and the Real Fuels of the Future*. His articles have appeared in numerous publications, including the *Wall Street Journal*, *New York Times*, *Austin Chronicle*, *Guardian*, and *National Review*. He has given over three hundred invited or keynote lectures to groups ranging from the Marines Corps War College to the Sydney Institute and has appeared on dozens of media outlets ranging from Fox News to Al Jazeera. Bryce is also the producer of a new feature-length documentary: *Juice: How Electricity Explains the World*. He lives in Austin, Texas, with his wife, Lorin.

www.robertbryce.com
Twitter: @pwrhungry

PublicAffairs is a publishing house founded in 1997. It is a tribute to the standards, values, and flair of three persons who have served as mentors to countless reporters, writers, editors, and book people of all kinds, including me.

I. F. STONE, proprietor of *I. F. Stone's Weekly*, combined a commitment to the First Amendment with entrepreneurial zeal and reporting skill and became one of the great independent journalists in American history. At the age of eighty, Izzy published *The Trial of Socrates*, which was a national bestseller. He wrote the book after he taught himself ancient Greek.

BENJAMIN C. BRADLEE was for nearly thirty years the charismatic editorial leader of *The Washington Post*. It was Ben who gave the *Post* the range and courage to pursue such historic issues as Watergate. He supported his reporters with a tenacity that made them fearless and it is no accident that so many became authors of influential, best-selling books.

ROBERT L. BERNSTEIN, the chief executive of Random House for more than a quarter century, guided one of the nation's premier publishing houses. Bob was personally responsible for many books of political dissent and argument that challenged tyranny around the globe. He is also the founder and longtime chair of Human Rights Watch, one of the most respected human rights organizations in the world.

• • •

For fifty years, the banner of Public Affairs Press was carried by its owner Morris B. Schnapper, who published Gandhi, Nasser, Toynbee, Truman, and about 1,500 other authors. In 1983, Schnapper was described by *The Washington Post* as "a redoubtable gadfly." His legacy will endure in the books to come.

Peter Osnos, *Founder*